TRIBOLOGY OF POLYMER COMPOSITES

TRIBOLOGY OF POLYMER COMPOSITES

Characterization, Properties, and Applications

Edited by

SANJAY MAVINKERE RANGAPPA

SUCHART SIENGCHIN

JYOTISHKUMAR PARAMESWARANPILLAI

KLAUS FRIEDRICH

ELSEVIER

Elsevier
Radarweg 29, PO Box 211, 1000 AE Amsterdam, Netherlands
The Boulevard, Langford Lane, Kidlington, Oxford OX5 1GB, United Kingdom
50 Hampshire Street, 5th Floor, Cambridge, MA 02139, United States

Notices
Knowledge and best practice in this field are constantly changing. As new research and experience broaden our understanding, changes in research methods, professional practices, or medical treatment may become necessary.

Practitioners and researchers must always rely on their own experience and knowledge in evaluating and using any information, methods, compounds, or experiments described herein. In using such information or methods they should be mindful of their own safety and the safety of others, including parties for whom they have a professional responsibility.

To the fullest extent of the law, neither the Publisher nor the authors, contributors, or editors, assume any liability for any injury and/or damage to persons or property as a matter of products liability, negligence or otherwise, or from any use or operation of any methods, products, instructions, or ideas contained in the material herein.

Library of Congress Cataloging-in-Publication Data
A catalog record for this book is available from the Library of Congress

British Library Cataloguing-in-Publication Data
A catalogue record for this book is available from the British Library

ISBN: 978-0-12-819767-7

Publisher: Matthew Deans
Acquisitions Editor: Dennis McGonagle
Editorial Project Manager: Ana Claudia A.Garcia
Production Project Manager: Kamesh Ramajogi
Cover Designer: Matthew Limbert

Typeset by SPi Global, India

Contents

Contributors

Ahmed Abdelbary
Egyptian Government, Alexandria, Egypt

Sleptsova Sardana Afanasyevna
Chemistry Department, Institute of Natural Sciences, North-Eastern Federal University, Yakutsk, Russia

Nikiforov Leonid Aleksandrovich
Chemistry Department, Institute of Natural Sciences, North-Eastern Federal University, Yakutsk, Russia

Okhlopkova Aitalina Alekseevna
Chemistry Department, Institute of Natural Sciences, North-Eastern Federal University, Yakutsk, Russia

Sandro C. Amico
Laboratory of Polymeric Materials, Federal University of Rio Grande do Sul, Porto Alegre, RS, Brazil

P. Arulraj
Department of Mechanical Engineering, Einstein College of Engineering, Tirunelveli, Tamil Nadu, India

R. Arun Ramnath
Department of Mechanical Engineering, PSG College of Technology, Coimbatore, Tamil Nadu, India

C. Bennet
Department of Mechanical Engineering, Kalasalingam Academy of Research and Education, Krishnankoil, Tamil Nadu, India

Muthukumar Chandrasekar
School of Aeronautical Sciences, Hindustan Institute of Technology & Science, Chennai, Tamil Nadu, India

Chatchapol Chungchoo
Department of Mechanical Engineering, Faculty of Engineering, Kasetsart University, Bangkok, Thailand

C. Deepa
Department of Computer Science and Engineering, KIT-Kalaignarkarunanidhi Institute of Technology, Coimbatore, Tamil Nadu, India

A. Deiva Ganesh
Department of Mechanical Engineering, KPR Institute of Engineering and Technology, Coimbatore, Tamil Nadu, India

Nazanin Emami
Polymer Composites and Biotribology Group, Division of Machine Elements, Luleå University of Technology, Luleå, Sweden

Klaus Friedrich
Institute for Composite Materials (IVW GmbH), Technical University of Kaiserslautern, Kaiserslautern, Germany

R. Hemanth
Department of Mechanical Engineering, NIE Institute of Technology, Mysore; Visvesvaraya Technological University, Belagavi, Karnataka, India

M.R. Ishak
Department of Aerospace Engineering; Aerospace Malaysia Research Centre, Tower Block, University of Putra Malaysia, Serdang, Selangor, Malaysia

A. Jumahat
Faculty of Mechanical Engineering, Universiti Teknologi MARA, Shah Alam, Selangor, Malaysia

Subramaniam Karthikeyan
Department of Mechanical Engineering, Kalasalingam Academy of Research and Education, Krishnankoil, Tamil Nadu, India

D. Lenin Singaravelu
Department of Production Engineering, National Institute of Technology, Tiruchirappalli, Tamil Nadu, India

Yucheng Liu
Key Laboratory of Bionic Engineering (Ministry of Education, People's Republic of China); College of Biological and Agricultural Engineering, Jilin University, Changchun, People's Republic of China

Yunhai Ma
Key Laboratory of Bionic Engineering (Ministry of Education, People's Republic of China); College of Biological and Agricultural Engineering, Jilin University, Changchun, People's Republic of China

PON. Maheshkumar
Department of Mechanical Engineering, Nandha College of Technology, Perundurai, Tamil Nadu, India

K. Mayandi
Department of Mechanical Engineering, Kalasalingam Academy of Research and Education, Krishnankoil, Tamil Nadu, India

Yasser S. Mohamed
Faculty of Engineering, Department of Mechanical Engineering, Alexandria University, Alexandria, Egypt

Lazareva Nadezhda Nikolaevna
Chemistry Department, Institute of Natural Sciences, North-Eastern Federal University, Yakutsk, Russia

Panadda Niranatlumpong
National Metal and Materials Technology Center, Pathum Thani, Thailand

Peerawatt Nunthavarawong
Tribo-Systems for Industrial Tools and Machinery Research Center, Department of Teacher Training in Mechanical Engineering, Faculty of Technical Education, King Mongkut's University of Technology North Bangkok, Bangkok, Thailand

Jyotishkumar Parameswaranpillai
Department of Mechanical and Process Engineering, The Sirindhorn International Thai-German Graduate School of Engineering (TGGS), King Mongkut's University of Technology North Bangkok, Bangkok, Thailand

Vasilev Andrey Petrovich
Chemistry Department, Institute of Natural Sciences, North-Eastern Federal University, Yakutsk, Russia

Matuphum Phuthotham
Department of Mechanical and Process Engineering, The Sirindhorn International Thai-German Graduate School of Engineering (TGGS), King Mongkut's University of Technology North Bangkok, Bangkok, Thailand

Nagarajan Rajini
Department of Mechanical Engineering; Centre for Composite Materials, International Research Centre, Kalasalingam Academy of Research and Education, Krishnankoil, Tamil Nadu, India

S. Ramakrishnan
Faculty of Mechanical Engineering, School of Building and Mechanical Sciences, Kongu Engineering College, Perundurai, Tamil Nadu, India

M. Ramesh
Department of Mechanical Engineering, KIT-Kalaignarkarunanidhi Institute of Technology, Coimbatore, Tamil Nadu, India

Sanjay Mavinkere Rangappa
Department of Mechanical and Process Engineering, The Sirindhorn International Thai German Graduate School of Engineering, King Mongkut's University of Technology North Bangkok, Bangkok, Thailand

M.R. Sanjay
Department of Mechanical and Process Engineering, The Sirindhorn International Thai-German Graduate School of Engineering (TGGS), King Mongkut's University of Technology North Bangkok, Bangkok, Thailand

Prabakaran Saravanan
Assistant Professor, Department of Mechanical Engineering, BITS-Pilani, Hyderabad, Telengana, India

T.P. Sathishkumar
Department of Mechanical Engineering, Kongu Engineering College, Perundurai, Tamil Nadu, India

M. Chithirai Pon Selvan
School of Science and Engineering, Curtin University Dubai, Dubai, United Arab Emirates

Struchkova Tatiana Semenovna
Chemistry Department, Institute of Natural Sciences, North-Eastern Federal University, Yakutsk, Russia

Thiagamani Senthil Muthu Kumar
Department of Mechanical Engineering; Centre for Composite Materials, International Research Centre, Kalasalingam Academy of Research and Education, Krishnankoil, Tamil Nadu, India; Department of Mechanical and Process Engineering, The Sirindhorn International Thai-German Graduate School of Engineering (TGGS), King Mongkut's University of Technology North Bangkok, Bangkok, Thailand

Krishasamy Senthilkumar
Department of Mechanical Engineering; Centre for Composite Materials, International Research Centre, Kalasalingam Academy of Research and Education, Krishnankoil, Tamil Nadu, India; Department of Mechanical and Process Engineering, The Sirindhorn International Thai-German Graduate School of Engineering (TGGS); Center of Innovation in Design and Engineering for Manufacturing (CoI-DEM), King Mongkut's University of Technology North Bangkok, Bangsue, Bangkok, Thailand

H. Sharudin
Faculty of Mechanical Engineering, Universiti Teknologi MARA Cawangan Pulau Pinang, Permatang Pauh, Pulau Pinang, Malaysia

Suchart Siengchin
Department of Mechanical and Process Engineering, The Sirindhorn International Thai-German Graduate School of Engineering (TGGS), King Mongkut's University of Technology North Bangkok, Bangkok, Thailand

I. Siva
Department of Mechanical Engineering; Centre for Composite Materials, International Research Centre, Kalasalingam Academy of Research and Education, Krishnankoil, Tamil Nadu, India

D. Sivakumar
Department of Mechanical Engineering, Technical University of Malaysia Melaka, Melaka, Malaysia

B. Suresha
Visvesvaraya Technological University, Belagavi; Department of Mechanical Engineering, The National Institute of Engineering, Mysore, Karnataka, India

A.A.A. Talib
Faculty of Mechanical Engineering, Universiti Teknologi MARA, Shah Alam, Selangor, Malaysia

Jin Tong
Key Laboratory of Bionic Engineering (Ministry of Education, People's Republic of China); College of Biological and Agricultural Engineering, Jilin University, Changchun, People's Republic of China

Thitinun Ungtrakul
Department of Mechanical and Process Engineering, The Sirindhorn International Thai-German Graduate School of Engineering (TGGS), King Mongkut's University of Technology North Bangkok, Bangkok, Thailand

P.S. Venkatanarayanan
School of Aeronautical Sciences, Hindustan Institute of Technology & Science, Chennai, Tamil Nadu, India

R. Vijay
Department of Production Engineering, National Institute of Technology, Tiruchirappalli, Tamil Nadu, India

J.T. Winowlin Jappes
Department of Mechanical Engineering, Kalasalingam Academy of Research and Education, Krishnankoil, Tamil Nadu, India

T.G. Yashas Gowda
Department of Mechanical and Process Engineering, The Sirindhorn International Thai-German Graduate School of Engineering (TGGS), King Mongkut's University of Technology North Bangkok, Bangkok, Thailand

Krittirash Yorseng
Department of Mechanical and Process Engineering, The Sirindhorn International Thai-German Graduate School of Engineering (TGGS), King Mongkut's University of Technology North Bangkok, Bangkok, Thailand

N.W.M. Zulfikli
Department of Mechanical Engineering, Faculty of Engineering, University of Malaya, Kuala Lumpur, Malaysia

Acknowledgment

The editors highly acknowledge the funding of the German Academic Exchange Organization (DAAD) (Project Number: 1700100000) and the Thailand Research Fund (TRF) on a common tribology research project through which this book collaboration was initiated.

CHAPTER 1

Outline to tribology of polymer composites

Sanjay Mavinkere Rangappa[a], Jyotishkumar Parameswaranpillai[a], Suchart Siengchin[a], and Klaus Friedrich[b]
[a]Department of Mechanical and Process Engineering, The Sirindhorn International Thai-German Graduate School of Engineering (TGGS), King Mongkut's University of Technology North Bangkok, Bangkok, Thailand
[b]Institute for Composite Materials (IVW GmbH), Technical University of Kaiserslautern, Kaiserslautern, Germany

Composites are structural materials that consist of two or more chemically distinct constituents that are combined at the macroscopic/nanoscopic level, with a distinct interface that separates them. In other words, composites contain two phases, i.e., a discontinuous phase and a continuous phase. The discontinuous phase is dispersed in the continuous phase. The discontinuous phase is usually harder and stronger so that it is termed the *reinforcement*, and the one in which it is embedded, i.e., the continuous phase, is termed the *matrix*. The reinforcement phase may be in the form of fibers, particles, or flakes. The matrix material can be metallic, polymeric, and ceramic. Some of the properties of composite materials are their high specific strength and modulus, thermal properties, corrosion resistance, wear resistance, and fatigue life [1–5].

Composite materials can be classified based on the geometry of the reinforcement or by the type of matrix. Depending on the types of reinforcement, composite materials can be classified as particulate composites, flake composites, and fiber composites. Particulate composites consist of a polymer matrix with very small particles, sometimes with a size of less than 0.25 μm. The particles can have any shape or size and may be oriented randomly in the polymer matrix. Polymer composites with metal particles or glass spheres are examples for particulate composites. They have higher physical, mechanical, and thermal properties compared to the pure matrix materials. Flake composites consist of flat reinforcements. Common examples of flake reinforcements are glass, mica, silicon carbide, graphite, etc.

Tribology of Polymer Composites
https://doi.org/10.1016/B978-0-12-819767-7.00001-3

Flake composites exhibit high modulus, strength, and thermal properties. Fiber composites consist of discontinuous or endless fibers in the continuous matrix. Fiber-reinforced composites are further classified into unidirectional fiber composites, bidirectional composites, discontinuous fiber composites, and woven fiber-reinforced composites. Recently, natural fibers are often preferred instead of synthetic fiber reinforcements due to their environmental friendliness, easy availability, cheapness, and lower density. The fiber-reinforced composites have more strength, modulus, and good elongation at break. In laminated composites, layers of preimpregnated fibers (so-called prepregs) are stacked one over the other at various angles to form a multilayer laminate [6–10].

Based on the matrices, the composites are classified into three categories such as polymer matrix composites (PMC), metal matrix composites (MMC), and ceramic matrix composites (CMC). The PMC can be further classified into thermosetting, thermoplastic, and rubber composites. Recent research on PMC surpasses MMC and CMC due to their lightweight, good thermomechanical properties, lower cost, and easy processability. The most commonly used thermoplastics are polypropylene, polystyrene, polyethylene, polyamide, high-impact polystyrene, polyetheretherketone, etc. Common examples of thermosets are phenol-formaldehyde resin, epoxy resins, polyesters, etc. Natural rubbers, polyisoprene, polybutadiene, chloroprene, butyl rubber, ethylene-propylene rubber, silicone rubbers, etc., are examples of elastomers.

For the advancement of environmentally friendly polymer composites, the development of new materials is always preferred. Natural fibers are one of such materials that are lightweight, environmentally friendly, and of low cost. Sometimes natural fibers are also preferred over synthetic fibers because of their nontoxicity. Some examples of natural fibers are sisal, jute, kenaf, banana, bamboo, hemp, flax, etc. Studies have shown that continuous fiber reinforcements usually provide better mechanical properties compared to discontinuous fiber or particle reinforcements [11–15]. The polymer-based composites are produced by various manufacturing processes depending on the types of polymer composites. The short fiber and nanoparticle-based polymer composites are developed by common manufacturing processes such as injection molding, compression molding, sheet forming, thermal stamping, and thermoplastic filament winding. The thermoset-based composites are fabricated by resin transfer molding, vacuum-assisted resin transfer molding, compression resin transfer molding, pultrusion, autoclave processing, and filament winding [2, 16, 17].

"Tribology" is the study of wear, friction, and lubrication of interacting surfaces in relative motion. Tribology is an interdisciplinary subject; it is connected with mechanical engineering, material science, physics, chemistry, biology, and mathematics [18]. Great attention was focused on the study of the tribology of polymeric composites due to the increased replacement of metals and alloys in tribological components. This is because polymer composites are lightweight, corrosion resistant, and they also have a lower friction coefficient compared with materials based on metals/alloys [19, 20]. Self-lubricating properties of polymeric composites in industrial machinery can prevent contamination of the use of lubricating oils or greases [21, 22]. Friction refers to the resistance to movement when a solid surface is in contact with another solid surface. This is due to electromagnetic interaction between the charged particles present in either of the solid surfaces. There are two types of friction, static and kinetic friction. Static friction is the force that keeps the object at rest or in other words it is the minimum force needed to move the object. On the other hand, kinetic friction is the friction that is generated between moving partners. Wear is the surface damage, surface deformation and/or removal of material from the surface, all generated between the moving surfaces [23–26]. Lubrication is the use of oil or grease to reduce friction and wear. The failures related to wear and friction lead to high costs for industries. However, wear seems to be more critical than friction, since it can cause catastrophic and operational failures that can negatively affect productivity and cost [27]. Various parameters of polymer composites, namely, surface roughness, surface tension, surface wettability, and physical and chemical interactions of lubricants have a significant influence on the tribological performance of the materials [28–30]. Therefore, the study of friction, wear, and lubrication is of high interest in industries as it reduces the waste of resources because of friction and wear losses. In the laboratory, a tribometer is used for the analysis of friction and wear of the mating components. The common types of tribometers are pin on disc, four ball testers, reciprocating sliding friction and wear test, falex pin and vee block tester, high-temperature tribometer, low-temperature tribometer, vacuum tribometer, micro/nano tribometer, air-jet erosion test, bio tribometer, etc. [31–34]. The main tribological applications of polymer composites include gears, piston parts, brakes, clutches, axle box, bearings, submarine components, turbines, textile machines, filament wound bushings, shock absorbers, pulleys, oil pump parts, door hinges, knee, hip joint replacements, etc. [35–39].

References

[1] S. Mazumdar, Composites Manufacturing: Materials, Product, and Process Engineering, CRC Press, 2001.
[2] S.T. Peters (Ed.), Handbook of Composites, Springer Science & Business Media, 2013.
[3] F.C. Campbell Jr. (Eds.), Manufacturing Processes for Advanced Composites, Elsevier, 2003.
[4] J.M. Henshaw, Recycling and disposal of polymer-matrix composites, in: ASM Handbook, vol. 21, 2001, pp. 1006–1012. https://doi.org/10.31399/asm.hb.v21.9781627081955.
[5] A.K. Kaw, Mechanics of Composite Materials, Taylor & Francis Group, Boca Raton, 2006.
[6] D. Gay, Composite Materials: Design and Applications, CRC Press, 2014.
[7] S.V. Hoa, Principles of the Manufacturing of Composite Materials, DEStech Publications, Inc., 2009.
[8] G. Staab, Laminar Composites, Butterworth-Heinemann, 2015.
[9] E.M. Sozer, S.G. Advani, Process Modeling in Composites Manufacturing, CRC Press, 2010.
[10] V.V. Vasiliev, E.V. Morozov, Mechanics and Analysis of Composite Materials, Elsevier, 2001.
[11] F.C. Campbell, Structural Composite Materials, ASM International, 2010.
[12] B.T. Astrom, Manufacturing of Polymer Composites, Routledge, 2018.
[13] S.K. Malhotra, K. Goda, M.S. Sreekala, Part one introduction to polymer composites, Polym. Compos. 1 (2012) 1–2.
[14] R. Wool, X.S. Sun, Bio-Based Polymers and Composites, Elsevier, 2011.
[15] D. Bhattacharyya, Fakirov, S. (Eds.), Synthetic Polymer-Polymer Composites, Carl Hanser Verlag GmbH Co KG, 2012.
[16] S.G. Advani, Hsiao, K.T. (Eds.), Manufcturing Techniques for Polymer Matrix Composites (PMCs), Elsevier, 2012.
[17] F. Hussain, M. Hojjati, M. Okamoto, R.E. Gorga, Polymer-matrix nanocomposites, processing, manufacturing, and application: an overview, J. Compos. Mater. 40 (17) (2006) 1511–1575.
[18] K. Friedrich, Polymer composites for tribological applications, Adv. Ind. Eng. Polym. Res. 1 (1) (2018) 3–39.
[19] K. Friedrich, L. Chang, F. Haupert, Current and future applications of polymer composites in the field of tribology, in: Composite Materials, Springer, London, 2011, pp. 129–167.
[20] E. Omrani, P.L. Menezes, P.K. Rohatgi, State of the art on tribological behavior of polymer matrix composites reinforced with natural fibers in the green materials world, Eng. Sci. Technol. Int. J. 19 (2) (2016) 717–736.
[21] W. Luo, Q. Liu, Y. Li, S. Zhou, H. Zou, M. Liang, Enhanced mechanical and tribological properties in polyphenylene sulfide/polytetrafluoroethylene composites reinforced by short carbon fiber, Compos. Part B: Eng. 91 (2016) 579–588.
[22] S.W. Zhang, State-of-the-art of polymer tribology, Tribol. Int. 31 (1–3) (1998) 49–60.
[23] A.D. Moghadam, E. Omrani, P.L. Menezes, P.K. Rohatgi, Mechanical and tribological properties of self-lubricating metal matrix nanocomposites reinforced by carbon nanotubes (CNTs) and graphene—a review, Compos. Part B: Eng. 77 (2015) 402–420.
[24] E. Omrani, A.D. Moghadam, P.L. Menezes, P.K. Rohatgi, Influences of graphite reinforcement on the tribological properties of self-lubricating aluminum matrix composites for green tribology, sustainability, and energy efficiency—a review, Int. J. Adv. Manuf. Technol. 83 (1–4) (2016) 325–346.
[25] A.A. El-Sayed, M.G. El-Sherbiny, A.S. Abo-El-Ezz, G.A. Aggag, Friction and wear properties of polymeric composite materials for bearing applications, Wear 184 (1) (1995) 45–53.

[26] W. Brostow, V. Kovačevic, D. Vrsaljko, J. Whitworth, Tribology of polymers and polymer-based composites, J. Mater. Educ. 32 (5) (2010) 273.
[27] K. Holmberg, A. Erdemir, Influence of tribology on global energy consumption, costs and emissions, Friction 5 (3) (2017) 263–284.
[28] I. Hutchings, P. Shipway, Tribology: Friction and Wear of Engineering Materials, Butterworth-Heinemann, 2017.
[29] K. Kato, Wear in relation to friction—a review, Wear 241 (2) (2000) 151–157.
[30] K. Friedrich (Ed.), Friction and Wear of Polymer Composites, vol. 1, Elsevier, 2012.
[31] G. Stachowiak, A.W. Batchelor, Experimental Methods in Tribology, vol. 44, Elsevier, 2004.
[32] P.L. Menezes, M. Nosonovsky, S.P. Ingole, S.V. Kailas, M.R. Lovell (Eds.), Tribology for Scientists and Engineers, Springer, New York, 2013, pp. 295–389.
[33] S. Wen, P. Huang, Principles of Tribology, John Wiley & Sons, Hoboken, NJ, 2012, pp. 135–142.
[34] V.A. Bely, A.I. Sviridenok, M.I. Petrokovets, Friction and Wear in Polymer-Based Materials, Elsevier, 2013.
[35] N.K. Myshkin, S.S. Pesetskii, A.Y. Grigoriev, Polymer tribology: current state and applications, Tribol. Ind. 37 (3) (2015).
[36] K.B.A.U.K. Friedrich, Z.B.A.U.K. Lu, A.M. Hager, Recent advances in polymer composites' tribology, Wear 190 (2) (1995) 139–144.
[37] D.L. Burris, B. Boesl, G.R. Bourne, W.G. Sawyer, Polymeric nanocomposites for tribological applications, Macromol. Mater. Eng. 292 (4) (2007) 387–402.
[38] Z. Rymuza, Tribology of polymers, Arch. Civ. Mech. Eng. 7 (4) (2007) 177–184.
[39] J.K. Lancaster, Polymer-based bearing materials: the role of fillers and fibre reinforcement, Tribology 5 (6) (1972) 249–255.

CHAPTER 2

Recent advances in tribology of hybrid polymer composites

M. Ramesh[a], A. Deiva Ganesh[b], and C. Deepa[c]

[a]Department of Mechanical Engineering, KIT-Kalaignarkarunanidhi Institute of Technology, Coimbatore, Tamil Nadu, India

[b]Department of Mechanical Engineering, KPR Institute of Engineering and Technology, Coimbatore, Tamil Nadu, India

[c]Department of Computer Science and Engineering, KIT-Kalaignarkarunanidhi Institute of Technology, Coimbatore, Tamil Nadu, India

1 Introduction

Tribological science is important to improve the performance of components and moving parts. It involves studying and applying friction, wear adhesion, surface fatigue, and lubrication principles among lubricating elements. It is interdisciplinary in nature and covers a number of fields, including physical-chemical sciences, materials science, and engineering. Failures associated with wear and friction lead to heavy industry costs. It is estimated that wear and friction losses will exceed 63% of the total cost incurred by the industries for any major component of a mechanical system [1]. The research on this topic is very important in order to evaluate the failure of many auto and industrial components. It is also necessary to replace the traditional material with composite materials in order to work with high load, high temperature, high corrosiveness, and high pressure. The researchers have tried to study the effects of tribology in sectors such as transport, manufacturing, power generation, etc. Around 23% of the total energy consumption in the world comes from tribological contacts. Due to wear and related failures, 20% is used to overcome friction and 3% is used to remanufacture worn out and spare parts. Advanced tribological technologies can reduce the global emissions of carbon dioxide by roughly 1460×10^6 kg [2].

The demand for hybrid composite materials is increasing because of their characteristics such as low weight, excellent strength-to-weight ratio, corrosion behavior, fatigue resistance, high self-lubrication, noiseless operation,

Tribology of Polymer Composites
https://doi.org/10.1016/B978-0-12-819767-7.00002-5

etc. [3–5]. A tribological system has operating inputs such as the type of motion, load, velocity, and temperature as well as functional outputs such as motion, power, materials, etc. [6]. Components made of fiber-reinforced polymer composites are in many circumstances under tribological conditions of loading [7]. The literature has shown that the tribological quality evaluation of synthetic fiber-based polymer composites has long been derived on the friction and wear properties of these composites [7, 8]. Valasek et al. [9] prepared alkali solution-treated coir fiber-reinforced composites and examined the influence of surface treatment on the composites' mechanical and abrasive wear properties. This study concluded that coir fiber chemistry could enhance interaction with the matrix and increase the fiber tensile strength. In contrast, the composite containing coir fibers that were treated had the highest resistance to wear as well as improved porosity and hardness. The hot pressing process was used to manufacture polymer composites reinforced by raw maize stalk fiber. The effect of the concentration of silane solution on the tribological and morphological properties of the polymer composites was investigated [10]. The results obtained will direct the use of agricultural waste from maize stalks and the development of new or improved composite manufacturing technology.

Because the composite selection and surface processing heavily affect the wear and tear, this is the important part of tribology. For their optimal functioning, the long-term reliability of components and devices, and their feasibility, a good grasp of interacting surfaces is needed. When a new composite is formed and used as a surface with more friction, wear, and tear, tribological testing plays an important role in determining the material's ability to withstand friction and wear. With tribological packed composites, the coefficient of friction, the load-carrying capacity, and the wear frequency are the important factors to consider. The working conditions, the quality of the material, the lubricants, and the atmosphere are some of the areas to be considered. Due to various types of loads during motion, many engineering components such as brake shoes, bearings, gears, wheels, clutches, and shafts are subject to friction and wear. Therefore, the material's tribological behavior is essential for the manufacture of components based on tribology. Fig. 1 shows the causes of friction and wear [11].

The performance, cost, and ecofriendliness are some of the factors for developing a frictional material. The use of asbestos fiber composites as the dominant frictional materials in brake pads, linings, brake couplings, etc., was banned due to adverse environmental and human health problems. The use of natural fibers as reinforcements in composites of thermoplastic and

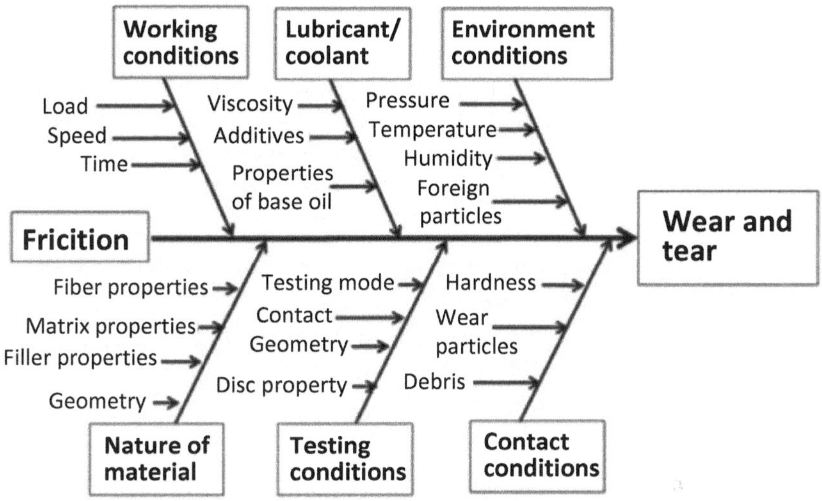

Fig. 1 Friction and wear causes.

thermoset matrices offers ecological benefits in terms of disposability and the efficient use of raw materials [12]. The key friction reduction methods are: (i) solid filler inclusion, (ii) internal lubrication, (iii) external lubrication, and (iv) generation of transfer film.

2 Tribological properties

Polymer matrix composites, metal matrix composites, ceramic matrix composites, and carbon-carbon composites are the various groups of tribological materials. Polymer matrix composites are commonly used in which tribological properties are of great value due to their properties of self-lubrication, wear resistance, and corrosion [13]. This is the one that uses polymer as the matrix and fiber as the reinforcement. The matrix materials distribute the applied load uniformly and transfer the loads to the fiber. Thermosetting polymers are used widely as adhesives, encapsulated materials, tooling compounds, and matrix materials for reinforced composites due to their high cross-linked molecular structures. The important considerations in the development of polymer composites are the fiber length, the orientation, the density of the fibers, and the fiber characteristics [14]. The fiber orientation plays a vital role in the development of composites. For example, fibers parallel to the tensile load may have good performance, but under tribological fiber loading, pullout take place. Hence, the orientation of the fibers at each loading condition needs to be studied for better composite performances.

Biocomposite stiffness and strength deficiencies can be overcome through structural configurations and better arrangement in the sense of placing fibers at specific locations for maximum strength performance. Some of the composite material properties depend on the matrix material's characteristics. Low wear resistance and high friction are the main drawbacks of polymer composite coatings [15]. When integrating nano- or microparticles and fibers into the matrix, these can be strengthened. In order to enhance tribological properties, filler materials such as glass, carbon, and other fibers are added. The choice of suitable fillers is critical in the design of composites used in tribological applications, as the higher friction coefficient does not only cause high friction resistivity. In tribological loading, the mechanical energy is transformed to thermal energy and part of this heat is conducted to the interface of the contacting solids. Poor thermal stability and thermal conductivity are the key constraints for polymer composites. This causes mechanical properties to change, even at 100°C. To eliminate this limitation, polymer matrix composites reinforced with ceramic powders are used. The ceramic powder reinforcement results in increased wear resistance and an extended temperature range [16].

Frictional heat from sliding surfaces is difficult to dissipate. During the rubbing process, it is also difficult to form a protective oxide film on contact surfaces. Because of these issues, adhesive wear can result in the failure of moving parts. When selecting the polymer matrix, it is also important to consider the working environment. Different polymers, when exposed to water or other liquids, have different resistances to air, sun, and swelling degradation. Another important parameter to consider is temperature. These composites play a vital role in bearing applications apart from structural applications. The impact of hybrid composites on tribological characteristics can be explored. Two or more reinforcements will be combined as hybrid composites through predetermined geometry and scale. As reinforcing elements, high-strength fibers such as glass, aramid, carbon, and kevlar fibers were used [17].

In response to environmental problems, the early part of the 21st century focused more on tribocomposite applications. For aerospace as well as marine and sports equipment, carbon fiber-reinforced composites are used extensively. These composites' sliding wear behavior depends on the matrix and reinforcement used in addition to operating conditions. It is found that the carbon fiber's self-lubricating property increases wear resistance by decreasing the friction coefficient. Carbon fiber is one of the latest reinforcement materials used in composites. It has high specific stiffness and strength.

As discussed, because of the self-lubricating property of the carbon fiber, no lubrication is needed. Hence, carbon polymeric materials are generally used in bearing materials [18].

Compared to untreated fibers, the tribological properties of natural fiber-reinforced composites treated on the surface are improved [19]. The efficient method of managing the interfacial adhesion between the fiber and the matrix is the combination of alkali and silane coupling agent treatments. The strength of the composites depends on the nature of the filler, the even distribution of filler, the spacing between the filler particles, and the interfacial interaction strength. The hybrid composite is the predominant material for the formation of biomaterial. Hybrid composites have two or more different reinforcements in a single matrix. In principle, several different fibers can be incorporated into a hybrid system, but in practice a combination of only two types of fibers would be more useful. Due to the high cost and brittleness of carbon fiber, one can use glass fibers along with carbon fiber to make a hybrid composite in order to overcome these limitations. Alumina (Al_2O_3) and titanium oxides (TiO_2) are used to improve the mechanical as well as tribological properties [20]. A minimal development effort, high rigidity, high strength, corrosion safety, and high temperature tolerances are important considerations for synthetic fiber-reinforced polymer composites as an appropriate material for modern applications. Testing is performed in many ways, and some of the important ones are: (i) dry sand rubber wheel test, (ii) pin-on-disc test, (iii) block-on-ring test, and (iv) pin-on-drum test.

2.1 Synthetic fiber-reinforced polymer composites

Polymer matrix composites made from conventional fibers can be made by the hand-layup technique, compression molding, and resin transfer molding. It has been studied that by the pultrusion process, the amount of the wear rate is controlled with an increase in fiber content. Also, the wear rate resistance varies based on the type of mode of wear and the fiber percentage. An experimental study was carried out among composites reinforced by glass, carbon, and aluminum particles. The results revealed that the tri-phase composite comprising all these fibers has the highest friction and wear resistance. It has also been stated that glass is better than steel in a comparative study of carbon and glass-reinforced composites [11]. The tribological properties of synthetic fiber-reinforced composites are presented in Table 1.

In a study, the effect of temperature on the friction and wear properties of phenolic composites was conducted. It was found that the rate of wear increases with the increase in temperature, irrespective of loading

Table 1 Tribological behavior of synthetic fiber reinforced composites [11].

S. no	Synthetic fiber	Matrix	Testing method	Major findings
1.	Potassium titanate	Polyether ether ketone/ Polytetrafluoroethylene	Pin-on-disc setup	The addition of potassium titanate above 5 wt% effectively reduces the friction and wear rate.
2.	Carbon/ionic liquid	Polyether ether ketone	Ring-on-disc tribometer	Carbon fiber creates effective holes to suck the ionic liquid, which results in reduction of friction and wear.
3.	Glass/Al_2O_3	Epoxy	Pin-on-disc setup	10 wt% of Al_2O_3 is found to be optimum proportion to enhance the resistance to friction and wear.
4.	Ferrous particles	Nylon/acrylonitrile butadiene-styrene (ABS)	Pin-on-disc setup	Fe-reinforced nylon offers higher wear resistance than ABS and the friction coefficient is higher for friction.
5.	Glass/polyethylene	Phenol	Block-on-ring test rig	The addition of polyethylene in microform at 5% improves the friction and wear resistance to a notable extent.
6.	Graphene	Polytetrafluoroethylene	Pin-on-disc setup	Wear rate of the composite starts to decline from 4% of fiber loading and a minimum wear rate is observed at 10% fiber loading.
7.	Glass/carbon, polytetrafluoroethylene, SiC, Al_2O_3 fillers	Trichloroethylene	Pin-on-disc setup	Wear rate elevates with a hike in the fiber/filler content but goes down with a hike in the grit size of SiC and sliding distance.
8.	SCF	Epoxy	Pin-on-disc setup	Addition of SCF in the matrix improves both the wear resistance and friction resistance significantly.
9.	Graphite, MoS_2	Polyethylene	Block-on-ring test rig	Addition of 3 wt% of graphite and 10 wt% of MoS_2 reduces the wear rate by two times and the friction coefficient by 1.5 times.

10.	Polytetrafluoroethylene	Polyethylene	Block-on-disc setup	Addition of 10 wt% of polytetrafluoroethylene enhances the wear resistance more than two times and alleviates the friction coefficient by two times.
11.	ZrB_2/Al_2O_3	Epoxy	Block-on-disc setup	Addition of ZrB_2 and Al_2O_3 fillers at 5 wt% reduces the friction coefficient, wear rate and frictional temperature.
12.	Glass	Polyetherimide	Pin-on-disc setup	30 wt% of glass reinforcement is the optimum proportion for reducing the wear rate.
13.	Woven glass/Al	Epoxy	Pin-on-disc setup	Addition of 10 wt% of Al particles and 60 wt% of glass gives minimum friction coefficient and maximum wear resistance.
14.	Glass/SiC, TiO_2, graphite	Epoxy	Pin-on-disc setup	A filler content of 5–7 wt% must be added to alleviate the wear rate

conditions. From the works of SiC fillers on aluminum composites, glass fiber epoxy composite with SiO_2 fillers, aluminum matrix with red mud fillers it is found that fillers play a vital role in controlling the wear. The operating conditions and material parameters are the main factors that affect the friction and wear properties [13]. The research in wear and friction studies of synthetic fiber composites is shown in Table 2.

Under three loads varying from 40 to 120 N and two sliding speeds of 2.51 to 3.14 m/s, the impact of the increase in load and sliding velocity of the glass fiber-reinforced polymer composites was found by using a pin-on-disc apparatus. There was weight loss, a rubbing coefficient, and changes in wear level due to the increase in both speed and load. The matrix wear and fiber debonding are the major reasons for the increase in wear rate. The tests were conducted to know the tribological properties of epoxy-based composites with fillers under dry sliding conditions. Abrasive tests on two distinct sorts of glass/epoxy composites were performed with the help of a tribometer. The wear behavior was observed by varying the sliding speeds and the connected load. The increase in the connected load and the size of the wear rate of the molecule of both composite increments were found. However, the wear rate decreased to 40% of the fiber loading with increasing sliding speed and then it began to rise. These results showed the effect of different sliding parameters on the tribological properties of polymer matrix composites reinforced by synthetic fiber.

The glass fiber examination was conducted on composites under three loads, three speeds, and three different surrounding conditions such as dry, oil, and inert. The results showed that the tribological properties are greater in the case of inert gas and less so in the condition of oil greased. The oil-greased condition prompts an improvement in tribological behavior due to the self-greasing property. The erosion and wear behavior of the polytetra-fluoroethylene composite was studied under ocean water as the surrounding condition. The carbon fiber and polyamide were used as the material for reinforcing and filling. Five distinct composites are produced by adjusting the volume ratio of the carbon fiber and polyamide. The addition of carbon fiber and polyamide strengthens the tribological properties and the volume fraction of 5% polyamide with 15% carbon fiber composite indicates a minimum coefficient of erosion and wear level among all the composites.

The study on the effects on the friction and wear behavior of different filler materials (short glass fiber, short carbon fiber, silicon oxide, and graphite) showed that incorporating filler materials has a positive effect on

Table 2 Wear and friction studies of synthetic fiber composites.

Matrix	Fiber	Filler	Variables	Manufacturing technique	Coefficient of friction	Wear rate	Observed wear mechanism
Polyester	Glass	–	FL, L	Molding	Decrease	Decrease	Plastic deformation
Polyester	Al	SiC	L, FVF	Powder	–	Decrease	–
PEI	Glass	Graphite, MoS_2,	L	–	–	Decrease	Adhesion
Epoxy	Carbon	–	L, SD	Resin transfer molding	Decrease	Decrease	Worn surfaces
Phenolic	Graphite	–	L, T	Compression molding	High T high FF	High T high wear	–
Epoxy	Carbon	–	L, SV, sliding time, FVF	Hand lay up	–	Decrease	–
Polyester	Glass	–	Speed, FO	Hand lay up	Decrease	Decrease	Back transfer polymer film
Epoxy	Rubber dust	–	Speed, L, time	Molding	–	Decrease	Abrasive
Epoxy	Glass	SiO_2	L, filler content	Cylindrical mold	Decrease	Decrease	Adhesion
Epoxy	Glass	SiC	SD	Hand lay up	Decrease	Decrease	Debonding, fiber breakage
Epoxy	Glass	Graphite and SiC	SV, L, SD	Hand lay up	–	Decrease	–

wear behavior and provides a lower friction coefficient. By increasing carbon black as a filler, the wear in the polymer composites was reduced at a specified impact velocity. Under the influence of electrical current and varying speeds from 10 to 60 m/s, the tribological behavior of carbon fiber-reinforced composites was studied. Three different fiber orientation types were taken: vertical, parallel, and regular. At high speed, the composite's normal orientation shows better wear resistance, but not at low speed. It was also found that the wear rate was due to high-speed erosion and low-speed matrix elimination. The wear resistance and friction coefficient of glass fiber-reinforced composites were studied for three different orientations, cross-laminar, regular, and parallel, under varying loads (4–22 N) and varying sliding speeds from 0.05 to 1.2 m/s. The results showed that the coefficient of friction is minimal in the direction of cross-laminar and peak in the direction of parallel. Because of the rapid exposure of new fibers on the counter surface with load variation and sliding speed, the friction coefficient in the parallel orientation is the highest.

On glass fiber-reinforced polymer composites under different glass fiber content and varying fiber length (1.5–3 mm), research on the effect of fiber volume fraction and fiber length on tribological properties was performed. The results showed that the wear rate of the longest fiber was lower than that of the short fiber. A composite with 5.7 wt% showed better tribological behavior at varying load and sliding distances. This results showed the effect of the surrounding environment, fiber orientation, fiber fraction, and length on the tribological properties of polymer composites reinforced by synthetic fiber [14].

Wieczorek et al. [16] carried out the experiments on an effective utilization of materials incorporated with silver nanopowders, ceramics, and polymer for the production of polytetrafluoro-ethylene composites. The utilization of the silver-bearing waste material Al_2O_3 or SiC ceramics as the polymer composite reinforcement at low cost creates a base for an efficient postproduction waste management system. The material is a mixture of silver powders (10%–50%), SiC and Al_2O_3 powders (20%–50%), and polymer powders (20%–50%). The disc-shaped specimen with a diameter of 30 mm and the circular rod specimen with a diameter of 6 mm were obtained by the press molding method. The tests were conducted under dry sliding conditions using the pin-on-block equipment. It was determined from the results that the tribological properties depend on the reinforcement quantity. The results showed a higher friction coefficient with an increment in reinforcement fraction. The results did not affect the friction coefficient for a further increase in reinforcement fraction. The alloy with a higher quantity of reinforcement results in greater resistance to abrasion. The increases in wear resistance are more distinct with increased reinforcement than improvements in the friction coefficient. It was therefore discovered that the main factor affecting the coefficient of friction is the matrix material, whereas the level of wear depends on the strength. The composites obtained may be used for producing tribo-systems that work under dry sliding conditions.

Based on the study of the effect of carbon nanotube inclusion, it was observed that due to the addition of nanoparticles, there is a decline in the tribological parameters. The introduction of nanoparticles also shifts the wear process from the wear mechanism found in the pure polymer to abrasive wear. The type of polymeric matrix material used in a composite plays a major role on the type of wear produced. It was reported that the composites produced from polycarbonate, polymethyl methacrylate, polyethylene terepthalate, polyamide, polyether ether ketone, polyvinylidene fluoride, polytetre fluroethylene, and polyethylene resins exhibited abrasive loss; otherwise, the composites produced from polypropylene and polyamide resins produced adhesive wear. The results of a comparative study among different types of polymer resin composites showed that a very low friction coefficient was provided by these composites. Polyoxymethylene composites were also stated to have the highest resistance to wear. The temperature of carbonization and the proportion of fiber also play important roles in wear and friction. The study revealed that the optimal wear rate conditions are a carbonization temperature of 950°C and a fiber content of 30%. The wear rate has been stated to be directly affected by the pressure applied and the sliding length, but as the pressure goes above 5 MPa, the wear rate begins to decline due to the formation of a transfer film at the interface. With the stress applied, the formation of transfer film increases and decreases wear. It has been stated from the investigations carried out on mica-filled polyamide and glass-reinforced polysulfone that the coefficient of friction is highly dependent on the material and not on the load applied.

2.2 Natural fiber-reinforced polymer composites

The best alternatives to synthetic fibers are organic or plant-based fibers to reduce the rise in dependence on fossil fuels. One of the main areas of research is the tribological characterization and application of natural components. Fig. 2 shows the number of research articles in synthetic and natural fiber tribology over the past decade.

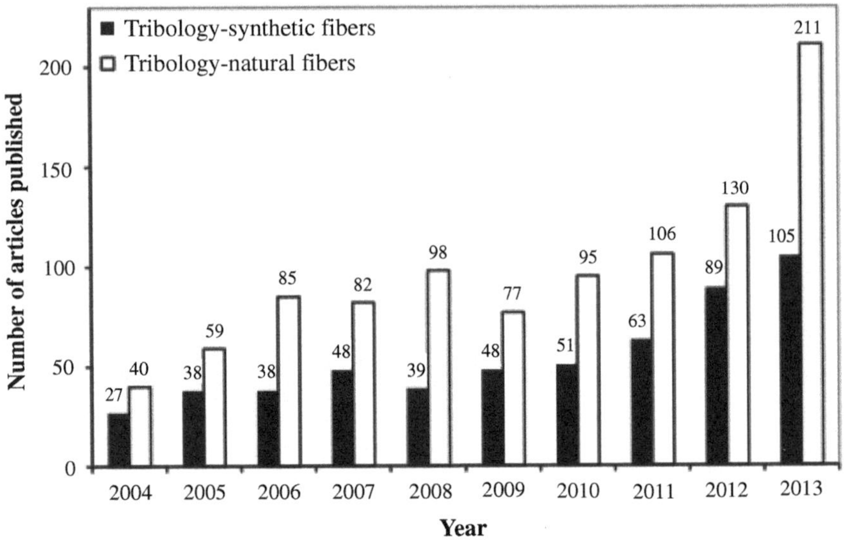

Fig. 2 Papers published on the tribological characteristics of synthetic and natural fibers.

The cellulose content plays a vital role in deciding the properties of the natural fibers. Also, a lower lignin content is responsible for the degradation of fibers, which takes place at a higher processing temperature. In the study of borassus fruit fibers and woven coconut sheath fibers, it was observed that the alkali-modified fibers improved the wear characteristics compared to the untreated fibers. The study was conducted on the effect of plasma treatment on the properties of phenolic composites reinforced by cotton fiber. For the fibers treated under air plasma and nitrogen plasma mode, the impact of treatment time was examined. It has been documented that fibers treated in air plasma mode showed better wear and friction than fibers treated with nitrogen. The friction properties of sisal fibers were found out based on different methods of NaOH, aminosilane, and borax treatments. The fibers treated with borax exhibited superior heat absorption with low wear rates and a consistent friction coefficient. It was also found that sisal fibers exhibited better wear and moderate friction performance [1].

The crop absorbs nitrogen and phosphorus from the soil, carbon dioxide, easily recyclable and light weight, based on a study of kenaf fibers. The wear resistance in kenaf fiber-reinforced composites was reported to be higher compared to plain polymer. Normal directed fiber composites also showed better wear quality in kenaf fiber. The wear mechanism was microcracking, with no fiber pullout suggesting a strong interfacial adhesion between the fiber and matrix, studied from the surface morphology. The tests on kenaf fiber-reinforced composites showed better friction wear quality than cotton, oil palm, jute, sugar cane, and even glass fiber composites.

Research on the wear and frictional properties of polymer composites reinforced by sugarcane fiber was performed to compare with composites reinforced by glass fiber. It has been concluded that composites of sugarcane have strong potential and are equivalent to composites of glass fiber. Wear of the unidirectional sugarcane fiber-reinforced polymer composite is decreased as load increased. In abrasive tests, wear for randomly oriented sugarcane fiber reinforced polymer composite increased with applied load due to the penetration of less abrasive particles on the composite surface at lower speeds. Compared to antiparallel sugarcane fiber composites, higher wear was observed on parallel-oriented sugarcane fiber composites because the debris produced is easily removed due to the free wear path in parallel-oriented fiber composites. The microcutting motion is limited in antiparallel-oriented fiber composites due to the movement of the abrasive particles through various fiber layers and the matrix. The results showed that polymer composites reinforced by unidirectional sugarcane fiber gave better abrasive wear than randomly oriented fiber composites [21].

Factors such as particle size, volume fraction, and surface treatment affect the tribological activity in particulate-reinforced composites. Particle size plays a crucial role in the composite's strength. This results in low wear levels and an improved friction coefficient with increased particle content and decreased particle size. Increasing reinforcement particles to more than 30% increases the composite friction coefficient. The volume fraction is the proportion of the number of particles in the whole volume. The banana peel particle experiments showed that with an increase in wt%, the mechanical and tribological properties increased. The samples are packed in phenolic resin and rice husk in epoxy with different formulations of banana peel powder. The results showed that interfacial reaction, uniform distribution, and no pores are observed and the wear rate decreases with a wt% increase in particulate fraction. In a composite material, surface treatment will strengthen the poor adhesion and low interlaminar shear strength between the particles and matrix. Particles immersed for lighter liquids are documented to give better results compared to liquids with high viscosity. Likewise, fiber orientation, length, volume fraction, surface treatment, and aspect ratio are the major factors responsible for the tribological properties of composites. Experiments on agricultural fibers such as cotton, maize, and sugarcane have been performed. The results show that there was a slight increase in the friction coefficient and a significant decrease in wear with the increase in fiber content.

In strengthening and toughening the polymer matrix, the function of fiber reinforcement is critical. It is stated that even small amounts of fibers will dramatically increase the strength of composites. The study on epoxy composites enhanced by borassus fiber indicated that the shorter length of the fiber gives better tribological properties. The fiber's geometric aspects depend on their length and diameter. The results of aramid fiber investigations showed that the friction and wear rate coefficient first increase and then decrease at defined speeds and loads with the increase in aspect ratio. The tests were carried out without taking humidity into account. Nonetheless, the main factor influencing the porosity of the composites is humidity, which affects tribological properties [22]. Chand and Dwivedi [23] documented the effect of adding maleic anhydride-grafted polypropylene (MA-g-PP) to the abrasive wear resistance of polypropylene composites reinforced by jute fiber. The treated and untreated specimens were tested on differential scanning calorimetry (DSC) for their melting point. Polymer wear proceeds in steps such as matrix wear, fiber wear, fiber fracture, and fiber-to-matrix interfacial debonding. Adding MA-g-PP as a coupling agent dominated the other mechanisms by interfacial debonding. With the introduction of the binding agent MA-g-PP, the wear properties of jute fiber-reinforced polypropylene composites are strengthened due to reduced fiber pullout and less fiber debonding.

Yosuke et al. [24] researched the effects of a silane coupling agent on the tribological properties of plant-derived polyamide composites reinforced with hemp fibers. The samples are treated with two types of surface treatments, sodium hydroxide alkali treatment and silane binding agent treatment. Three types of silane coupling agents, aminosilane, epoxysilane, and ureidosilane, have been used. A ring-on-plate sliding wear tester was used to measure the tribological properties of composites. The constant normal load and constant sliding velocity showed the best effect on the tribological properties of all the composites. The effect of the volume fraction of the ureidosilane coupling agent-treated hemp fiber composites was different due to the change in the wear mechanism by orientation and fiber dispersion. It was concluded that the ureidosilane coupling agent had the best effect on properties such as friction coefficient, specific wear rates, and limits of pressure and speed.

Due to the incorporation of reinforcement, the tribological properties of polymers can improve or deteriorate [8, 25]. Different studies have been carried out on the tribological analysis of natural fiber-reinforced polymer composites such as betelnut fiber-reinforced polyester composites [25, 26], polyester-reinforced sugarcane fiber composites [22], kenaf fiber-based epoxy composites [27], cotton with polyester resin composites [28], sisal fiber-based phenolic resin composites [29], etc. El-Tayeb [30] researched the possible use of polymer composites for tribological applications reinforced by sugarcane fiber. The composites are prepared using the hand-lay-up method, unsaturated polyester resin is used as a matrix, and methyl ethyl ketone peroxide is the base for processing composites. Using a pin-on-disc tribological test device, adhesive wear tests were performed. It was found that sugarcane fiber composites give strong wear resistance with increased load and friction coefficient compared to glass fiber composites. It was also found that the unidirectional sugarcane fiber-reinforced polymer composite had the lowest friction coefficient when measured in an antiparallel orientation and the highest in parallel orientation, indicating that the sugarcane fiber in an antiparallel orientation could be competitive with glass fiber. From the microscopic observation, it was shown that the sugarcane fibers are capable of having a good bond with the polyester matrix. This made it more difficult to separate the fibers from the composite and thus improve the wear resistance of sugarcane fiber composites.

The investigations on the plant fibers such as sisal, kenaf, cotton, palm tree fiber, bamboo, and jute addressed the following issues [31]: (i) operating parameters influence the frictional and wear behavior, (ii) though the polymer reinforcement with plant fibers improves the wear resistance, this is not true for the composites, (iii) the addition of solid lubricants such as graphite can reduce the friction coefficient and maintain high wear resistance, and (iv) the wear performance of natural polymer composites is affected by using water as a lubricant due to the transfer of adhesive mode into three body abrasion wear modes. Because they can provide good interfacial bonding between the fiber and the matrix, fiber treatments are important. Fiber orientations also play a vital role in improving the composites' wear and friction properties. Normal orientation of the fibers is the optimal orientation against the sliding condition for the greatest tribological properties [32]. It was found that when used as a reinforcement in polyester composites, surface modification on coir fibers showed better interfacial adhesion. Compared to untreated carbon fiber-reinforced composites, they displayed greater wear resistance and low friction [33]. Table 3 shows the different studies in the field of wear and friction analysis.

Table 3 Analysis of friction and wear of composites reinforced with natural fibers [13].

Polymer	Fiber	Fillers	Variables	Manu. technique	Friction coefficient	Wear rate	Observed wear
Epoxy	*Lantana camara*	–	FVF, L, SD	Molding	⇩	⇩	Abrasion
Epoxy	Kenaf	–	FO, L, SD, SV	Molding	⇧	⇧	Adhesion
Epoxy	Pin wood dust	–	FVF, SD, SV, L	–	⇔	⇩	–
Polyester	Betelnut fruit fiber	–	SD, L	Hand lay up	⇧	⇩	Macro and microcracks, debonding and fiber pullout
Epoxy	sugarcane	–	FO, L	Hand lay up	⇔	⇩	Microcutting and microploughing
Polyester	Linen and jute	–	FVF	Molding	⇧	⇩	Fiber debonding
Polyester	Coir	Alkaline	SD, L	Hand lay up	⇧	⇩	Adhesion
Epoxy	Borassus fruit fiber	Alkaline	L, SV, FL	Cylindrical mold	⇩	⇩	Microcracks
Polyester	Untreated and treated oil palm	Alkaline	SD, L	Cylindrical mold	⇩	⇩	Debonding
Epoxy	Bamboo	–	L, FO, SD	Hand lay up	⇩	⇩	Back film transfer
Epoxy	Rice husk	Benzolylation	L, FVF	Cylindrical mold	⇔	⇩	Abrasive
Polyester	Bamboo powder	–	FVF, SD	Cylindrical mold	⇔	⇩	Abrasive
Epoxy	Coir	Benzolylation	FVF, L, speed	Cylindrical mold	⇔	⇩	Abrasive
Polyurethane	Kenaf	Alkaline	FO, L, SD	Hand lay up	⇩	⇩	Plastic deformation
Epoxy	Recycled cellulose fiber	Nano SiC nano clay	–	–	⇔	⇩	–

Devadas et al. [34] investigated the tribological properties of epoxy resin enhanced by kenaf fiber particles. A dry sliding test was performed on a pin-on-disc system at 2.8 m/s for 6.72 km sliding distance under various loads between 5 and 20 N. The results showed that composites reinforced by kenaf fiber particles show better wear performance compared to plain epoxy. It was also found that the coefficient of friction was significantly improved.

3 Tribological properties of hybrid polymer composites

Hybrid composites are developed to display combined component properties. Adding a reinforcing agent can be in the form of fillers, short fibers, or fabrics to formulate the hybrid composites with various forms of reinforcing pairs. The reinforcement selection is an important factor for addressing the suitability of hybrid composites in tribological applications. The hybrid composites developed in combination with natural and synthetic fibers have enhanced mechanical properties and can be suitable for tribological applications with higher loading. Wear debris, fiber breakage, voids, matrix loss, debonding, and transfer surface are the primary wear mechanisms in the rubbing action of hybrid composites. Including fillers is the important factor influencing their tribological characteristics. The improved mechanical strength with increased tribological efficiency, however, depends on the uniform dispersion of the fillers. The transfer layer plays an important role in assessing the tribological quality of composites based on the polymer. The existence of the transfer layer can be achieved by maintaining low friction and reduced wear through protective shielding.

It is expected that the design of hybrid composites using natural and synthetic fibers will achieve better interfacial strength between the fiber and matrix and minimize the water absorption characteristics. Jesthi et al. [20] researched the dry abrasive wear behavior of hybrid polymer composites reinforced by glass and carbon fibers. The hybrid composites are made up of six glass fiber layers and four carbon fiber layers by using a manual lay-up method and an epoxy polymer. Three body abrasion wear tests were conducted, and it was found that the specific wear rate decreases with the increase in applied load and sliding distance. It was observed that composite hardness, applied load, and sliding distance are the three independent variables that contribute to the prediction of wear rate.

From the investigation into hybrid composites fabricated by using bamboo and glass fibers, it was found that the layer sequence affects the hybrid composite wear behavior. The increased weight percentage of glass fiber showed improvement in mechanical properties while the increase in bamboo fiber weight percentage showed better wear resistance. A study of the erosion behavior of jute and SiC hybrid composites with respect to different parameters such as velocity of impact, fiber content, filler content, and angle of impingement was conducted. It was found that these are the primary factors involved in reducing wear rate. Several researchers used the Taguchi optimization approach to minimize wear loss. The results of the investigation with the combination of cotton fiber and graphite-modified polyester on the hybrid composites show a significant decrease in the common wear rate. It was also observed that the frictional force decreases with the inclusion of a graphite filler and rises in hybrid composites with the introduction of cotton fiber. In addition to ceramic fillers such as SiC and Al_2O_3, a study on the performance of jute fiber-reinforced epoxy composites was carried out. It was found that composites filled with Al_2O_3 have increased wear resistance compared to composites filled with SiC. In the case of SiC-added hybrid composites, damage such as large cracks, fiber breakage, and pores was also noted. With the introduction of multiwalled carbon nanotubes (MWCNTs), hybrid cotton fiber composites in a phenolic matrix exhibit improved wear resistance. Because of their excellent properties such as thermal conductivity, good mechanical strength, aspect ratio, and low thermal degradation limits, MWCNTs are considered an effective nanofiller for polymer matrix composite reinforcement [35].

Madhusudhan et al. [36] reviewed the mechanical properties and tribological characteristics of natural and glass fiber-reinforced hybrid composites. They reported that better flexural and impact properties as well as an increase in thermal properties can be achieved by hybridization. The wear resistance was better in bidirectional glass fiber composites compared to unidirectional reinforced composites. Natural and glass fiber-reinforced polymer composites may be utilized as an alternative material for pure synthetic fiber composites. Betelnut-treated fiber-reinforced polyester composites were manufactured, and the tribological properties were investigated under dry/wet contact conditions against the counterface of stainless steel [25]. The results revealed that the composite's wear and frictional quality were improved by about 54% and 95% compared to the dry condition under wet contact conditions and the composite showed high wear output in an antiparallel orientation under both dry/wet contact conditions.

Hong et al. [37] investigated the tribological properties of core shell structure SiO_2 hybrid filler composites and observed that the strengths can be enhanced by the suitable inclusion of SiO_2 fillers. The friction coefficient and wear were reduced nearly 88.4% when the addition of filler was about 0.5%. Mohammed et al. [38] investigated the tirbological and corrosion behavior of hybrid composites with fillers for biomedical applications. The packing of linear chains and the branching levels of high-density polyethelene materials resulted in enhanced mechanical and tribological properties. High-density polyethelene is used as a polymer matrix and the properties can be improved by using TiO_2 as a binding agent and Al_2O_3 as a ceramic reinforcement. The wear test was performed using a pin-on-disc tester at a constant speed and under different loads. It was observed that the wear loss decreased with an increase of 10% of TiO_2 and alumina. Also, the alumina particles offer wear resistance, a low friction coefficient, and uniform dispersion.

The performance of hybrid polytetrafluoroethylene composites filled with nanoparticles such as nano-β-carborundum, nanocopper, and nano-TiO_2 was studied and sliding friction tests were performed by Jia et al. [39]. It was found that the addition of nanoparticles decreases the polytetrafluoroethylene composite friction coefficient and differs within the range of 0.095–0.106. The rolling effect of the nanoparticles between the material pairs reduces the frictional resistance. This increases the properties of wear resistance and composite antifriction. During sliding contact, the transfer film formation is an important factor in reducing the composite wear rate. Lingaraju et al. [40] investigated the impact of nanoparticles on the epoxy composite based on glass fiber epoxy composite systems. A halloysite nanotube was used as a base medium for reinforcing clay and epoxy resin. The dry slide wear test results showed that hybrid nanocomposite wear rates are lower than pure composites. The addition of 2% nanopowders showed an increase of 4.2% in hardness and a decrease of 49% in wear level. The addition of 1% showed an increase of 2% in the hardness and 75% in the wear level.

Sandeep et al. [41] used Grewia optiva as the main reinforcement and rice husk/wheat as particulates for enhancing the wear and mechanical characteristics of composites. Three different types of composites were fabricated by the manual method. The wear test was conducted for different loads (5, 10, and 15 N) at the sliding velocity of 12.5 m/s. It was observed that in Grewia optiva composites incorporated with particulates, the water absorption increased as compared to the unfilled composites. The results revealed that the rice husk and wheat straw content in Grewia optiva composites gave better hardness and wear properties. From the evaluation, it was also found that the erosion behavior of the materials depends on the surface hardness. The hybridization of Grewia optiva composites with agro-waste particulates has potential applications in the areas that require better strength and wear resistance.

Akash and Srivastava [42] discussed the properties of randomly distributed synthetic fiber composites filled with alumina nanoparticles through wear and a microhardness test. The composite specimens were prepared by adding alumina particles, glass, and carbon fibers. The microhardness was tested and the specific wear rate was measured in relation to the addition of alumina, which increased gradually. It was also found that the coefficient of friction corresponding to the loading of alumina particles gradually decreased. The results revealed that the wear and friction behaviors of randomly oriented glass/carbon fiber-reinforced composites increased by adding filler materials. The loss of wear is also least for carbon fiber composites and it decreases when the quantity of the fillers in the composite is improved. It was found that the wear rate is more efficient in hybrid composites.

An examination of dry sliding and abrasive wear analysis on two body abrasive wears of carbon fibers and their hybrid composites showed that the optimal range for short carbon fibers in a matrix is 60%. The study of carbon and glass fiber friction and wear behavior under dry sliding conditions was conducted. For all the tested composites, the wear rate and the friction coefficient increased with the increase of sliding velocity. Among all the tested composites, polyether ether ketone/glass fiber composites exhibited excellent wear resistance while polyether ether ketone/carbon fiber composites exhibited the best tribological behavior. The investigation on bagasse/glass fiber hybrid composites showed that the inclusion of bagasse fibers increased the modulus of elasticity. The bending strength decreased with the addition of bagasse fibers while it improved by adding glass fibers. The wear and friction characteristics of carbon, bagasse, and their hybrid composites were studied. It was observed that the volume of the fiber is an important factor for tribological applications. Also, applied pressure and velocity play vital roles in sliding wear characteristics [43].

4 Conclusion

The recent research articles on the tribological behavior of polymer composites have been reviewed in this chapter. From the chapter, it is confirmed that the natural fiber-reinforced polymer composites exhibited less abrasive damage than their synthetic counterparts due to high cellulose content. Based on the review, the following points are listed for developments in the future.

- The adhesive wear mechanism that was incurred and the reliability of different types of composites under different test parameters on wear, friction, temperature, and contact condition are highlighted.
- The fiber size, orientation, volume fraction, and properties are important considerations in the development of fiber-reinforced polymer composites.
- Due to the rapid emergence of new fibers on the counter layer, the friction coefficient is minimal in the cross-laminar orientation and peak in the parallel orientation. The increase in the proportion of reinforcement also indicated a higher friction coefficient.
- The major setbacks such as poor wear and a high coefficient of friction can be improved by incorporating filler materials. The performances of composites depends on the nature of the filler, the even distribution of the filler, the spacing between the filler particles, and the interfacial interaction strength.
- It is found that the carbon fiber's self-lubricating property improves the wear resistance by reducing the friction coefficient. As a result, polymeric carbon materials are generally used in bearing materials.
- With the increase in temperature, the wear rate increases regardless of the loading conditions. The fillers play a major role in wear rate control. Operating parameters and material parameters are the main factors affecting friction and wear properties.
- Tribological properties are highest in the state of inert gas and less so in the case of oil-grained. The oil-greased condition prompts improvement in tribological properties due to the self-greasing property.
- The friction coefficient, pressure, temperature, and brittleness of the fibers are the contributing factors for better wear properties.
- Because they can provide good interfacial bonding between the fiber and the matrix, fiber treatments are important. A surface treatment will improve the poor adhesion and low interlaminar shear strength between the particles and the matrix. Particles in lighter liquids have been reported to have better results compared to high-viscosity liquids.
- There are more opportunities on new studies on tribology involving polymer composites that would bring further research in this field.

References

[1] U. Nirmal, H. Jamil, M.M.H.M. Ahmad, A review on tribological performance of natural fibre polymeric composites, Tribol. Int. 83 (2015) 77–104.

[2] K. Friedrich, Polymer composites for tribological applications, Adv. Ind. Eng. Polym. Res. 1 (2018) 3–39.

[3] M. Ramesh, R. Logesh, M. Manikandan, N.S. Kumar, D.V. Pratap, Mechanical and water intake properties of banana-carbon hybrid fiber reinforced polymer composites, Mater. Res. 20 (2017) 365–376.
[4] R. Bhoopathi, C. Deepa, G. Sasikala, M. Ramesh, Experimental investigation on mechanical properties of hemp-banana-glass fiber reinforced composites, Appl. Mech. Mater. 766 (2015) 167–172.
[5] M. Ramesh, P. Sudharsan, Experimental investigation on mechanical and morphological properties of flax-glass fiber reinforced hybrid composite using finite element analysis, SILICON 10 (2018) 747–757.
[6] S.Y. Bardekar, M.B. Kulkarni, Tribology of hybrid polymer composites, Int. J. Mech. Prod. Eng. 2321-2071, 6 (2018).
[7] N.S.M. El-Tayeb, B.F. Yousif, T.C. Yap, An investigation on worn surfaces of chopped glass fibre reinforced polyester through SEM observations, Tribol. Int. 41 (2008) 331–340.
[8] N.S.M. El-Tayeb, B.F. Yousif, T.C. Yap, Tribological studies of polyester reinforced with CSM 450-R-glass fiber sliding against smooth stainless steel counterface, Wear 261 (2006) 443–452.
[9] P. Valasek, R. D'Amato, M. Muller, A. Ruggiero, Mechanical properties and abrasive wear of white/brown coir epoxy composites, Compos. B Eng. 146 (2018) 88–97.
[10] Y. Liu, J. Xie, N. Wu, L. Wang, Y. Ma, J. Tong, Influence of silane treatment on the mechanical, tribological and morphological properties of corn stalk fiber reinforced polymer composites, Tribol. Int. 131 (2019) 398–405.
[11] R. Vinayagamoorthy, Friction and wear characteristics of fibre-reinforced plastic composites, J. Thermoplast. Compos. Mater. (2018), https://doi.org/10.1177/0892705718815529.
[12] P.L. Menezes, P.K. Rohatgi, M.R. Lovell, Studies on the tribological behavior of natural fiber reinforced polymer composite, in: Green Tribology, Springer, Berlin, Heidelberg, 2012, pp. 329 345.
[13] H.H. Parikh, P.P. Gohil, Tribology of fiber reinforced polymer matrix composites—a review, J. Reinf. Plast. Compos. 34 (2015) 1340–1346.
[14] J. JyotiKalita, K.K. Singh, Tribological properties of different synthetic fiber reinforced polymer matrix composites—a review, IOP Conf. Ser. Mater. Sci. Eng. 455 (2018) 012134.
[15] Y. Zhang, D. Zhang, X. Wei, S. Zhong, J. Wang, Enhanced tribological properties of polymer composite coating containing graphene at room and elevated temperatures, Coatings 8 (2018) 91.
[16] J. Wieczorek, A.B. Grzechnik, J. Labaj, G. Siwiec, Tribological properties of polymer composites reinforced with silver particles, Adv. Mater. Res. 1036 (2014) 65–70.
[17] K. Prabhakar, S. Debnath, R. Ganesan, K. Palanikumar, A review of mechanical and tribological behaviour of polymer composite materials, IOP Conf. Ser. Mater. Sci. Eng. 344 (2018) 012015.
[18] B. Shivamurthy, K. Murthy, S. Anandhan, Tribology and mechanical properties of carbon fabric/MWCNT/epoxy composites, Adv. Tribol. 2018 (2018) 1508145.
[19] P.K. Bajpai, I. Singh, J. Madaan, Tribological behavior of natural fiber reinforced PLA composites, Wear 297 (2013) 829–840.
[20] B.C. Routara, A. Nayak, D. Jesti, R. Nayak, Evaluation of mechanical and tribological properties of glass/carbon fiber reinforced polymer hybrid composite, Int. J. Eng. 31 (2018) 1088–1094.
[21] D. Chandramohan, L. Ravikumar, C. Sivakandhan, G. Murali, A. Senthilathiban, Review on tribological performance of natural fibre-reinforced polymer composites, J. Bio. Tribocorros. 4 (2018) 55.

[22] H. Sumithra, B.S. Reddy, A review on tribological behaviour of natural reinforced composites, J. Reinf. Plast. Compos. 37 (2018) 349–353.
[23] N. Chand, U.K. Dwivedi, Effect of coupling agent on abrasive wear behaviour of chopped jute fibre-reinforced polypropylene composites, Wear 261 (2006) 1057–1063.
[24] Y. Nishitani, K. Tetsuto, Y. Toshiyuki, Effect of silane coupling agent on tribological properties of hemp fiber-reinforced plant-derived polyamide 1010 biomass composites, Materials 10 (2017) 1040.
[25] U. Nirmal, B.F. Yousif, D. Rilling, P.V. Brevern, Effect of betelnut fibres treatment and contact conditions on adhesive wear and frictional performance of polyester composites, Wear 268 (2010) 1354–1370.
[26] N.S. Gill, B.F. Yousif, Wear and frictional performance of betelnut fibre-reinforced polyester composite, Proc. Inst. Mech. Eng. Part J 223 (2009) 183–194.
[27] C.W. Chin, B.F. Yousif, Potential of kenaf fibres as reinforcement for tribological applications, Wear 267 (2009) 1550–1557.
[28] S.A.R. Hashmi, U.K. Dwivedi, N. Chand, Graphite modified cotton fibre reinforced polyester composites under sliding wear conditions, Wear 262 (2007) 1426–1432.
[29] X. Xin, C.G. Xu, L.F. Qing, Friction properties of sisal fibre reinforced resin brake composites, Wear 262 (2007) 736–741.
[30] N.S.M. El-Tayeb, A study on the potential of sugarcane fibers/polyester composite for tribological applications, Wear 265 (2008) 223–235.
[31] A. Shalwan, B.F. Yousif, In state of art: mechanical and tribological behaviour of polymeric composites based on natural fibres, Mater. Des. 48 (2013) 14–24.
[32] E. Omrani, P.L. Menezes, P.K. Rohatgi, State of the art on tribological behavior of polymer matrix composites reinforced with natural fibers in the green materials world, Eng. Sci. Technol. Int. J. 19 (2016) 717–736.
[33] P.L. Menezes, P.K. Rohatgi, M.R. Lovell, Tribology of natural fiber reinforced polymer composites, in: Proc. of the International Joint Tribology Conference ASME/STLE 2011, American Society of Mechanical Engineers, 2011.
[34] A. Devadas, U. Nirmal, J. Hossen, Investigation into mechanical and tribological performance of kenaf fibre particle reinforced composite, Cogent Eng. 5 (2018) 1479210.
[35] S. Karthikeyan, N. Rajini, M. Jawaid, J.T.W. Jappes, M.T.H. Thariq, S. Siengchin, J. Sukumaran, A review on tribological properties of natural fiber based sustainable hybrid composite, Proc. Inst. Mech. Eng. Part J 231 (2017) 1616–1634.
[36] T. Madhusudhan, M.S. Kumar, D. Athith, Mechanical characteristics and tribological behavior study on natural-glass fiber reinforced polymer hybrid composites: a review, Int. Res. J. Eng. Technol. 3 (2016) 2243–2246.
[37] J. Hong, C. Liu, X. Deng, T. Jiang, L. Gan, J. Huang, Enhanced tribological properties in core–shell structured SiO_2@GO hybrid fillers for epoxy nanocomposites, RSC Adv. 6 (92) (2016) 89221–89230.
[38] M. Yunus, M.S. Alsoufi, Experimental investigations into the mechanical, tribological, and corrosion properties of hybrid polymer matrix composites comprising ceramic reinforcement for biomedical applications, Int. J. Biomater. (2018), https://doi.org/10.1155/2018/9283291.
[39] Z.N. Jia, C.Z. Hao, Y.L. Yang, Tribological performance of hybrid PTFE/serpentine composites reinforced with nanoparticles, Tribol. Mater. Surf. Interf. 8 (2014) 139–145.
[40] D. Lingaraju, K. Ramji, M.P. Devi, U.R. Lakshmi, Mechanical and tribological studies of polymer hybrid nanocomposites with nano reinforcements, Bull. Mater. Sci. 34 (2011) 705–712.
[41] K. Sandeep, B. Gangil, V.K. Patel, Physico-mechanical and tribological properties of Grewia Optiva fiber/bio-particulates hybrid polymer composites, in: AIP Conference Proceedings, vol. 1728(1), AIP Publishing, 2016.

[42] M. Akash, V.K. Srivastava, Tribological behavior of particles and fiber-reinforced hybrid nanocomposites, Tribol. Trans. 58 (2015) 1142–1150.
[43] K. Supriya, S.S. Mer, A. Joshi, Mechanical and tribological properties of carbon, bagasse and their hybrid polymer composite: a literature review, in: Proc. of 2nd National Conference on Recent Advances in Mechanical Engineering, 26–27, Sep, 2014.

CHAPTER 3

Development and tribological properties of polymer composites

Vasilev Andrey Petrovich, Lazareva Nadezhda Nikolaevna, Nikiforov Leonid Aleksandrovich, Okhlopkova Aitalina Alekseevna, Sleptsova Sardana Afanasyevna, and Struchkova Tatiana Semenovna
Chemistry Department, Institute of Natural Sciences, North-Eastern Federal University, Yakutsk, Russia

Abbreviations

BF	basalt fibers
CF	carbon fibers
EP	epoxy polymer
GF	glass fibers
GO	graphene oxide
Limiting PV	the limiting pressure-velocity
LLDPE	linear low-density polyethylene
PCM	polymer composite materials
PEEK	polyetheretherketone
POD	polyoxadiazole fibers
PI	polyimide
PTFE	polytetrafluoroethylene
RWR	a ratio of wear rates
UHMWPE	ultra-high-molecular-weight polyethylene
UPTFE	ultrafine polytetrafluoroethylene

1 Introduction

Currently, polymers and polymer composite materials (PCM) are widely used as parts of friction units of machines and equipment for their high specific strength; excellent ability to absorb noise and vibration; high resistance to wear, abrasion, impact, and corrosion; and ease of manufacturing [1]. Polymer composites with self-lubricating properties attract a high interest due to their excellent tribological performance to avoid the contamination of using lubricating oil or grease in industrial communities [2]. Polyimides, polyoxymethylene, fluoroplastics, polyaryletherketones, and UHMWPE are widely used polymers with self-lubricating properties and their industrial

Tribology of Polymer Composites
https://doi.org/10.1016/B978-0-12-819767-7.00003-7

production is growing every year. However, the properties of industrial polymers are not always sufficient to meet the increased technical requirements for tribological materials. The primary approach in the improvement of the mechanical and tribological properties of polymers is the introduction of various modifiers—dispersed particles and fibers as well as nanosized particles.

However, the features of various types of fillers and their combination with polymer matrices lead to the need to use specific approaches to the development of materials for the tribological application. For example, it is necessary to prevent the agglomeration of nanoparticles or exfoliate layered fillers in the bulk of the polymer. Thus in this chapter, various approaches and technological methods used in the development of PCM using multiple types of fillers will be considered.

2 The polymer matrix

In contrast to metals and ceramics, the use of polymer materials and PCM in friction units is convenient as they can operate in sliding conditions without any lubricants (dry friction). It is crucial especially when lubricants cannot be used due to possible contamination of the product (oils and lubricants) or the environment, under extreme operating conditions (vacuum, chemically aggressive environments, and high humidity) or in the absence of the possibility of maintenance of the unit. Despite a large number of publications on the development and research of polymers and PCM of tribological purpose, this topic remains relevant with the increasing requirements for the operational characteristics of materials.

Currently, the industry produces a comprehensive range of polymer matrices of engineering applications. However, as is known from the literature, a limited number of polymer matrices are used for the manufacture of tribological parts: polytetrafluoroethylene (PTFE), ultra-high molecular weight polyethylene (UHMWPE), polyamides (PA), polyoxymethylene (POM), polysulfone (PSU), epoxy polymer (EP), polyethersulfone (PES), polyphthalamide (PPA), poly(*p*-vinylidene) fluoride (PVDF), polyetherimide (PEI), poly(*p*-phenylene) sulfide (PPS), polyetheretherketone (PEEK), polybenzimidazole (PBI), poly(ether ketones) (PEK), etc.

The choice of the polymer matrices as discussed earlier, which belong to thermoplastics, is due to such characteristics as mechanical properties, resistance to vibration, shock impact, resistance to aggressive environments and temperature difference, as well as ease of manufacturing into products. However, the use of unfilled polymers is usually limited by several reasons, such as low wear resistance, poor thermal conductivity, high linear expansion coefficient, high friction coefficient, and low frost resistance of the initial polymer matrix. Thus the development of tribological materials involves, as a rule, modification of the polymer matrix, which initially has the complex of necessary parameters, with various fillers. Therefore the primary strategy for introducing fillers in the polymer matrix is not only to improve the tribological and mechanical properties but also to preserve the own advantages of the polymer matrix: resistance to corrosion and aggressive environments, shock and vibration resistance, frost and heat resistance, and the other properties. Besides their processing methods, moulding methods and process parameters strongly influence the properties of polymer materials [3].

The manufacturing of products from the polymer composite mainly depends on the polymer matrix. Moreover, the moulding methods and parameters of the manufacturing process of the polymer composite are mainly determined by the polymer matrix. Therefore it is important to study and understand the composition, role, and performance of matrix materials and their impact on operational properties.

3 Polymer composites modified with layered fillers

One of the most promising approaches to the development of polymer composites is the use of layered silicates as fillers for polymers. Layered silicates are widespread and well known as clay minerals. Natural layered silicates are more accessible than synthetic nanoparticles [4]. The particle sizes of clay minerals in the volume of the polymer can vary in diameter from 250 to 1000 nm, a thickness of about 1 nm. Due to the size achieved by dispersing silicates in the polymer, nanocomposites based on them exhibit significantly improved physicochemical, optical, and thermal properties, for example, heat resistance and increased modulus. The structure of the layered silicates can be of two main types: 1:1 and 2:1. The layered silicates with the structure 1:1 consist of the repetition of one tetrahedral and one octahedral grid. The layered silicates with the structure 2:1 consist of one octahedral grid sandwiched between two tetrahedral grids [5].

Attempts to create composite materials based on the layered silicates have been known since the 1950s of the 20th century [6]. However, the developers faced the problem of agglomeration of layered silicate particles, which sharply reduced the exploitation characteristics of PCM. In 1976, researchers at Unitika Ltd. created a material based on polyamide-6 and layered silicate using in situ polymerization of ε-caprolactam with montmorillonite [7]. Using this technology, the researchers expected to eliminate the agglomeration of the particles of the layered silicate. Fukushima and Inagaki made the next attempt to create PCM based on the layered silicates at Toyota Central Research & Development Co. Int. in 1987 [8]. The developed method of substitution of the cations in the interplanar space of the original clay cation surfactant significantly increased the compatibility of the layered silicate and a nonpolar polymer matrix. Such modified clays are called lipophilic, organically modified, or organoclay. The organomodification leads to the expansion of the interlayer space of layered silicates due to the surfactant molecules penetration. Further, it may lead to exfoliation of clays—delamination of the layered silicate into individual plates with a thickness of 1 nm. Usuki, Fukushima, and their colleagues in 2006 for the first time obtained exfoliated nanocomposites based on polyamide-6 and montmorillonite using organically modified montmorillonite and in situ polymerization [9].

The introduction of layered silicates in the polymer matrix significantly improves the mechanical, optical properties as well as increases the wear resistance of the material. Concerning the development of antifriction materials, interest in layered silicates as fillers for a polymer matrix arises in connection with the prospect of creating self-lubricating polymer composites. It is also attractive when even a small amount of clay minerals (2–3 wt.%) in a polymer matrix, the coefficient of thermal expansion of the materials, decreases significantly [10, 11]. Achieving significant improvement in the characteristics of the material occurs in the case of intercalation of polymer macromolecules into the interlayer space of silicates with their subsequent exfoliation to individual nanolayers [12].

Thus the layered silicates are very advantageous as nanomodifiers of the polymer matrix. In the formation of the polymer-silicate system, the property of layered silicates is used to intercalate various molecules into the interlayer space with further expansion of the interlayer space, i.e., exfoliation, of the filler particles into monolayers. Thus three types of polymer-silicate composites with different distribution of layered silicates in the polymer matrix are shown in Fig. 1. In this case, good dispersion of nanoparticles is achieved in the polymer matrix [13].

In the literature, the researchers with varying degrees of success have achieved some success in unleashing the potential of layered silicates as

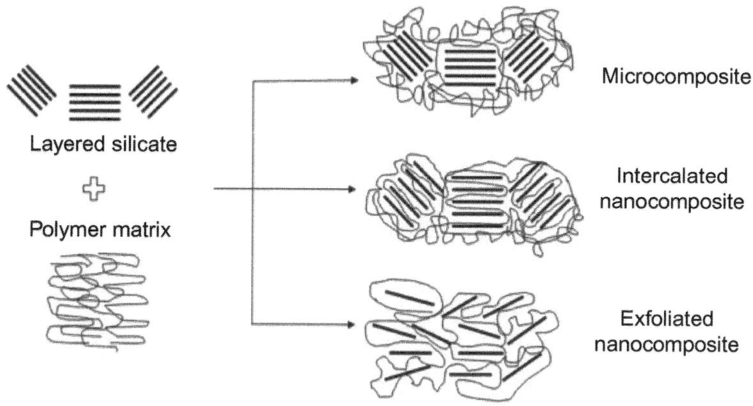

Fig. 1 Scheme of the formation of three types of composites "polymer-layered silicate."

fillers in a polymer matrix. For example, A. Pegoretti, A. Dorigato, and A. Penaty studied PCM based on high-density polyethylene filled with synthetic layered silicates [14]. In this work, the authors used the method of mixing in the polymer melt. They used two types of polyethylene with different melt viscosity as a polymer matrix, two types of organically modified layered silicate as fillers, and polyethylene with grafted maleic anhydride as a compatibilizer. It was found that intercalation proceeds most intensively in a polymer with low viscosity, and a more considerable expansion of the interplanar space is achieved with an increase in the content of the compatibilizer. With an increase in the silicate content, PCM hardness increases; however, filling with layered silicate increased the yield strength slightly.

However, to date, the developed polymer-layered composites based on nonpolar or weakly polar polymers have not been characterized by a significant change in operational characteristics compared to the composites based on polar polymers [15]. The problem persists due to the hydrophilicity of the silicates used and so it is difficult to combine them with a polymer matrix. One way to solve this problem is to modify layered silicates with surfactants to replace inorganic cations inside silicate layers with organic cations. Substitution with cationic surfactants, such as bulk ammonium and phosphonium ions, increases the space between the layers, reduces the surface energy of clay minerals, and makes the surfaces of the particles hydrophobic [16–19]. In the general case, the organomodification of layered silicates occurs according to the following scheme: when polar centers adsorb surfactants, organophilic layers are formed on the inner and outer surfaces of the layered silicate, which ensures the thermodynamic compatibility of clay with polymer molecules. It is the properties of the organophilic layers that determine the ability of layered silicates to disperse in the polymer. It was shown in Ref. [15] that the use of "optimized" surfactants for clay modification allows one to achieve a significant increase in the mechanical characteristics of a composite based on polyethylene (elastic modulus by 50%, tensile strength by 20%) with a clay silicate content of only 5 wt.% while maintaining plasticity at the level of the original unmodified polymer.

In the well-known work of Park [20], the authors studied PCM based on UHMWPE and organomodified montmorillonite. X-ray diffraction (XRD) studies showed the formation of an exfoliated nanocomposite (Fig. 2).

The obtained PCM are characterized by higher strength compared to the initial UHMWPE. Such change in the properties was explained by the formation of a system with nanometer-sized layered silicate uniformly distributed in the polymer matrix.

However, the modification methods discussed earlier have several disadvantages, such as the complexity of the technological process for creating composites and the use of organic solvents, catalysts, and reaction stabilizers.

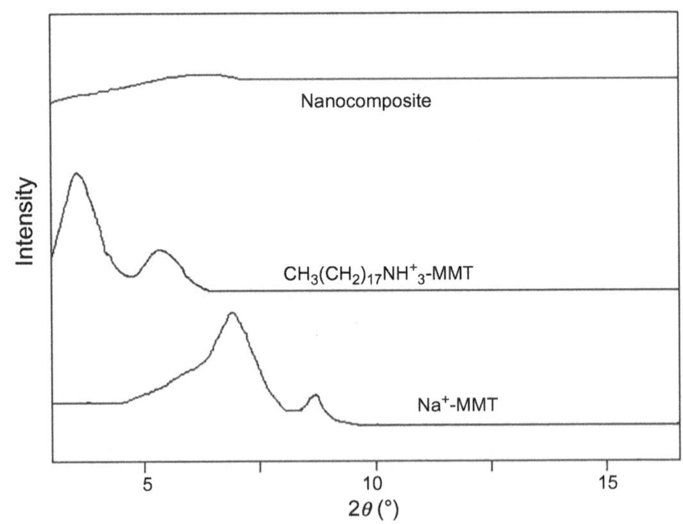

Fig. 2 Diffraction patterns of PCM based on ultra-high-molecular-weight polyethylene and organomodified montmorillonite. *(Representation in work from S.-J. Park, K. Li, S.-K. Hong, Preparation and characterization of layered silicate-modified ultrahigh-molecular-weight polyethylene nanocomposites, J. Ind. Eng. Chem. 11 (2005) 561–566.)*

It should be noted that under real conditions, nanocomposites based on polymer and clay, which have a mixed morphology, are often obtained; such materials are all three types of filler distribution in the bulk of the polymer material.

In the works [21–23], layered silicates were modified by such surfactants as CTAB and ADMBAH. A positive effect during processing was proved by the fact that the tensile strength of composites based on UHMWPE containing 1.0 wt.% Biotite treated with CTAB was increased by 30% compared with compositions with untreated biotite [23].

In the work of Kirillina et al. [21], the results of a study of the structure and properties of composites based on PTFE containing bentonite and vermiculite. In this article, the results given of the comparison of the mechanical and tribological properties of composites based on PTFE filled with layered silicates treated with surfactants and without treatment. A comparison showed that surfactant treatment slightly affects the characteristics of composites. The reason is the excess of the temperature of PTFE processing above the temperature of the destruction of used organic surfactants. However, the resulting composites, regardless of the processing of surfactants, were characterized by significantly increased wear resistance (up to 2000 times) compared with the initial PTFE. The mechanical properties remain at the initial polymer level when the content of the filler is under the 5.0 wt.%. For example, as a rule, the introduction of a filler contributes to a rather sharp decrease in these properties. Since the layered silicates used in the work were mechanically activated, the main explanation of the obtained results is an increase in the reactivity of silicate particles as a result of their dispersion in a planetary mill, which contributes to an increase in interfacial interaction at the phase boundary.

Several works [21, 22, 24–26] have shown that minerals of the subclass of silicates—serpentinite, vermiculite, kaolinite, and bentonite—previously mechanically activated, were used as fillers in the polymer matrix and confirm the effectiveness of this method as a way to improve the compatibility of fillers with the polymer matrix. Since PTFE belongs to rigid-chain polymers, the mechanism of composites formation with layered silicates is unlikely to include the stage of intercalation of polymer macromolecules, although we do not exclude the fact that single plates of layered silicates with a structure type (1:1) such as kaolinite and serpentine exfoliate in the polymer volume. In Ref. [22], it was shown that in composites UHMWPE/kaolinite and PTFE/serpentinite + magnesium nanospinel, the disappearance of basal reflexes was detected by the XRD (Fig. 3).

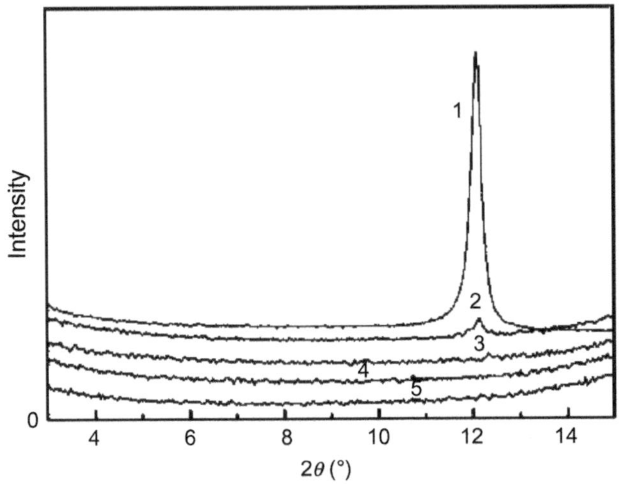

Fig. 3 X-ray diffraction patterns of PTFE/layered silicate nanocomposites: (1) serpentinite processed in a planetary mill, (2) PTFE containing 5 wt.% serpentinite, (3) PTFE containing 4.5 wt.% serpentinite and 0.5 wt.% MAS, (4) PTFE containing 4 wt.% serpentinite and 1 wt.% MAS, and (5) initial PTFE. *(Representation in work from I.V. Kirillina, L.A. Nikiforov, A.A. Okhlopkova, S.A. Sleptsova, C. Yoon, J.-H. Cho, Nanocomposites based on polytetrafluoroethylene and ultrahigh molecular weight polyethylene: a brief review, Bull. Kor. Chem. Soc. 35 (2014) 3411–3420.)*

An analysis of studies of the mechanical activation of layered silicates shows that, upon activation, the octahedral grids are destroyed first. There are many works [27–30] in which after mechanical activation of layered silicates, cations of octahedral grids are released and become active centers, which are, in fact, coordinatively unsaturated ions.

The study on the properties and structure of silicates before and after activation explain their effects on the mechanical and tribological properties of PCM. It was shown in Refs. [26, 31] using capillary electrophoresis that the mechanical activation of bentonite (montmorillonite, which is part of bentonite, has a 1:2 structure), serpentine and kaolin leads to the appearance of mobile cations, and, in particular, central cations of octahedral networks of silicates, Al^{3+} and Mg^{2+}. A comparison of the IR spectra of the layered silicates before and after activation revealed a shift in the absorption band of the Si–O bond, which confirms the destruction of the octahedral grids of silicates. Of these two metals, aluminum is characterized by a pronounced ability to complexation based on the ability of atoms to form coordination bonds by the donor-acceptor mechanism. Furthermore, this ability of aluminum is widely used in practice for binding fluorine atoms [32, 33]. The disappearance of the basal peak during the combined modification of PTFE with serpentinite with the introduction of magnesium spinel, shown in Ref. [22], can also be explained from this point of view. A more detailed explanation is given in Ref. [34]. The authors tried to explain the mechanism of the improvement of the adhesion at the PTFE/serpentinite/magnesium nanospinel interphase. They found the octahedral grid is destroyed in the process of dispersing serpentinite in a planetary mill exposing uncompensated bonds on the plate surface. As a result, it becomes possible to distribute single plates in the polymer volume (as we believe, this is typically only for layered silicates with a 1:1 structure).

However, in the absence of nanospinel, the uncompensated bonds of the surface atoms of the activated silicate plates reconnect with each other. The introduction of activated nanospinel adds aluminum ions into the mixture. It is assumed that aluminum ions, due to their coordination ability and their exceptional affinity for fluorine atoms, participate in interfacial interaction by the donor-acceptor mechanism with the PTFE macromolecule by the one side and with the surface of mechanically activated serpentinite by the other side.

The structural studies of the friction surface of composites can help to explain a significant increase in the wear resistance of PTFE/layered silicates composites.

A feature of composites containing layered silicates is the formation of a thin surface layer on the friction surface that even differs visually from the material itself (Fig. 4).

The morphology of the layer (Fig. 4B) plays the role of a solid lubricant that protects the surface of the material from abrasion. The layer is formed mainly from wear debris that appears in large quantities at the beginning of abrasion and deposited on the friction surface [34].

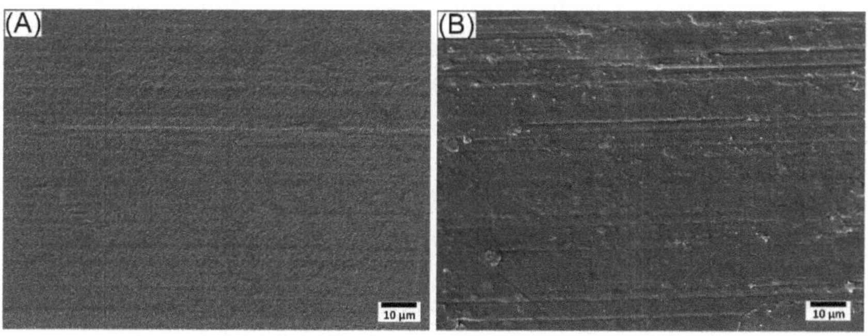

Fig. 4 Micrograph of the friction surface of the initial PTFE (A), a composite containing 2.0 wt.% mechanically activated bentonite (B).

In Ref. [35], a diagram of the formation of a protective layer on the surface of PTFE modified with aluminum oxide is presented. It was noted that the circulation of wear debris between the transfer film on the counterbody and the layer on the friction surface of the composite is a critical factor in maintaining the ultra-low wear of composites based on PTFE and nanoalumina.

Thus wear debris repeatedly coming again onto the friction surface undergoes mechanochemical processes of dispersion, defragmentation, activation, and oxidation. Simultaneously, the friction surface is accumulated with silicates particles oriented along the direction of friction. When moving of rubbing surfaces under a load, layered compounds, for example, graphite, are able to orient themselves so that the main axes become parallel to the sliding direction [36]. The forming protective layer, due to the particles of silicates contained in them, has low shear resistance, and it can slip relative to the surface of the counter-body and the surface of the composite in the direction of shear. Owing to its plasticity, the protective layer smoothes the friction surface, filling cavities, and irregularities as well as levels microprotrusions and microroughness of the friction surfaces.

4 Polymer composites modified with dispersed fillers

Nanotechnology is a set of methods for producing nanomaterials and as an interdisciplinary direction formed as a result of the practical application of fundamental knowledge from different fields of science and technology. A filling of a polymer with nanosized particles with average sizes from 0.1 to 100 nm forms a dispersion-strengthened nanostructured material or, in other words, a polymer nanocomposite [37].

The main physical reasons for the structural features and properties of nanosized materials include quantum-size effects, when the geometric dimensions are comparable with the characteristic lengths of a physical phenomenon, an abnormally high value of specific surface energy, which provokes the predominance of surface phenomena and processes, as well as extreme synthesis conditions leading to metastable structure. For these reasons, nanomaterials are one of the types of nonequilibrium state of a substance in which energy-saturated metastable structures, high-temperature phases, structural, concentration, or phase inhomogeneity are formed [38]. Macroscopic properties of the nanomaterials: mechanical, electrical, magnetic (especially superparamagnetism), thermal, optical, chemical, etc. manifest their features noted earlier [39].

Despite the vast number of publications stating the possibility of a sharp improvement in specific functional properties with a significant decrease in the content of the amount of filler compared to traditional micron-sized fillers, the physical foundations of "nanofilling" are fragmented, and sometimes contradictory. In review papers [40, 41], various approaches are given to explain the influence of nanofillers based on the chemical nature and their effect on the tribological behavior of polymers in the process of friction with metal. Thus in the review of Krasnov [42], a system for classifying nanofillers by their type of influence on the friction process is given:

- fillers that directly affect friction through active tribochemical processes or nano- and ultrafine fillers which, due to their physical characteristics, come into discrete contact with a metal counterbody;
- indirectly affect friction through the formation in the polymer binder of a more wear-resistant structure;
- intercalation compounds (intercalates).

Among nanoscale polymer structure modifiers, oxide ceramics occupy a special place. SiO_2, Al_2O_3, ZnO, TiO_2, and Fe_2O_3 nanopowders are produced on an industrial scale [43].

However, the features of nanoparticles cause problems in the development of nanocomposites based on them: difficulties in their uniform distribution in the polymer volume, agglomeration of nanoparticles due to their high surface energy, low interfacial interaction at the polymer-filler interface [44]. A variety of approaches are used to solve these problems.

Melt mixing is most acceptable for thermoplastic polymers and therefore is most often used. So in Ref. [45], metal-polymer composites based on LLDPE were obtained by melt mixing. A quasicrystal powder of the composition $Al_{65}Cu_{22}Fe_{13}$ with a particle size in the range of 0.01–3 μm was used as fillers. The components were mixed in an extruder mixer with which subsequently test samples were obtained. As a result, wear resistance increased by 57% at 1.0 wt.% of filler. The authors explain this increase in wear resistance by the peculiarity of LLDPE wearing, which consists in the melting of the polymer on the friction surface. The authors found that when filling the polymer with quasicrystals, the melting temperature rises, which leads to the resistance of the composites to heating during friction. Also, a decrease in the coefficient of friction was recorded compared to the initial LLDPE for composites filled with quasicrystals. However, at 147 N of the load, the samples showed the instability of the friction coefficient and were catastrophically worn. Only some concentrations of the filler showed stable friction. Such composites formed a protective antifriction film on the friction surface.

Another well-known method of mixing components is mechanical mixing using various high-energy methods such as planetary mills, ball mills, high-speed paddle mixers and the other equipment [42]. In Ref. [46], UHMWPE and TiO_2 powders in various ratios were mixed in a ball mill and then made samples for testing by hot pressing. Some samples were treated with gamma radiation at doses of 120, 250, and 500 kGy. Darkening of the samples was noted after irradiation relative to untreated. The tests were carried out according to the scheme simulating the knee joint in saline. The study of unirradiated samples showed that the wear resistance when filling UHMWPE 6 wt.% TiO_2 is reduced by 33% compared to the original UHMWPE. When processing filled composites with gamma radiation, an increase in wear resistance by 50% is observed compared to the unirradiated composites. Besides, when treated with radiation, filled composites showed a decrease in the oxidation index, which is vital for composites intended for use in joint prostheses. The authors suggested that the nanoparticles crosslink UHMWPE macromolecules by Van der Waals forces and hydrogen bonds and thereby strengthen the polymer. The use of gamma radiation leads to chemical crosslinking of macromolecules among themselves, improving wear resistance. Also, the oxidation index decrease was explained by the inhibiting effect of titanium oxide nanoparticles on the oxidation of the polymer during irradiation. Thus nanoparticles can act not only as a reinforcing component but also as functional additives.

Moreover in Ref. [47], UHMWPE was combined with Al_2O_3, AlN, SiO_2, and Si_3N_4 nanoparticles by their combined mechanical activation in a planetary mill. The filling of UHMWPE with nanoparticles led to significant decrease in the coefficient of friction and the wear rate. The authors suggested antifriction properties increased because of a cross-linked structure formed on the friction surface. During friction, nanoparticles are concentrated on the friction surface of the composite and form secondary structures as a result of the dissipation of thermal energy. The methods of IRS, EDS, and SEM confirmed the formation of a wear-resistant layer on the composite surface. Presumably, such layers consist of filler particles, wear debris, and fragments of UHMWPE macromolecules. Both chemical and adsorption interactions bind the components. Besides, the action of the friction process as an initiator of mechanochemical reactions was revealed. Thus in the processing and operation, it is necessary to take into account the participation of nanoparticles in various chemical interactions with PCM components.

This type of technology should modify the surface of nanoparticles to prevent their agglomeration and achieve uniform distribution during mixing with the polymer, especially at low degrees of filling [48].

Another effective method is combining with a liquid phase. Okhlopkova et al. [49] combined nanoparticles of metal compounds and UHMWPE microparticles under the action of ultrasound in an alcohol medium. Moreover, ultrasound also acts during the removal of a liquid medium, which prevents the agglomeration of nanoparticles. Studies have shown that, as a result, nanoparticles are uniformly deposited on the surface of UHMWPE particles. One of the most obvious criteria for evaluating the effectiveness of the chosen combination method is the uniform distribution of the nanoparticles (NP) filler in the polymer volume. The uniformity was evaluated by surface mapping of UHMWPE/Al_2O_3 nanocomposite using the EDS method. As a result, it was possible to increase the wear resistance by five times compared with the initial UHMWPE in the case of 0.5–1.0 wt.% CuO.

There is another approach consisting of the synthesis of components of composites in the process of combining. The in situ polymerization method is widely known [50]; however, the opposite approach, in which the synthesis of nanoparticles proceeds, has not been sufficiently studied. This solution is most often implemented by impregnating UHMWPE reactor powder in an organic solvent in which suspended nanoparticles are stabilized by various chemical reagents [51]; organic solutions of nanoparticle precursors, for example, Pd(II) organocomplexes [52], Ag salts of carboxylic and alkyl sulfonic acids [53], and inorganic fullerenes WS_2 [54]. In Ref. [55], CuO nanoparticles were synthesized in the presence of UHMWPE particles by the interaction of $CuCl_2$ with KBH_4 in micelles formed by cetyltrimethylammonium bromide. As a result, wear resistance increased by 33%, and the friction coefficient decreased by 33%.

The problem of enhancing interfacial interaction between components in PCM for various purposes is solved mainly by optimizing technological factors. Various methods are used to modify the surface structure of not only the polymer itself but also of the NP. Among the existing methods, one of the most effective is the chemical modification of the polymer matrix. The authors of Ref. [56] performed the bromination of the surface of UHMWPE powder particles in a CCl_4 medium. As a result, the material was obtained, which adhesion to nanoparticles in some cases increased up to 3 times compared with the initial UHMWPE. In Ref. [57], brominated UHMWPE was used as an agent that changes the interfacial interaction between PCM components. In this case, the wear resistance for certain types of nanoparticles increased by 5–6 times compared to the initial UHMWPE and by 2–3 times compared to PCM without modified UHMWPE. Concurrently, the authors concluded that such agents should be selected individually for each pair of polymer matrix-filler.

However, the authors in Ref. [58] revealed that the use of NP does not lead to a significant increase in wear resistance under conditions of abrasive wear. It is related to the ratio of the sizes of nanoparticles and abrasive particles. Microparticles showed a significant increase in wear resistance under similar conditions. Moreover, the maximum effect was noted with comparable sizes of filler microparticles and abrasive grains.

5 Polymer composites modified fibers fillers

Fibers for the PCM manufacture are used as threads and bundles, as well as short fibers. An essential characteristic of fibers is the shape factor—the ratio of length to diameter [59]. Thus the modification of polymers by the introduction of fibers leads to an increase in wear-resistance and mechanical properties, such as stiffness, creep resistance, impact resistance, and thermal conductivity [60, 61].

The most common fibers for the development of tribological polymer composites are the following fillers: carbon fibers (CF), glass fibers (GF), basalt fibers (BF), and aramid fibers (AF).

The characteristics of CF are high strength and modulus, heat resistance, wear-resistance and minimal impact on the counterbody, which makes them excellent material for obtaining PCM for tribological purposes [62]. However, CF is chemically inert material due to the stage of high-temperature graphitization in their manufacture and the smoothness of the crystallized surface [63]. In this regard, in polymer composites, in which CF is used without surface treatment, materials with lower shear strength are obtained, which, in turn, effect on the other properties [64, 65].

Qian-Qian et al. [66, 67] investigated the tribological characteristics of PTFE/CF composites. The authors used CF both without and with surface treatment according to the following method:

(1) Oxidation in the air: CF was placed in a furnace and heated at 450°C for 40 min, followed by cooling to room temperature inside the furnace.

(2) CF was immersed in an alcoholic solution of $LaCl_3$ (0.3 wt.%) for 3 h: in the first method, the treated fibers were dried in an oven at 800°C for 4 h, and in the second method, drying was carried out in inert atmosphere at 80°C for 4 h.

(3) CF was oxidized in the air, followed by treatment with $LaCl_3$.

Polymer composites samples for research were prepared as follows: dry mixing and moulding at room temperature (35 MPa with a soaking time of 5 min). Sintering of the samples was carried out in a high-temperature furnace at 380°C for 2.0 h, followed by cooling to room temperature inside the furnace. In work, the CF content in all PCM was 20 vol.% [67]. Surface treatment of CF in $LaCl_3$ solution leads to an improvement in the tribological characteristics of PCM in comparison with composites containing oxidized CF in the air and with CF without treatment. It was found that the treatment of carbon fibers with the use of a rare-earth element is preferable in comparison with fibers modified by oxidation in the air. An improvement in the mechanical and tribological properties of PCM was associated with an improvement in interfacial adhesion between CF and PTFE after treatment in $LaCl_3$ solution. Similar results were obtained in Ref. [68], where the authors of the work carried out the treatment of CF using HNO_3, KH-530 with a solution of La_2O_3 and La_2O_3 sol and studied their effect on the tribological properties of PTFE. Composites were prepared similarly, as shown in previous work. The authors found that PTFE/CF modified with La_2O_3 sol is characterized by the high wear resistance, which is 35% lower than that of PTFE filled with untreated CF.

In the works of Refs. [69–71], the effect of modified CF on the properties and structure of PTFE is studied. Modification of CF was carried out by plasma chemical treatment. The authors polymerized fluorinated compounds on the CF surface to form a fluoropolymer coating, which is characterized by a bound, branched, amorphous structure and the presence of unsaturated chemical bonds. The substrate's surface energy decreases and reaches a value close to the surface energy of PTFE, which ensures the compatibility of the components and adhesion between the filler and the matrix. Tribological tests have shown that in PCM with treatment CF, the wear rate is $(1.2–1.6) \times 10^{-7}$ mm^3/(N m), and composites without modification $(2.2–4.5) \times 10^{-7}$ mm^3/(N m). Similar results were obtained in the study of PTFE filled with polyoxadiazole (POD) fibers [72]. The authors used the RF plasma treatment technology of POD fibers according to the following parameters: 5.28 MHz RF-plasma was excited at 0.5 kV in the medium of tetrafluoroethylene at a pressure of ~0.1 kPa. Samples PCM for the study were obtained using dry mixing of the components. The resulting mix was pressed at 50 MPa at room temperature and sintered at (375 ± 5)°C with a heating-cooling rate of 40°C/h. The authors found that the organofluorine polymer coating on the POD surface enhances the wetting of the fibers with a PTFE melt, which reduces the defectives by the boundary layer and enhances the adhesion of the matrix to the fibers. Tribological tests showed that the wear rate of the composite with treated fibers is reduced compared to the original composite only at high-pressure tests.

It was shown in Ref. [73] that the introduction of CF into PTFE with graphene oxide (GO) leads to an improvement in the strength characteristics and wear resistance of PCM. Samples for research were obtained as follows: CF and GO were treated in a KH-550 solution with ultrasonic treatment for 2 min and dried, then the components were mixed in a mill and pressed, and then sintering was carried out. The authors showed that PTFE/CF/GO composites achieve the lowest wear and improve strength characteristics, at GO content of 1 wt.% and CF of 10 wt.%. When the CF and GO contents increase the given concentrations, PTFE strength decreases.

Glass fibers (GF) have a relatively high tensile strength and toughness, as well as high chemical resistance, but have a low modulus and low fatigue resistance. Nevertheless, due to its lightweight and low cost, GF is widely used in the development of composite materials [74]. It is known that the introduction of GF in polymers leads to a significant reduction in PCM wear rate [75–77]. In work [78], the authors compared the effects of GF and CF and their combination on the tribological properties of UHMWPE. Samples PCM for research were obtained in the following way: UHMWPE and fibers were mechanically mixed, and then the cold moulding and sintering were used, followed by hot pressing. Wear tests showed that UHMWPE/GF has better antiwear characteristics during dry friction compared to UHMWPE/CF and UHMWPE/GF/CF. However, some studies indicate an abrasiveness of the GF to a steel counterbody which can limit the GF for the manufacture of tribological materials [79, 80]. In this regard, for the development of tribological polymer composites, short GF are used as fillers in conjunction with nanosized particles or solid lubricants [81, 82].

Conte and Igartua [83] investigated commercial PCM with varying contents of fibrous and dispersed fillers and their combinations (Fig. 5).

Fig. 5 shows that the introduction in PTFE of 15 wt.% GF and 3.0 wt.% of MoS_2 reduce the coefficient of friction and wear compared to composites with a high content of fillers: PTFE + 60 wt.% bronze, PTFE + 15 wt.% graphite, PTFE + 25 wt.% GF and PTFE + 20 wt.% GF + 5 wt.% graphite.

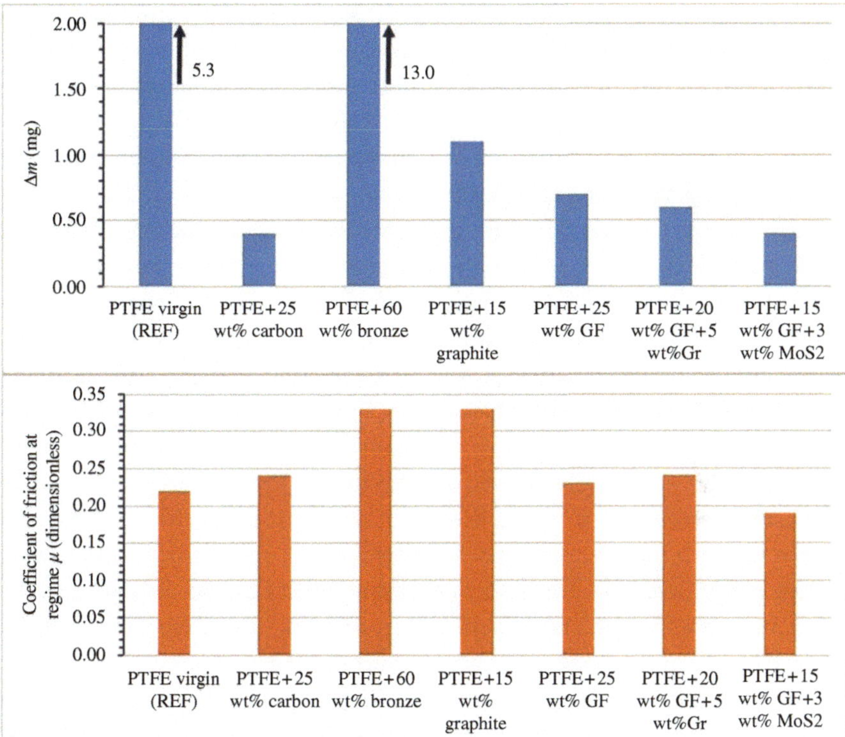

Fig. 5 Mass wear and coefficient of friction of PTFE and PCM.

The authors of the work suggested that the improvement of the tribological characteristics of PCM with additional MoS_2 content is since molybdenum disulfide particles reduce the contact between GF and the counterbody, which leads to the formation of stable transfer film.

Song et al. [84] studied the effect of the combining of fibrous fillers CF and GF with additional MoS_2 on the tribological behavior of PTFE. Samples were prepared in the following way: components were mixed in an electric mixer for 2 min in dry condition and then dried at 120°C; the resulting mix was processed by cold moulding at 40 MPa for 20 min, followed by sintering at 375°C for 120 min. It was shown in the work that the PTFE + 20 vol.% CF + 5 vol.% GF has weak antifriction property due to severe abrasive wear. However, when 20 vol.% CF and 5 vol.% GF was added, additional MoS_2 (5 vol.%) leads to a significant improvement in the tribological characteristics and the limit of PV as well. The limiting value of PV of this composite reaches 9.5 MPa m/s at sliding speed of 1 m/s and 15 MPa m/s at sliding speed of 2 m/s, respectively.

Basalt fibers (BF) are environmentally friendly materials and have excellent mechanical, chemical, and thermal characteristics [85]. Polymer composites with BF are attracting increasing attention as a reinforcing material, especially in comparison with traditional GF [86]. In work [61], the mechanical and tribological characteristics of epoxy resin/BF and epoxy resin/GF were compared. The composites were prepared by hand layup technique. It was shown that BF is effective in increasing the wear resistance of composites because of the strong bond between the polymer matrix and fibers. However, during modifying thermoplastics with BF, various surface treatment methods are used to improve the interfacial interaction between the polymer matrix and BF. In work [86], the surface of BF was modified using graphene in a polyurethane solution under ultrasonic treatment and mixing until obtaining a homogeneous consistency of the solution, followed by drying at 80°C for 12h. Subsequently, samples for studies were obtained using a twin-screw extruder. Studies of tribological properties showed that composites with modification have a low friction coefficient and high wear resistance compared to composites without modification, which may be due to improved interfacial adhesion by providing an active stress transfer pathway between the polymer matrices and the BF.

Wang et al. [87] investigated composites based on PTFE with CF and BF in water lubrication conditions. Samples were prepared as follows: PTFE powders were first mixed with milled fibers at a volume ratio of 4:1; the mixed powders were pressed at 50MPa, followed by sintering at 375°C for 2h. During friction under water lubrication conditions, the PTFE + CF composite showed better wear resistance compared to PTFE + BF. The authors of the work suggested that the high wear resistance of PTFE + CF due to the high resistance to water plasticization and friction stress as a result of better interfacial adhesion between the fibers and the matrix; on the contrary, the PTFE + BF composite has a high wear rate due to severe plasticization in water of the matrix and failure of fiber reinforcement.

Zhang et al. [88] studied about polymer composites based on polyimide (PI) and short BF. Samples were obtained by hot pressing; previously, BF was purified in acetone using ultrasound. The results of tribological studies have shown that the introduction of BF has a positive effect on the coefficient of friction and PCM wear rate compared to the initial polymer matrix. It was found that the optimal mass content of BF in composites is 10.0wt.%. In work [87], PI/BF composites with the addition of MoS_2 and graphite were studied. It was shown that the addition of solid lubricants leads to a significant improvement in tribological properties. The most optimal contents of MoS_2 and graphite in PI/BF are 40 and 35 vol.%, respectively.

The tribological properties of PTFE filled with CF with layered silicates (vermiculite and kaolin) and ultrafine PTFE (UPTFE) were studied in works [89–91]. Previously, vermiculite and kaolin were mechanically activated in a planetary mill, then mixed with PTFE in a mixer; obtained mix were moulded at 50 MPa at room temperature and sintered in an electric furnace at 375°C. Conducted researches showed that the additional introduction of layered silicates and UPTFE in the PTFE/CF composite leads to a significant increase in the wear resistance of PCM. Such enhancement authors associate with the formation of secondary structures on the friction surface containing layered silicates (vermiculite and kaolin) and ultrafine PTFE.

6 Factors affecting the adaptation of PCM to friction conditions

Currently, according to the fundamental paradigms of materials science, the relationship between the processing-structure-properties relationships of the material is undeniable. Studies conducted with PTFE and nanosized particles reveal that the control of the structure by the introduction of fillers allows us to achieve the optimal combination of properties in PCM [91–94]. However, the process of restructuring the PCM structure during various external influences, such as mechanical deformation, heating, and other factors, is rarely studied. The process of formation of the PCM structure depends on the chemical nature, size, structure, and content of various fillers. It is evident that under the action of external influences, the PCM rearrange its structure, and this rearrangement should largely depend on the nature of the external influence and the parameters of the fillers. One of the hardest loads experienced by PCM is frictional loading because the material undergoes plastic and shear deformations, heating, and various mechanochemical processes during friction. The response of the structure to such effects is to a large extent of fundamental and applied interest.

Two main components can be distinguished in adapting materials to friction conditions: restructuring of the polymer matrix and changing the chemical composition of the PCM surface layer. Moreover, a change in composition can also cause a change in the structure of PCM. At the initial stage of attrition, which includes running-in, conditions are created for the transition of the system to processes leading to a minimum of energy costs for friction under steady conditions. It is known that this stage is characterized by the smoothing of microroughness and the establishment of optimal surface roughness, accompanied by plastic deformation of the surface layers, loosening and amorphization of the material structure [26, 90]. These processes, along with high temperature, lead to the formation of an intermediate phase separating the rubbing surfaces.

PCM studies were carried out to determine the adaptability of secondary structures during friction (Figs. 6 and 7). For this, PCM was tested for running-in at a sliding speed (0.2 m/s) and specific load (2 MPa) for (1.0 + 3.0) and (1.0 + 4.0) h, as a control we used the initial results of the corresponding composites. After each test, the weight of the samples was determined.

As can be seen from Figs. 6 and 7, for all the PCM studied, the technological running-in increases the PCM wear resistance compared to the control sample (PCM without running-in). A significant improvement in PCM wear resistance after 1 and 4 h of running-in is observed at a CF content of 1–3 wt.%, where the wear resistance of composites after running-in increased by 2–3 times compared to the control sample. The improvement in the wear resistance of composites does not depend on the time of pre-running. Studies on friction and wear have shown that the wear resistance of composites with a CF content of 1–3 wt.% depends on the content of mechanically activated kaolin and UPTFE. Besides, tribological tests of

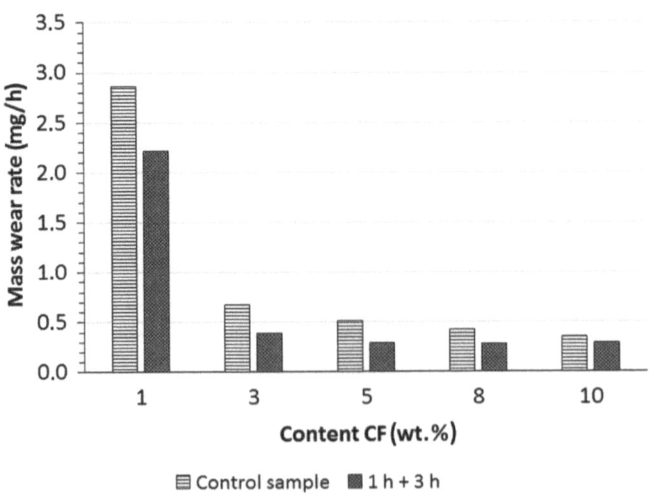

Fig. 6 Comparison of the mass wear rate of PCM composition of PTFE + CF.

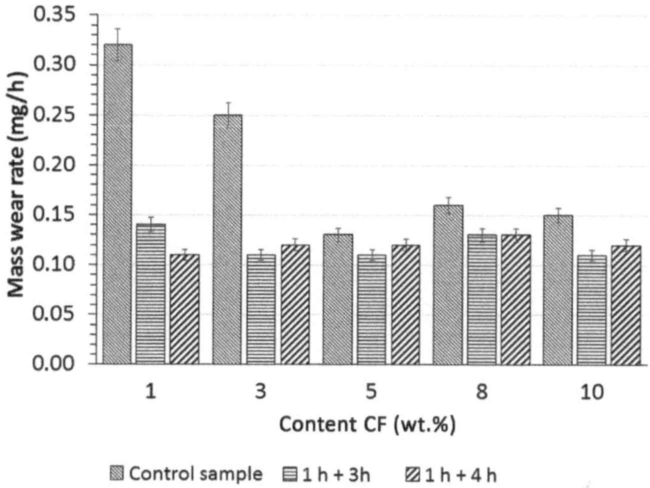

Fig. 7 Comparison of the mass wear rate of the composites PTFE + CF + 1 wt.% kaolinite + 1 wt.% UPTFE.

PCM filled with mechanically activated and nonactivated bentonite for technological running-in were carried out (Fig. 8).

PCM test results based on PTFE and bentonite also showed a positive impact of running-in on the wear resistance of materials. To identify the significance of adaptation processes for various fillers, we constructed graphs of

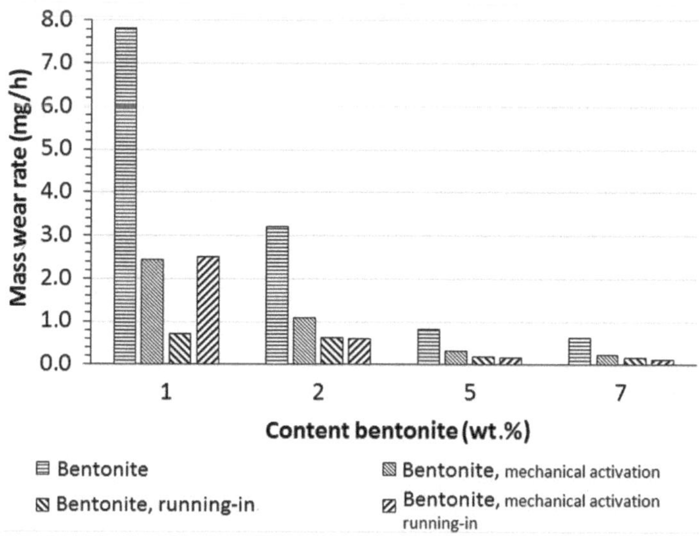

Fig. 8 Comparison of PCM mass wear rate of PTFE + bentonite composites.

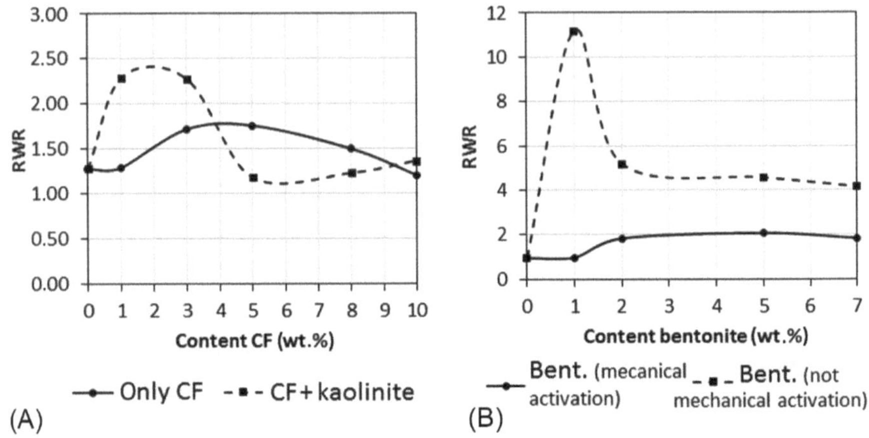

Fig. 9 Comparison of the ratios of mass wear rate (RWR) of PCM without running-in and running-in: (A) PTFE+CF and PTFE+CF+1 wt.% kaolinite+1 wt.% UPTFE; (B) PTFE+bentonite and PTFE+mechanically activated bentonite.

the dependences of the ratio of wear rates (RWR) on the content of fillers (Fig. 9). The RWR indicator can characterize the depth and significance of adaptation processes on the near-surface layers of PCM during friction.

It was found that running-in affects not only the PCM wear resistance but also the wear resistance of the unfilled polymer, which confirms the hypothesis that the adaptation of the material occurs due to the restructuring of the near-surface PCM layers. However, the addition of fillers to the PCM leads to a change-like adaptation, as shown in Fig. 9. Such changes are likely to depend on the on-going physicochemical processes during friction.

To study the processes occurring during friction, we considered the IR spectra of composites containing the same amount of bentonite before and after friction (Fig. 10).

In contrast to the IR-spectra of composites before friction (Fig. 10A), spectra after friction are characterized by new peaks corresponding to the stable products of tribooxidation of the PTFE carbon chain. The appearance of broad bands in the region of 2800–3600 cm^{-1} indicates the presence in the surface layer of associates of OH groups, interconnected by hydrogen bonds, moreover of an intermolecular nature by the shift toward lower frequencies and the intensity of these bands [95]. The absorption band of the unassociated OH group (for comparison) is at 3600 cm^{-1} in the spectrum of the initial bentonite (Fig. 10B).

The following intense bands at 1656 and 1435 cm^{-1} are characteristic of anions of salts of carboxylic acids, in this case, perfluorocarboxylic acids.

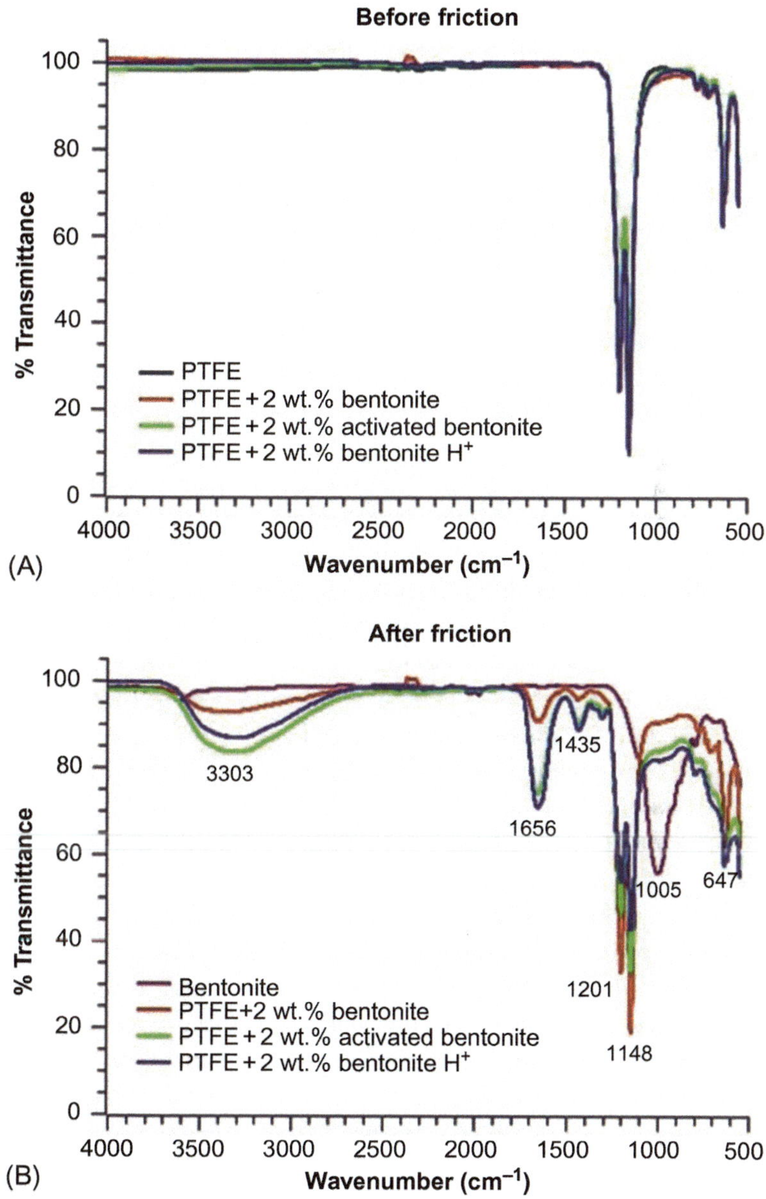

Fig. 10 IR-spectra of composites based on PTFE and bentonite depending on modification: (A) before friction and (B) after friction.

The results of XPS were presented in earlier work [25] to confirm the likelihood of the formation of carboxyl groups as a result of tribooxidation of PTFE-based composites. The presence of carboxyl groups on the transfer films and the friction surface of composites is also indicated by the authors of Refs. [35, 96], who studied the wear of nanocomposites based on PTFE and alumina nanoparticle.

In the process of friction, the polymer matrix in PCM is involved in various tribochemical reactions, as evidenced by many studies. For example, in the friction process of PTFE-based PCM, carboxyl and hydroxyl functional groups are formed, which is confirmed by IR spectra of friction surfaces. Oxidation of perfluoroalcohols leads to the formation of perfluorocarboxylic acids, which then react with cations of mineral fillers and form a salt. The process of salt formation is inhibitory and leads to the structuring of the surface layer, contributing to the formation of a secondary structure adapted to external influences. At this stage, metal cations released from the filler structure, due to the pronounced tendency to complexation, may be more active in the formation of perfluorocarboxylic acid salts. Complexation is a thermodynamically beneficial process and the result of the natural desire of a complex system to stabilize. Therefore we can assume that the formation of complex salts of metal cations from the structure of fillers with perfluorocarboxylic acids through a donor-acceptor interaction is the most likely process that allows one to obtain mutually coordinated labile structures on the friction surface. Formation so-called "cross-linked" structures from fragments of oxidized PTFE macromolecules described in the work of some authors can occur by the mechanism of complexation. According to the authors in Ref. [35], the formation of carboxylic acids in the process of PTFE/alumina tribodestruction causes the formation of a transfer film on the counterbody, since carboxyl groups at the end of PTFE macromolecules react with and chelate to the surface of the metal and alumina particles. Thus the formation of the terminal carboxyl group may be a critical point, in addition to structural rearrangement, which determines the mechanism of adaptation of the PTFE-metal system under friction conditions.

Thus the sharp decrease in RWR (Fig. 9A) in composites with kaolinite, when CF content increases, can be explained by a decrease in the content of active layered silicates relative to CF, which can participate in complexation and the formation of cross-linked structures on the PCM surface. It also explains the generally higher values of RWR for PCM filled with bentonite. Although higher RWR values for PCM containing nonactivated bentonite may raise questions, a comparison of wear resistance of PCM with activated and nonactivated bentonite after running-in showed a small difference. It indicates more profound changes in the PCM structure under the action of frictional loading, i.e., more profound adaptation changes, which explains the high values of RWR.

Based on studies of tribological characteristics, friction coefficient, the temperature in the friction zone, structure of friction surfaces, tribochemical processes using IR spectroscopy, SEM, X-ray spectral analysis, the PCM wear mechanism with carbon fibers, and combined fillers can be represented as follows (Fig. 11). In the surface layer of PCM, the fibers act as microprotrusions, which undergo significant loads during wear due to shear stresses, which leads to a weakening of the interaction at the polymer-fiber interface (see Fig. 11A). The formation of secondary structure causes an improvement of the wear resistance of PCM with combined fillers during friction. Such secondary structure protects the surface from destruction, reducing the value of contact pressure (see Fig. 11B). The effect of self-organization of the developed composites under friction is caused by several processes, due to which a new, more wear-resistant surface is formed. It is shown that the restructuring occurs under the influence of temperature and shear deformations, selective wear of less stable polymer regions and the gradual enrichment of the surface with more stable thermostable filler particles. As a result, not only an oriented structure is formed, but also a discrete surface, where the filler particles are elements that protrude on the surface and absorb the load and decrease stress during frictional contact. Due to an increase in the actual contact area, respectively, this increases the load capacity of the material and reduces PCM wear.

The next factor for providing the materials with adaptive properties is the use of components characterized by reactive activity in the processes of

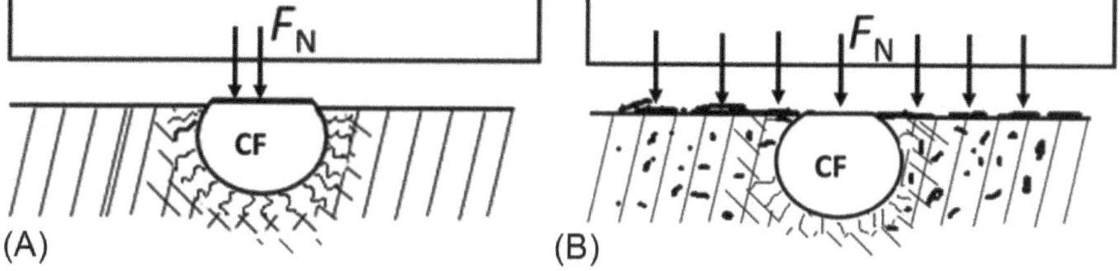

Fig. 11 Models of the PCM wear mechanism depending on the content of fillers: (A) PTFE/CF; (B) PTFE/CF/layered silicates (UPTFE).

tribodestruction and subsequent tribostructuring. Based on the identified products of tribooxidation by the analysis of the composites IR spectra, the most probable schemes of tribochemical reactions are substantiated. Cations with pronounced coordinating activity are capable of complexation reactions with fragments of a PTFE macromolecule with terminal carboxyl groups. Using PCM based on PTFE and mechanically activated bentonite and PCM with combined fillers, as an example, it is shown that metal cations that are part of layered silicates actively participate in creating a strong cross-linked surface and form complex compounds as a result of mechanochemical processes.

The reactions of tribooxidation, destruction, and structuring of surface layers under the action of significant shear stress with the participation of nanofillers initiate the formation of secondary structures from the products of chemical and physicochemical interaction, not only in the case of PTFE but also in the case of UHMWPE nanocomposites. It has been shown that tribooxidative degradation and structuring processes occur on the friction surface with an increase in crystallinity and orientation of the surface layers in the direction of friction, as well as the development and formation of complex ordered secondary structures that contribute to a significant increase in the wear resistance of UHMWPE upon modification with nanosized ceramics.

Based on the research and analysis of the results, it is possible to identify the main factors that contribute to the adaptation of the structure of materials to friction conditions:

1. Structural factors
2. Chemical factors

Structural factors:

- the molecular weight of the polymer matrix;
- ambient temperature;
- supramolecular structure of PCM before friction;
- the particle size of the filler;
- the form factor of filler particles.

Chemical factors:

- composition and structure of fillers;
- the reactivity of fillers and their components with respect to wear products under conditions of mechanochemical reactions;
- composition of the environment.

Ensuring the liability of the structural elements of the polymer by varying the parameters of the technology and formulation allows us to achieve the optimal set of structural factors, which was demonstrated by using CF as fillers.

The addition of components chemically active with respect to the oxidized forms of polymer wear products under friction conditions, such as bentonite or kaolinite, makes it possible to achieve an optimal set of chemical factors. The superposition of both components of the adaptation to friction conditions leads to a synergistic effect, which was demonstrated using combined fillers.

7 Conclusion

The paper considers the approaches and technological methods used in the development of PCM. One of the key factors influencing the properties of PCM during their development is the processing method: modification of the filler and polymer, a combination of components, production method (extrusion, compacting, injection moulding and sintering).

Fiber-reinforced PCM are of particular interest owing to their excellent mechanical and tribological properties. In PCM filled with fibers, the main factor determining the mechanical and tribological properties is the interfacial interaction between fibers and polymers.

The use of layered silicates as polymer modifiers leads to the creation of materials characterized by high wear resistance and mechanical properties, low friction coefficient, and self-lubricating during friction.

Such modification methods as ultrasonic treatment, mechanical activation, and the use of a compatibilizer are used to improve the interaction of nanosized fillers with a polymer matrix.

Polymer composite materials based on PTFE and UHMWPE can operate in dry sliding conditions without the use of lubricants. The studies of PCM with self-lubricating properties have shown that tribooxidative degradation and structuring processes occur on the friction surface with an increase in crystallinity and orientation of the surface layers in the direction of friction. The wear resistance of UHMWPE with nanoceramics increases due to the formation of complex-ordered secondary structures on a friction surface.

Acknowledgment

This research was funded by the Ministry of Science and Higher Education of the Russian Federation (Grant Number [FSRG-2020-0017]).

References

[1] S. Sharma, J. Bijwe, S. Panier, Exploration of potential of Zylon and Aramid fibers to enhance the abrasive wear performance of polymers, Wear 422–423 (2019) 180–190.

[2] W. Luo, Q. Liu, Y. Li, S. Zhou, H. Zou, M. Liang, Enhanced mechanical and tribological properties in polyphenylene sulfide/polytetrafluoroethylene composites reinforced by short carbon fiber, Compos. Part B: Eng. 91 (2016) 579–588.
[3] R.-M. Wang, S.-R. Zheng, Y.-P. Zheng, Polymer Matrix Composites and Technology, Woodhead Publishing in Materials, WP, Woodhead Publ., Oxford, 2011.
[4] V. Mittal, Polymer layered silicate nanocomposites: a review, Materials 2 (2009) 992–1057.
[5] M.F. Brigatti, D. Malferrari, A. Laurora, C. Elmi, Structure and mineralogy of layer silicates: recent perspectives and new trends, in: M.F. Brigatti, A. Mottana (Eds.), Layered Mineral Structures and Their Application in Advanced Technologies, Mineralogical Society of Great Britain and Ireland, London, 2011, pp. 1–71.
[6] L.W. Carter, J.G. Hendricks, D.S. Bolley, Elastomer reinforced with a modified clay, 1950. US Patent 2531396.
[7] S. Fujiwara, T. Sakamoto, Polyamide-organoclay composites, 1976. Japanese Patent 109998.
[8] Y. Fukushima, S. Inagaki, Synthesis of an intercalated compound of montmorillonite and 6-polyamide, in: J.L. Atwood, J.E.D. Davies (Eds.), Inclusion Phenomena in Inorganic, Organic, and Organometallic Hosts, Springer Netherlands, Dordrecht, 1987, pp. 365–374.
[9] M. Okamoto, Polymer/layered silicate nano-composites: structure development and processing operations, Int. Polym. Process. 21 (2006) 487–496.
[10] T. Lan, P.D. Kaviratna, T.J. Pinnavaia, On the nature of polyimide-clay hybrid composites, Chem. Mater. 6 (1994) 573–575.
[11] K. Yano, A. Usuki, A. Okada, Synthesis and properties of polyimide-clay hybrid films, J. Polym. Sci. A: Polym. Chem. 35 (1997) 2289–2294.
[12] S. Sinha Ray, M. Okamoto, Polymer/layered silicate nanocomposites: a review from preparation to processing, Prog. Polym. Sci. 28 (2003) 1539–1641.
[13] R.A. Vaia, R.K. Teukolsky, E.P. Giannelis, Interlayer structure and molecular environment of alkylammonium layered silicates, Chem. Mater. 6 (1994) 1017–1022.
[14] A. Pegoretti, A. Dorigato, A. Penati, Tensile mechanical response of polyethylene–clay nanocomposites, Express Polym. Lett. 1 (2007) 123–131.
[15] V.A. Gerasin, M.A. Guseva, A.V. Rebrov, Y.M. Korolev, E.M. Antipov, The effect of the physicomechanical characteristics of the polymer matrix and structure of a filler on the stress-strain behavior of polymer-montmorrilonite nanocomposites, Polym. Sci., Ser. A 51 (2009) 317–330.
[16] A. Olad, Polymer/clay nanocomposites, in: B. Reddy (Ed.), Advances in Diverse Industrial Applications of Nanocomposites, InTech, 2011.
[17] M. Tortora, G. Gorrasi, V. Vittoria, G. Galli, S. Ritrovati, E. Chiellini, Structural characterization and transport properties of organically modified montmorillonite/polyurethane nanocomposites, Polymer 43 (2002) 6147–6157.
[18] N.A. Rakhimova, S.V. Kudashev, Organophilization of Na^+-montmorillonite with polyfluorinated alcohols, Russ. J. Appl. Chem. 83 (2010) 2035–2040.
[19] V.N. Moraru, Structure formation of alkylammonium montmorillonites in organic media, Appl. Clay Sci. 19 (2001) 11–26.
[20] S.-J. Park, K. Li, S.-K. Hong, Preparation and characterization of layered silicate-modified ultrahigh-molecular-weight polyethylene nanocomposites, J. Ind. Eng. Chem. 11 (2005) 561–566.
[21] Y.V. Kirillina, N.N. Lazareva, S.A. Sleptsova, A.A. Okhlopkova, Effect of organomodified layered silicates on the properties and structure of polytetrafluoroethylene, Polym. Sci., Ser. A 58 (2016) 95–101.
[22] I.V. Kirillina, L.A. Nikiforov, A.A. Okhlopkova, S.A. Sleptsova, C. Yoon, J.-H. Cho, Nanocomposites based on polytetrafluoroethylene and ultrahigh molecular weight polyethylene: a brief review, Bull. Kor. Chem. Soc. 35 (2014) 3411–3420.

[23] L. Nikiforov, T. Okhlopkova, I. Kapitonova, S. Sleptsova, A. Okhlopkova, E. Shim, J.-H. Cho, Surfactant effects on structure and mechanical properties of ultrahigh-molecular-weight polyethylene/layered silicate composites, Molecules 22 (2017) 2149.
[24] G. Wang, Y. Chen, Q. Wang, Structure and properties of poly(ethylene terephthalate)/ Na^+-montmorillonite nanocomposites prepared by solid state shear milling (S^3M) method, J Polym Sci B 46 (2008) 807–817.
[25] S.A. Sleptsova, A.A. Okhlopkova, I.V. Kapitonova, N.N. Lazareva, M.M. Makarov, L.A. Nikiforov, Spectroscopic study of tribooxidation processes in modified PTFE, J. Frict. Wear 37 (2016) 129–135.
[26] S.A. Sleptsova, N.N. Lazareva, V.I. Fedoseeva, Y.V. Kapitonova, A.A. Okhlopkova, The influence of metal cations of mechanoactivated bentonite on tribochemical processes in PTFE, J. Frict. Wear 39 (2018) 469–475.
[27] A.M. Kalinkin, A.A. Politov, V.V. Boldyrev, E.V. Kalinkina, V.N. Makarov, V.T. Kalinnikov, Study of mechanical activation of diopside in a CO_2 atmosphere, J. Mater. Synth. Process. 10 (2002) 61–67.
[28] S. Yariv, W. Bodenheimer, Specific and sensitive reactions with the aid of montmorillonite II: pyrocatechol and its derivatives, Isr. J. Chem. 2 (1964) 197–200.
[29] P. Baláž, Mechanochemistry in Nanoscience and Minerals Engineering, Springer, Berlin, 2008.
[30] I.A. Vorsina, T.F. Grigor'eva, E.T. Devyatkina, S.V. Vosmerikov, T.A. Udalova, N.Z. Lyakhov, Interaction of organic acids with silicates in the course of joint mechanical activation, Russ. J. Appl. Chem. 90 (2017) 334–341.
[31] S. Laukkanen, S.A. Sleptsova, P.N. Tarasov, V.I. Fedoseeva, E.V. Abakunov, S. Laukkanen, Influence of layered silicates 1:1 on PTFE properties, in: Proceedings of the International Symposium "Engineering and Earth Sciences: Applied and Fundamental Research" Dedicated to the 85th Anniversary of H.I. Ibragimov (ISEES 2019), Atlantis Press, Grozny, Russia, 2019.
[32] A. Teutli-Sequeira, M. Solache-Ríos, P. Balderas-Hernández, Modification effects of hematite with aluminum hydroxide on the removal of fluoride ions from water, Water Air Soil Pollut. 223 (2012) 319–327.
[33] O.S. Ayanda, F.A. Adekola, A review of niobium-tantalum separation in hydrometallurgy, J. Miner. Mater. Charact. Eng. 10 (2011) 245–256.
[34] S.A. Sleptsova, Y.V. Kapitonova, N.N. Lazareva, A.A. Okhlopkova, L.A. Nikiforov, V.I. Fedoseeva, L.A. Grigoryeva, U. Gohs, Effect of Magnesium Spinel on the Properties and Structure of Composites PTFE/Layered Silicate. Preprints 2018 https://doi.org/10.20944/preprints201811.0258.v1. 2018110258.
[35] K.L. Harris, A.A. Pitenis, W.G. Sawyer, B.A. Krick, G.S. Blackman, D.J. Kasprzak, C.P. Junk, PTFE tribology and the role of mechanochemistry in the development of protective surface films, Macromolecules 48 (2015) 3739–3745.
[36] B.I. Kostetsky, The structural-energetic concept in the theory of friction and wear (synergism and self-organization), Wear 159 (1992) 1–15.
[37] D.R. Paul, L.M. Robeson, Polymer nanotechnology: nanocomposites, Polymer 49 (2008) 3187–3204.
[38] C.P. Poole, F.J. Owens, Introduction to Nanotechnology, J. Wiley, Hoboken, NJ, 2003.
[39] B. Bhushan (Ed.), Springer Handbook of Nanotechnology, fourth ed., Springer HandbooksSpringer, Berlin, 2017.
[40] B.J. Briscoe, S.K. Sinha, Tribological Applications of Polymers and Their Composites: Past, Present and Future Prospects, Tribology and Interface Engineering Series, Elsevier, 2008, pp. 1–14.
[41] D.L. Burris, B. Boesl, G.R. Bourne, W.G. Sawyer, Polymeric nanocomposites for tribological applications, Macromol. Mater. Eng. 292 (2007) 387–402.

[42] A.P. Krasnov, V.N. Aderikha, O.V. Afonicheva, V.A. Mit', N.N. Tikhonov, A.Y. Vasil'kov, E.E. Said-Galiev, A.V. Naumkin, A.Y. Nikolaev, Categorization system of nanofillers to polymer composites, J. Frict. Wear 31 (2010) 68–80.
[43] Y. Zhang, Y. Chen, P. Westerhoff, K. Hristovski, J.C. Crittenden, Stability of commercial metal oxide nanoparticles in water, Water Res. 42 (2008) 2204–2212.
[44] F. Hussain, M. Hojjati, M. Okamoto, R.E. Gorga, Review article: polymer-matrix nanocomposites, processing, manufacturing, and application: an overview, J. Compos. Mater. 40 (2006) 1511–1575.
[45] I.E. Uflyand, E.G. Drogan, V.E. Burlakova, K.A. Kydralieva, I.N. Shershneva, G.I. Dzhardimalieva, Testing the mechanical and tribological properties of new metal-polymer nanocomposite materials based on linear low-density polyethylene and $Al_{65}Cu_{22}Fe_{13}$ quasicrystals, Polym. Test. 74 (2019) 178–186.
[46] D. Xiong, J. Lin, D. Fan, Z. Jin, Wear of nano-TiO_2/UHMWPE composites radiated by gamma ray under physiological saline water lubrication, J. Mater. Sci. Mater. Med. 18 (2007) 2131–2135.
[47] A.A. Okhlopkova, R.V. Borisova, T.A. Okhlopkova, L.A. Nikiforov, Structure and friction behavior of UHMWPE/inorganic nanoparticles, Key Eng. Mater. 670 (2015) 69–75.
[48] W.G. Sawyer, K.D. Freudenberg, P. Bhimaraj, L.S. Schadler, A study on the friction and wear behavior of PTFE filled with alumina nanoparticles, Wear 254 (2003) 573–580.
[49] T.A. Okhlopkova, R.V. Borisova, L.A. Nikiforov, A.M. Spiridonov, P.P. Sharin, A.A. Okhlopkova, Technology of liquid-phase compounding of ultra-high-molecular-weight polyethylene with nanoparticles of inorganic compounds under the action of ultrasonic vibrations, Russ. J. Appl. Chem. 89 (2016) 1469–1476.
[50] J.S. Bergman, G.W. Coates, H. Chen, E.P. Giannelis, M.G. Thomas, Synthesis and characterization of polyolefin–silicate nanocomposites: a catalyst intercalation and in situ polymerization approach, Chem. Commun. (1999) 2179–2180.
[51] A.Y. Vasil'kov, D.A. Migulin, A.V. Naumkin, Y.V. Zubavichus, A.V. Budnikov, O.G. Ellert, Y.V. Maksimov, A.M. Muzafarov, Hybrid materials based on polymethylsilsesquioxanes containing Fe, Pt, and Fe–Pt metallic nanoparticles, Russ. J. Phys. Chem. A 91 (2017) 2188–2194.
[52] A.M. Nemeryuk, M.M. Lylina, Synthesis of dichloride-diphenylacetonitrile palladium and metal-polymer composite based on UHMWPE, Orient. J. Chem. 32 (2016) 927–932.
[53] V. Glushko, L. Blokhina, E. Anisimova, M. Bogdanovskaya, V. Kozhukhov, T. Cherdyntseva, Silver nanoparticles modification of ultra high molecular weight polyethylene in non-aqueous medium, Orient. J. Chem. 32 (2016) 3089–3097.
[54] T. Ma, T. Zhang, P. Gao, J. Zhang, Synthesis and properties of ultrahigh molecular weight polyethylene/WS2 nanoparticle fiber for bullet-proof materials, Chin. Sci. Bull. 58 (2013) 945–948.
[55] Z. Cao, G. Shi, X. Yan, Q. Wang, *In situ* fabrication of CuO/UHMWPE nanocomposites and their tribological performance, J. Appl. Polym. Sci. 136 (2019) 47925.
[56] R.V. Borisova, A.M. Spiridonov, T.A. Okhlopkova, L.A. Nikiforov, A.N. Golikov, N.V. Shadrinov, A.A. Okhlopkova, Bromination of UHMWPE surface as a method of changing adhesion to nanoparticles, Mater. Today Commun. 14 (2018) 65–71.
[57] R.V. Borisova, L.A. Nikiforov, A.M. Spiridonov, T.A. Okhlopkova, A.A. Okhlopkova, N.S. Koryakina, The influence of brominated UHMWPE on the tribological characteristics and wear of polymeric nanocomposites based on UHMWPE and nanoparticles, J. Frict. Wear 40 (2019) 27–32.
[58] S. Panin, L.A. Kornienko, N.X. Thuc, L.R. Ivanova, S.V. Shilko, Role of micro- and nanofillers in abrasive wear of composites based on ultra-high molecular weight polyethylene, Adv. Mater. Res. 1040 (2014) 148–154.

[59] J. Karger-Kocsis, H. Mahmood, A. Pegoretti, Recent advances in fiber/matrix interphase engineering for polymer composites, Prog. Mater. Sci. 73 (2015) 1–43.
[60] K. Friedrich, Z. Lu, A.M. Hager, Recent advances in polymer composites' tribology, Wear 190 (1995) 139–144.
[61] C.A. Chairman, S.P. Kumaresh Babu, Mechanical and abrasive wear behavior of glass and basalt fabric-reinforced epoxy composites, J. Appl. Polym. Sci. 130 (2013) 120–130.
[62] P.W.R. Beaumont, C. Soutis, A. Hodzic, The Structural Integrity of Carbon Fiber Composites—Fifty Years of Progress and Achievement of the Science, Development, and Applications. Springer, Cham, (2017), https://doi.org/10.1007/978-3-319-46120-5.
[63] M.C. Paiva, C.A. Bernardo, M. Nardin, Mechanical, surface and interfacial characterisation of pitch and PAN-based carbon fibres, Carbon 38 (2000) 1323–1337.
[64] I.A. Rashkovan, Y.G. Korabel'nikov, The effect of fiber surface treatment on its strength and adhesion to the matrix, Compos. Sci. Technol. 57 (1997) 1017–1022.
[65] P.C. Varelidis, R.L. McCullough, C.D. Papaspyrides, The effect on the mechanical properties of carbon/epoxy composites of polyamide coatings on the fibers, Compos. Sci. Technol. 59 (1999) 1813–1823.
[66] S. Qian-Qian, C. Xian-hua, On the friction and wear behavior of PTFE composite filled with rare earths treated carbon fibers under oil-lubricated condition, Wear 260 (2006) 1243–1247.
[67] X.H. Cheng, Q.Q. Shang-guan, Effect of rare earths on mechanical and tribological properties of carbon fibers reinforced PTFE composite, Tribol. Lett. 21 (2006) 153–160.
[68] Y. Shi, X. Feng, H. Wang, X. Lu, Tribological properties of PTFE composites filled with surface-treated carbon fiber, J. Mater. Sci. 42 (2007) 8465–8469.
[69] V.A. Shelestova, P.N. Grakovich, S.G. Danchenko, V.A. Smirnov, New antifriction materials of the Fluvis group based on modified carbon fibers, Chem. Pet. Eng. 42 (2006) 663–666.
[70] V.A. Shelestova, P.N. Grakovich, S.F. Zhandarov, A fluoropolymer coating on carbon fibers improves their adhesive interaction with PTFE matrix, Compos. Interfaces 18 (2011) 419–440.
[71] V.A. Shelestova, S.F. Zhandarov, S.G. Danchenko, P.N. Grakovich, Surface modification of carbon fiber by fluoropolymer in a low-temperature plasma, Inorg. Mater. Appl. Res. 6 (2015) 219–224.
[72] V.N. Aderikha, V.A. Shapovalov, Mechanical and tribological behavior of PTFE–polyoxadiazole fiber composites. Effect of filler treatment, Wear 271 (2011) 970–976.
[73] L. Wang, S. Wang, D. Wen, G. Wang, L. Yang, The study of the composite material Go/CF/PTFE tribological property, in: MATEC Web of Conferences, vol. 130, 2017, p. 09005.
[74] T. Sathishkumar, S. Satheeshkumar, J. Naveen, Glass fiber-reinforced polymer composites—a review, J. Reinf. Plast. Compos. 33 (2014) 1258–1275.
[75] K. Tanaka, Friction and wear of glass and carbon fiber-filled thermoplastic polymers, J. Lubr. Technol. 99 (1977) 408.
[76] H. Voss, K. Friedrich, On the wear behaviour of short-fibre-reinforced peek composites, Wear 116 (1987) 1–18.
[77] S. Bahadur, Y. Zheng, Mechanical and tribological behavior of polyester reinforced with short glass fibers, Wear 137 (1990) 251–266.
[78] Y. Wang, Z. Yin, H. Li, G. Gao, X. Zhang, Friction and wear characteristics of ultrahigh molecular weight polyethylene (UHMWPE) composites containing glass fibers and carbon fibers under dry and water-lubricated conditions, Wear 380–381 (2017) 42–51.
[79] B. Suresha, K. Shiva Kumar, S. Seetharamu, P. Sampath Kumaran, Friction and dry sliding wear behavior of carbon and glass fabric reinforced vinyl ester composites, Tribol. Int. 43 (2010) 602–609.

[80] H.C.Y. Cartledge, C. Baillie, Y.-W. Mai, Friction and wear mechanisms of a thermoplastic composite GF/PA6 subjected to different thermal histories, Wear 194 (1996) 178–184.
[81] L. Zhang, G. Zhang, L. Chang, B. Wetzel, B. Jim, Q. Wang, Distinct tribological mechanisms of silica nanoparticles in epoxy composites reinforced with carbon nanotubes, carbon fibers and glass fibers, Tribol. Int. 104 (2016) 225–236.
[82] T. Wan, S. Liao, K. Wang, P. Yan, M. Clifford, Multi-scale hybrid polyamide 6 composites reinforced with nano-scale clay and micro-scale short glass fibre, Compos. A: Appl. Sci. Manuf. 50 (2013) 31–38.
[83] M. Conte, A. Igartua, Study of PTFE composites tribological behavior, Wear 296 (2012) 568–574.
[84] F. Song, Q. Wang, T. Wang, Effects of glass fiber and molybdenum disulfide on tribological behaviors and PV limit of chopped carbon fiber reinforced polytetrafluoroethylene composites, Tribol. Int. 104 (2016) 392–401.
[85] V. Dhand, G. Mittal, K.Y. Rhee, S.-J. Park, D. Hui, A short review on basalt fiber reinforced polymer composites, Compos. Part B: Eng. 73 (2015) 166–180.
[86] V. Fiore, T. Scalici, G. Di Bella, A. Valenza, A review on basalt fibre and its composites, Compos. Part B: Eng. 74 (2015) 74–94.
[87] J. Wang, B. Chen, N. Liu, G. Han, F. Yan, Combined effects of fiber/matrix interface and water absorption on the tribological behaviors of water-lubricated polytetrafluoroethylene-based composites reinforced with carbon and basalt fibers, Compos. A: Appl. Sci. Manuf. 59 (2014) 85–92.
[88] X. Zhang, X. Pei, Q. Wang, Friction and wear properties of polyimide matrix composites reinforced with short basalt fibers, J. Appl. Polym. Sci. 111 (2009) 2980–2985.
[89] A.A. Ohlopkova, T.S. Struchkova, A.P. Vasilev, A.G. Alexeev, Studying the properties and structure of polytetrafluoroethylene filled with Belum modified carbon fibers, J. Frict. Wear 37 (2016) 529–534.
[90] A.P. Vasilev, A.A. Okhlopkova, T.S. Struchkova, P.N. Grakovich, A.L. Bashlakova, Investigation of the influence of complex fillers on the properties and structure of polytetrafluoroethylene, J. Frict. Wear 39 (2018) 427–432.
[91] A.P. Vasilev, T.S. Struchkova, L.A. Nikiforov, A.A. Okhlopkova, P.N. Grakovich, E.L. Shim, J.-H. Cho, Mechanical and tribological properties of polytetrafluoroethylene composites with carbon fiber and layered silicate fillers, Molecules 24 (2019) 224.
[92] A.A. Okhlopkova, P.N. Petrova, S.N. Popov, E.G. Avvakumov, Composites based on polymers doped with disperse particles of inorganic compounds, Chem. Sustain. Dev. 12 (2004) 609–617.
[93] A.A. Okhlopkova, P.N. Petrova, O.V. Gogoleva, A.L. Fedorov, Fluoropolymer composites of triboengineering purposes, J. Frict. Wear 28 (2007) 551–556.
[94] A.A. Okhlopkova, S.A. Sleptsova, Influence of nanoceramics on the properties of polytetrafluoroethylene, Mech. Compos. Mater. 39 (2003) 123–128.
[95] J.W. Nibler, Infrared absorption spectroscopy. Second edition (Nakanishi, Koji), J. Chem. Educ. 55 (1978) A316.
[96] J. Ye, D. Burris, T. Xie, A review of transfer films and their role in ultra-low-wear sliding of polymers, Lubricants 4 (2016) 4.

CHAPTER 4

Tribological behavior of fiber-reinforced polymer composites

Ahmed Abdelbary[a] and Yasser S. Mohamed[b]
[a]Egyptian Government, Alexandria, Egypt
[b]Faculty of Engineering, Department of Mechanical Engineering, Alexandria University, Alexandria, Egypt

1 Introduction

Over the last few years, composites have attracted prolonged interest for tribological applications because of their extraordinary overall performance. In particular, the success of introducing polymeric substances in different applications has led to using polymers as matrices in composites, wherein the modifications to their characteristics are probably performed by including one or more filler reinforcements. These reinforcements can be fibers, solid particles, or any other forms to achieve the excessive strength, hardness, elasticity, or self-lubrication of the matrix. Generally, the natures of fibers are synthetic or natural. Most fiber-strengthened polymer composites are carbon, glass, or aramid fibers in a thermoset matrix or a thermoplastic matrix.

The fibers, of about 5–10 μm in diameter, are notably stronger than the polymer matrix. However, they will be predisposed to be brittle. Therefore, the construction of a composite allows taking benefit of the outstanding toughness and strength properties of carbon fiber or glass fiber by embedding these fibers in a more compliant polymer matrix. The reinforcement polymer fibers can be organized in a variety of orientations:

- Short randomly or aligned fibers.
- Continuous randomly or aligned fibers.
- Woven or unidirectional continuous fibers.

Fiber-reinforced polymer composites (FRPCs) are now quite common in many applications, including automotive, aerospace, sports, commercial, and many others. Yet, to date, there were many extensive studies performed to investigate the impact of reinforced fibers on the tribological conduct of polymeric composites.

Tribology of Polymer Composites
https://doi.org/10.1016/B978-0-12-819767-7.00004-9

A communal conclusion of the majority of the tribological investigations of polymer composites is that the friction and wear cannot be intrinsic properties. In fact, it depends upon the essential variables, which include the fiber material, the fiber orientation, the matrix polymer, and so forth. Furthermore, it relies upon constituent variables that include applied load, counter face material, sliding velocity, and others [1, 2]. Although reinforcing a polymer with fibers improves the tribological performance of the majority polymer, in some cases it could make that worse.

In most polymeric composites, extraordinary heating has been shown at the interfacial contact during sliding that arises from low thermal conductivity and high stiffness. It is postulated that the rate of wear enlarges very abruptly after a specific critical temperature [3].

2 Polymer matrix

As mentioned above, polymers will often comprise other reinforcements to shape a composite. Such reinforcements include synthetic, natural, metallic, organic, or inorganic materials. The binder (polymer) is normally referred to as the *matrix* or *resin*. The function of the matrix is to support the fiber as well as to protect it. Usually, the polymer resin (matrix) is of significantly lower density, stiffness, and strength than reinforcements, even though the mixture of matrix and reinforcements may have excessive stiffness and strength.

Fibers, as a reinforcement, are collected (bounded) in order to shape a desired element that can support loads. In polymeric composites, the introduction of proper fibers as reinforcements can alter the tribological and mechanical behavior compared to net polymers [4]. It is worth mentioning that the mechanical properties of polymer composites not only depend on the reinforcing fibers, but also on the polymer matrix. Usually, polymer materials can be further categorized into three categories: thermoplastics, thermosets, and elastomers.

2.1 Thermosetting matrices

Thermoset polymers are extremely cross-linked plastics that are dried by heat only, or by pressure and heat, or in some cases light irradiation may be applied. The most frequently used thermosetting matrices are polyester and epoxy. These thermosets cover a wide range of mechanical, physical, and chemical properties. Generally, thermosetting resins are isotropic. Besides, at the heat distortion temperature, they lose their stiffness properties.

This property defines an upper limit for their tribological applications in engineering systems. However, in this respect, polyesters are generally lower than epoxy resins.

2.2 Thermoplastic matrices

The structures of thermoplastics comprise one- or two-dimensional molecules. Therefore, at elevated temperatures, they have a tendency to become softer, and they roll back their properties by cooling. The most frequently used thermoplastic polymers include polyamides (PA), polyethylene (PE), polypropylene (PP), poly vinyl chloride (PVC), and polycarbonate (PC). Dissimilar to thermosets, thermoplastics can have anisotropic characteristics according to the curing cycle. They are generally used in mechanical products in which fiber reinforcement is achieved by injection molding. Moreover, an important property is that they *creep* under constant load. Under such condition of constant applied load, the strain increases with time. This means that during deformation, a load redistribution between the resin and fibers will be expected.

3 Fibers

Fibers in nature are solid elongated substances with an ultrahigh aspect ratio and an exceptional directional asset. Natural or synthetic fibers are combined with polymer matrices to achieve one or more tribological properties, which can be briefed in the next points:

(1) Superior load support by filler with high wear resistance.
(2) Inhibition of bulky film formation.
(3) Improved counter face adhesion/transfer film.
(4) Dimensional control of wear debris.

However, the discrepancy ascends because the improvement in one property possibly degrades other properties. Consequently, experimental laboratory investigations are obligatory to study the impact of the fiber on the polymer matrix.

3.1 Natural fibers

A simple definition of natural fibers (NF) is that they are fibers that aren't man-made or synthetic. They can be obtained from any natural source, including plants and animals [5]. Natural FRPCs (or green composites) have become popular in various applications due to the superior properties as well as the advanced benefits of NF over synthetic fibers. In particular, they have several properties that include being inexpensive, light weight, less damaging to processing devices, appropriate mechanical properties such as tensile and flexural modulus, renewable resources, improved surface finish of the molded part composite, flexibility in processing, biodegradability, and negligible health dangers [6].

On the other hand, there are some disadvantages of using NFs, such as weak bindings between the fiber and polymer matrix because of the fiber structure, which permits the absorption of moisture. This absorption ability is due to the existence of lignin, cellulose, hemicelluloses, and waxy and pectin substances. Accordingly, specific treatments are performed using reagent functional groups, which have ability for responding of the fiber structures and changing their composition [7]. Consequently, fiber modification results in a reduction of the absorbing moisture of the NFs that, in turn, results in an exceptional enhancement incompatibility between the matrix and the fiber. The diverse kinds of natural FRPCs are particularly significant in automotive, aerospace, sports, and other tribological applications [8, 9]. Tables 1 and 2 provide the chemical formulation and physicomechanical properties of some collective NFs [10].

Table 1 Chemical composition of some collective NFs.

Fiber	Cellulose (wt.%)	Hemicellulose (wt.%)	Lignin (wt.%)	Waxes (wt.%)
Abaca	56–63	20–25	7–9	3
Bagasse	55.2	16.8	25.3	–
Bamboo	26–43	30	21–31	–
Coir	32–43	0.15–0.25	40–45	–
Curaua	73.6	9.9	7.5	–
Flax	71	18.6–20.6	2.2	1.5
Hemp	68	15	10	0.8
Jute	61–71	14–20	12–13	0.5
Kenaf	72	20.3	9	–
Oil palm	65	–	29	–
Pineapple	81	–	12.7	–
Ramie	68.6–76.2	13–16	0.6–0.7	0.3
Rice husk	35–45	19–25	20	–
Rice straw	41–57	33	8–19	8–38
Sisal	65	12	9.9	2
Wheat straw	38–45	15–31	12–20	–

From O. Faruk, A.K. Bledzki, H.-P. Fink, M. Sain, Biocomposites reinforced with natural fibers: 2000–2010, Prog. Polym. Sci. 37(11) (2012) 1552–1596, https://doi.org/10.1016/j.progpolymsci.2012.04.003, with permission from Elsevier.

Table 2 Physicomechanical properties of some common natural fibers.

Fiber	Density (g/cm^3)	Tensile strength (MPa)	Young's modulus (GPa)	Elongation at break (%)
OPEFB	0.7–1.55	248	3.2	2.5
Flax	1.4	88–1500	60–80	1.2–1.6
Hemp	1.48	550–900	70	1.6
Jute	1.46	400–800	10–30	1.8
Ramie	1.5	500	44	2
Coir	1.25	220	6	15–25
Sisal	1.33	600–700	38	2–3
Abaca	1.5	980	–	–
Cotton	1.51	400	12	3–10
Kenaf (bast)	1.2	295	–	2.7–6.9
Kenaf (core)	0.21	–	–	–
Bagasse	1.2	20–290	19.7–27.1	1.1
Henequen	1.4	430–580	–	3–4.7
Pineapple	1.5	170–1672	82	1–3
Banana	1.35	335	33.8	53

From O. Faruk, A.K. Bledzki, H.-P. Fink, M. Sain, Biocomposites reinforced with natural fibers: 2000–2010, Prog. Polym. Sci. 37(11) (2012) 1552–1596, https://doi.org/10.1016/j.progpolymsci.2012.04.003, with permission from Elsevier.

3.2 Synthetic fibers

There are many forms of synthetic (or man-made) fibers used for reinforcing polymeric matrices. In general, glass fibers (GF) are broadly utilized in thermoplastic matrices as they provide excessive stiffness, chemical and impact resistance, and thermal stabilization. On the other hand, if the highest stiffness is required, carbon fibers (CF) can be used instead. They are designed in diameters analogous to GF with diameters alternating from 9 to 17 μm. Carbon fibers can be categorized into numerous classes based on their strength, modulus, and the temperature of the final heat treatment. Furthermore, there are other categories of synthetic fibers that can be applied such as basalt (BF), aramid (AF), polyethylene terephthalate (PET-F), polyacrylonitrile (PAN-F), and polypropylene fibers (PP-F). Figs. 1 and 2 show the Ashby plots of some mechanical properties of different fiber kinds [11].

4 Tribological properties of short fiber composites

The applications of different sorts of short fibers deliver prospects to enhance the tribological behavior of polymer composites. For instance, graphite flakes and polytetrafluoroethylene (PTFE) are often integrated to reduce

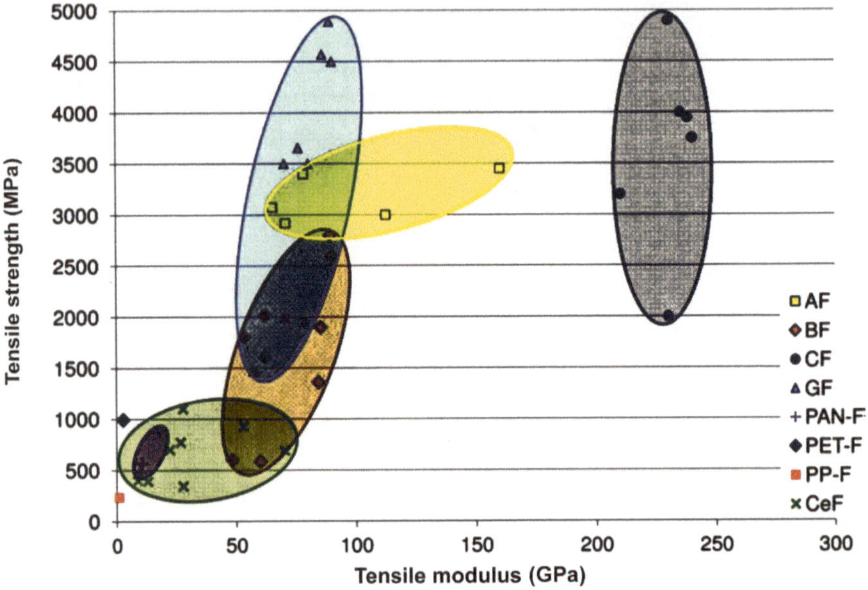

Fig. 1 Ashby plot presenting absolute tensile strength versus tensile modulus for various fiber types. *Note*: The term "CeF" includes plant-based fibers. *(From C. Unterweger, O. Brüggemann, C. Fürst, Synthetic fibers and thermoplastic short-fiber-reinforced polymers: properties and characterization, Polym. Compos. 35 (2) (2014) 227–236, with permission from Elsevier.)*

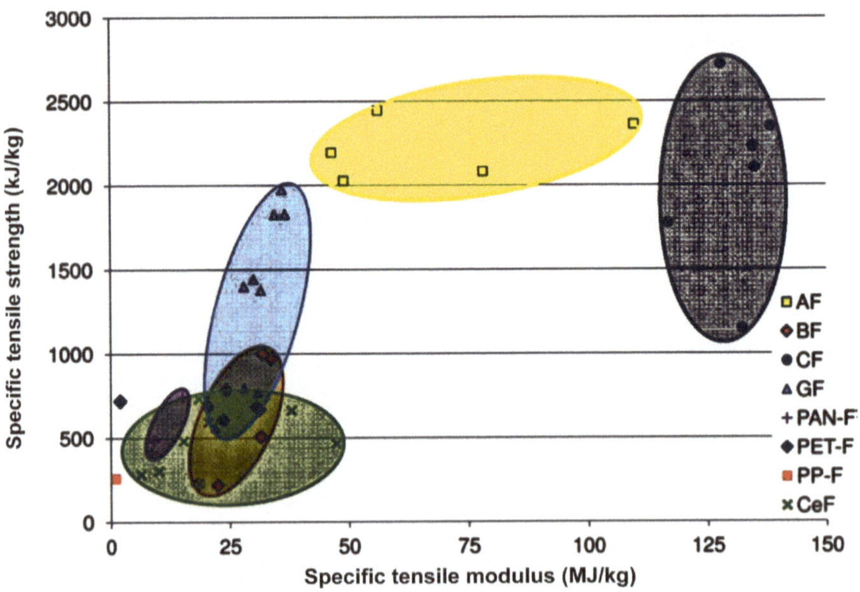

Fig. 2 Ashby plot presenting specific tensile strength versus specific tensile modulus for various fiber types. *Note*: The term "CeF" includes plant-based fibers. *(From C. Unterweger, O. Brüggemann, C. Fürst, Synthetic fibers and thermoplastic short-fiber-reinforced polymers: properties and characterization, Polym. Compos. 35 (2) (2014) 227–236, with permission from Elsevier.)*

the adhesion and internal lubricants. Carbon fibers (CF), glass fibers (GF), and aramid fibers (AF) are used to improve the compressive strength and the creep resistance of the polymeric matrix [12]. The choice of appropriate fibers is usually a compromise between the polymer properties and their tribological behaviors. Carbon, glass, and aramid fiber-reinforced polymer composites are 1–4 times stiffer than pure polymers [13]. Moreover, the rate of wear increases by about a factor of 10 compared to unreinforced polymers.

Glass and carbon fibers considerably decrease the coefficient of friction of polyamide composites up to 50% compared to unreinforced polyamide. It needs to be stated that the mechanical and chemical interactions of the aforesaid difficulties are very complex. As a result, advanced studies to understand these relationships in more detail are a subject of current and future investigations.

In order to improve the tribological properties of polymers, one traditional idea is to improve their hardness and compressive strength while also decreasing their adhesion to the counter face. This can be done correctly through using several fibers. Table 3 reviews the wear rate of various modified fillers and/or short fiber-reinforced thermoplastic composite systems.

Currently, there is potential in utilizing NF polymer composites, as they demonstrate the notable development of friction and wear behavior. Nonetheless, due to the high cellulose content of fibers, as they are soft by nature, NF composites showed less abrasive damage to processing equipment [4].

Table 4 summarizes the tribological properties of various natural fiber polymeric composites at diverse operating conditions. Compared to neat synthetic resins, the tribological behavior of NF composites is highly favorable. Besides, the friction and wear properties of NF composites can be further enhanced by utilizing proper fiber treatment and fiber loading during fabrication of the composite.

4.1 Effect of fiber length

Fiber length is considered one of the most critical elements affecting the tribological conduct of short fibers. It performs a critical characteristic to create interfacial bonding among the matrix and the fiber. Theoretically, comparatively longer fibers demonstrated an effective improvement to the abrasive wear resistance of thermosetting composites [12]. However, under wear situations, the abrasive particles locally destruct the longer fibers, where the rest of the fibers contribute to the wear resistance of the composites.

Table 3 Samples of different newly developed, filled polymer systems with excellent tribological properties under various loading conditions (block-on-ring tests, $p = 2$ MPa; $v = 1$ m/s; $t = 8$ h).

Material compositions (vol.%)							Specific wear rate
PTFE	PPS	PEEK	Graphite	CF	Bronz	Al_2O_3	
51.6	31.8	–	4.8	11.8	–	–	1.53
51	31.6	–	3.9	13.5	–	–	1.40
51.9	32.4	–	2.6	13.1	–	–	1.20
12.4	–	61.8	11.7	14.1	–	–	4.31
9.7	–	49.7	12.5	28.1	–	–	6.33
76.8	–	–	19.8	–	–	3.4	22.4
84.1	–	–	–	12.6	–	3.3	1.25
52.5	28	–	–	19.5	–	–	1.69
78.6	–	–	–	21.4	–	–	1.75
80	–	–	–	10	10	–	0.565

Where: p = contact pressure, MPa; v = sliding velocity, m/s; t = test duration, h.

From K. Friedrich, Z. Lu, A.M. Hager, Recent advances in polymer composites' tribology, Wear 190(2) (1995) 139–144, https://doi.org/10.1016/0043-1648(96)80012-3, with permission from Elsevier.

Table 4 Tribological properties of some polymeric composites based on NF at different working situations.

Type of composite	Test parameters	Imp. in W_s compared to NP (%)	Imp. in friction compared to NP (%)	Imp. in temp, compared to NP (%)	Increment in roughness compared to the virgin one (%)	Ref.
Oil palm fiber composite	POD, CC: dry, AL: 50–70 N, SD: 0–5 km, SV: 2.8 m/s, FL: 45–48 vol.%	79	14.7	4.7	ND	[14]
	BOR, CC: dry, AL: 30–100 N, SD: 0.85–5 km, SV: 1.7–3.9 m/s, FL: 48 vol.%	60	ND	ND	794.7	[15]
	POD, CC: wet, AL: 50 N, SD: 3–14 km, SV: 2.8–3.9 m/s, FL: 48 vol.%	ND	ND	ND	119	[16]
	BOR, CC: wet, AL: 50 N, SD: 5–27 km, SV: 5.6–7.8 m/s, FL: 48 vol.%	ND	ND	ND	85.7	[16]
Coir fiber composite	BOD, CC: dry, AL: 10–30 N, SD: 0.84–4.2 km, SV: 2.8 m/s, FL: 11.5 vol.% for 3 layer, 17.3 vol % for 4 layer	21.4	34.4	47.3	181	[17]
Bamboo fiber composite	POD, CC: dry, AL: 30 N, SD: 1–4 km, SV: 2.22 m/s, FL: 4 wt.%	60 compared to NE	46.4 compared to NE	34.7 compared to NE	ND	[18]
Sugarcane fiber composite	POD, CC: dry, AL: 20–80 N, SD: 2.25 km, SV: 2.5 m/s, FL: 3–4 wt.%	20–50 compared to C-SCRP	ND	ND	ND	[19]
Betelnut fiber composite	POD, CC: dry, AL: 5–30 N, SD: 0–6.72 km, SV: 2.8 m/s, FL: 48 vol.%	98	73	ND	ND	[20]
Sisal fiber composite	POD, CC: dry, AL: 10–100 N, SD: 6 km, SV: 1.75 m/s, FL: 10, 27, 30, 42 wt.%	16.7	120	60	ND	[21]
Jute fiber composites	POD, CC: dry, AL: 30–50 N, SD: 1.8 km, SV: 3–6 m/s, FL: 5–10 wt.%	90.2	17.7	ND	ND	[22]

Continued

Table 4 Tribological properties of some polymeric composites based on NF at different working situations—cont'd

Type of composite	Test parameters	Imp. in W_s compared to NP (%)	Imp. in friction compared to NP (%)	Imp. in temp, compared to NP (%)	Increment in roughness compared to the virgin one (%)	Ref.
Cotton fiber composite	POD, CC: dry, AL: 10–80 N, SD: 4 km, SV: 2.22 m/s	89.7	47.3	31.5	ND	[23]
	BOR, CC: dry, AL: 245 N, SD: 3.024 km, SV: 0.42 m/s	70	31.3	ND	ND	[24]
	POD, CC: dry, AL: 156.80–250.88 N, duration: 2 h, SV: 0.26 m/s	72.1	8.5	ND	ND	[25]
Basalt fiber composite	BOR, CC: dry, AL: 240 N, duration: 120 min, SV: 0.42 m/s. FL: 0–20 wt.%	98.9	2.1	ND	ND	[26]
Coconut fiber composite	POD, CC: dry, AL: 40 N, SD: 2.1 km, SV: 3.5 m/s, FL: 15–49 wt.%	61.1	16.7	ND	ND	[27]
	POD, CC: dry, AL: 40–60 N, SD: 1–6 km, SV: 1.66–3.33 m/s, FL: 40 wt.%, clay loading: 1–5 wt.%	31.4 and 45.2 (at 1.66 and 3.33 m/s)	ND	ND	ND	[28]
Borassus fiber composite	POD, CC: dry, AL: 15–30 N, Sliding time: 15 min, SV: 1.4–2.4 m/s, fiber length: 5 mm	69.3	22.5	ND	ND	[29]

Remark: *POD*, pin on disc; *BOR*, block on ring; *CC*, contact condition; *AL*, applied load; *SD*, sliding distance; *SV*, sliding velocity; *FL*, fiber loading; *ND*, not determined.

From U. Nirmal, J. Hashim, M.M.H. Megat Ahmad, A review on tribological performance of natural fibre polymeric composites, Tribol. Int. 83 (2015) 77–104, https://doi.org/10.1016/j.triboint.2014.11.003, with permission from Elsevier.

On the other hand, too short fibers decrease the capacity of carrying the load and are accountable for the excessive wear rate while too long fibers result in the easy pull out of the fibers.

Regarding natural fibers, several studies were conducted to investigate this issue. A composite prepared from Borassus fruit fibers and epoxy resin was tested by Boopathi et al. [29] and Parikh and Gohil [30] to investigate the effect of fiber length on the tribological properties of the composite. The impact of the fiber length of sugar cane fibers was also discussed by Mahapatra and Chaturvedi [31]. All these experimental examinations showed that fiber with a medium length provides the best wear and frictional properties as compared to extraordinary fiber length (too short or too long).

Recently, Kumar et al. [32] investigated the impact of fiber length and weight percentage fiber loading of Himalayan on physical and mechanical properties of agave fiber-reinforced polyester composites. They applied Taguchi analysis on the sliding wear experiments to define the ideal parameter to assess the damage during the wear process. They used the analysis of variance (ANOVA) technique to investigate the importance of selected parameters on the specific wear rate of the composite. The tribological experiments were performed on fibers with lengths of 3, 5, and 7 mm. It is postulated that the fiber length has a dominant impact on the mechanical properties of the composite. Ideal hardness, tensile, flexural strength, impact strength, and strength values were realized at fiber lengths of 7 and 5 mm. This work reveals that fiber length can be assumed as one of the essential control parameters that affects the wear rate.

During sliding in dry conditions of short carbon fiber-reinforced epoxy (SCF/EP) composites, the friction and wear behavior were found to be related to the fiber length [33]. The experimental investigations were conducted on composite samples using nominal fiber lengths of 90 and 400 μm. The polymer matrix (resin) was a bisphenol-A type epoxy. Additional fillers of graphite flakes (size of 20 μm) and TiO_2 particles (average diameter of 300 nm) were also applied. The results suggested that the fiber length has a distinct influence on the wear rate of the composite. The rate of wear was considerably decreased by approximately three times when the average fiber length was enlarged from 67 to 236 μm, as illustrated in Fig. 3.

Relatively, the composites with longer fibers showed enhanced resistance to wear than those with shorter fibers. This observation revealed that the relatively long fiber composites possessed a higher load-carrying functionality. The reason is attributed to their better stiffness and strength.

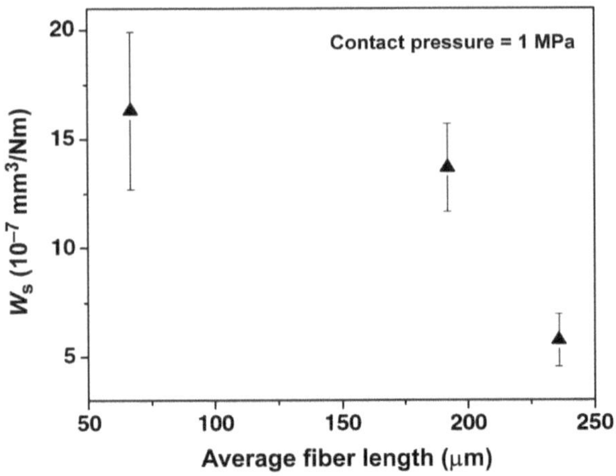

Fig. 3 Effect of fiber length on the specific wear rate of epoxy composites. *(From H. Zhang, Z. Zhang, K. Friedrich, Effect of fiber length on the wear resistance of short carbon fiber reinforced epoxy composites, Compos. Sci. Technol. 67 (2) (2007) 222–230, https://doi.org/10.1016/j.compscitech.2006.08.001, with permission from Elsevier.)*

Correspondingly, the fiber length was found to have a distinguished effect on the frictional coefficients of the wear samples after running-in wear periods. Regarding the effect on the friction coefficient of the polymeric composites, composites with longer fibers have a slightly lower frictional coefficient compared to shorter ones. It is suggested that the wear of the short fiber composites was attributed to fiber peeling-off. Also, the robust force of friction may "dig out" individual fibers directly without a recognizable fracture. In contrast, the peeling-off process of long fibers would not take place simply. Most of the fibers would remain embedded in the worn surface while small parts of the fibers would separate from the worn surface. Therefore, the worn surface becomes relatively smooth. Consequently, the coefficient of friction and the rate of wear dropped to some extent.

4.2 Effect of fiber orientation

Fiber orientation can be deliberated as one of the significant factors that influences the tribological behavior of short fiber polymer composites. Typically, the fiber orientation in short fiber composites is achieved by means of the injection molding technique. Subsequently, along the different fiber directions (normal, perpendicular, or parallel), the isotropic properties of the composites are obtained. The molding technique, or three-layer molded plates, is introduced in several scientific studies [34, 35]. Lhymn et al. [36] investigated the wear of PEEK/polyphenylene sulfide (PPS) thermoplastic polymers reinforced with short carbon/glass fiber. The study showed that the rate of wear was correlated to the orientation of the fiber axis with respect to the direction of sliding. A lower wear rate was detected in the normal oriented specimen compared to the parallel or antiparallel specimen.

Generally, the effect of the short fiber-filled polymers is to improve the tribological and mechanical performances. The friction and wear behaviors of PEEK composites filled with short carbon fibers/PTFE/graphite were examined by Rasheva et al. [37], considering various fiber orientations. A block-on-ring tribometer was used to perform the sliding wear tests. The test specimens were loaded along the fiber axis perpendicular and parallel to it, as illustrated in Fig. 4. The results established from the wear experiments are summarized in Table 5.

Generally, the results indicated that the rate of wear was relatively higher in specimens with parallel orientation fibers compared to perpendicular orientation ones. However, MB composites showed nearly no change independent from fiber orientation and applied loads. MB reinforced with parallel fibers showed the lowest wear rate as a result of no fiber breakage and fiber removal within the friction zone, as presented in Fig. 5B. On the other hand, extensive fiber fractures were detected in MC and MA composites, and they reached the maximum in MA, as shown in Fig. 5A.

It is postulated that fibers peeling off along the direction of sliding result in a high wear rate. Comparing wear rates of MB and MA composites, MB shows a lower wear rate and a more stable worn surface. The same conclusion can be found if MC and MB composites are considered. According to Fig. 5C, it can be assumed that a higher wear rate could result from a parallel fiber orientation. Also, a combination of more fiber content and less solid lubricant content (MC) results in low wear resistance. Consequently, the content of parallel short carbon fibers and solid lubricants, as in the MB

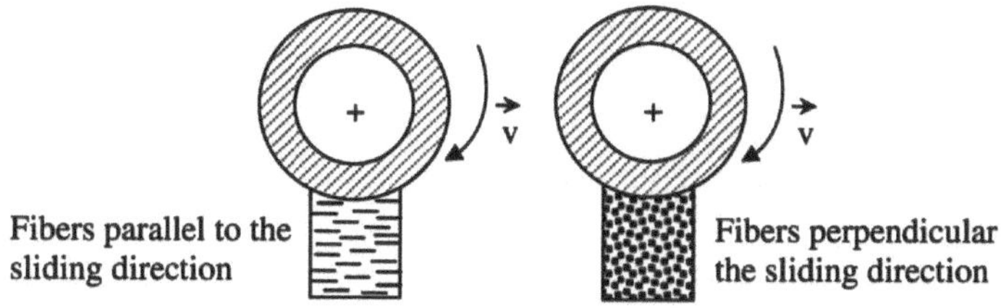

Fig. 4 Fiber orientation during the wear experiments. *(From Z. Rasheva, G. Zhang, T. Burkhart, A correlation between the tribological and mechanical properties of short carbon fibers reinforced PEEK materials with different fiber orientations, Tribol. Int. 43 (8) (2010) 1430–1437, https://doi.org/10.1016/j.triboint.2010.01.020, with permission from Elsevier.)*

Table 5 Specific wear rate perpendicular and parallel to the fiber direction.

Material code PEEK/PTFE/Graphite/SCF	Apparent pressure (MPa)	Spec. wear rate parallel (10^{-6} $mm^3/N\,m$)	Relative error (%)	Spec. wear rate perpendicular (10^{-6} $mm^3/N\,m$)	Relative error (%)
MA	1	0.554	8.66	0.410	5.85
70/10/10/10	4	0.494	6.88	0.367	2.39
MB	1	0.413	15.74	0.424	17.68
65/10/10/15	4	0.420	5.00	0.468	8.33
MC	1	0.501	13.77	0.397	10.30
75/05/05/15	4	0.454	11.23	0.525	15.12

From Z. Rasheva, G. Zhang, T. Burkhart, A correlation between the tribological and mechanical properties of short carbon fibers reinforced PEEK materials with different fiber orientations, Tribol. Int. 43(8) (2010) 1430–1437, https://doi.org/10.1016/j.triboint.2010.01.020, with permission from Elsevier.

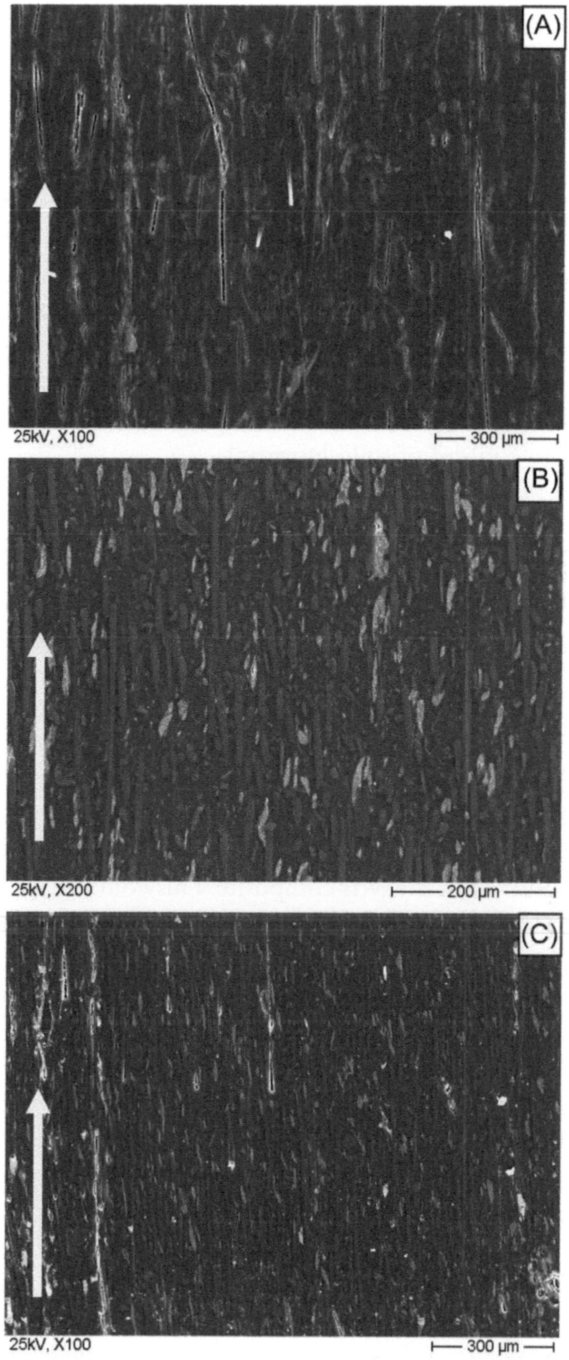

Fig. 5 (A) Fibers parallel to the sliding direction at high pressure (MA); (B) fibers parallel to the sliding direction at high pressure (MB); and (C) fibers parallel to the sliding direction at high pressure (MC). *(From Z. Rasheva, G. Zhang, T. Burkhart, A correlation between the tribological and mechanical properties of short carbon fibers reinforced PEEK materials with different fiber orientations, Tribol. Int. 43 (8) (2010) 1430–1437, https://doi.org/10.1016/j.triboint.2010.01.020, with permission from Elsevier.)*

composite, seems to be equally important. The MB composite, shown in Fig. 5B, shows a smooth worn surface attributed to a low wear rate.

On the other hand, the test results presented in Table 5 show that the low wear rate is attributed to fiber orientation, which is perpendicular to the direction of sliding. However, we should consider some factors. The MA composite at a perpendicular fiber orientation shows the lowest rate of wear, lower than the MB composite. The probable reasons could be acquired from Fig. 6A: Assuming that the MA composite is a composite with the lowest content of short carbon fibers, the potential reason for this consequence could be the combination of low content of short carbon fibers with much more content of solid lubricants.

The metallic counter face was covered by a transfer film of solid lubricant that, in turn, inhibits the distortion of the polymer matrix and fiber fracture. Therefore, a smooth surface results due to the accumulation of solid lubricant near the fibers [38]. In MC composites with fiber orientation perpendicular to the sliding direction, the high rate of wear is found in the material with less content of solid lubricants and high content of fiber.

It is also detected that the wear rate of the MC composites is higher than the wear rate of the MB composites. Fig. 6A and C suggests that higher destruction in the worn surface is detected compared to the MA composite. Wear fragments as a consequence of fiber fractures are higher compared to the MA and MB composites. However, no fiber fracture and a smooth worn surface were observed in the MB composite. This could be attributed to the effect of a combination of a certain content of solid lubricants and short carbon fibers that totally covers the fibers, forming a stable tribological surface. Despite that fact, the MA composite presents the lowest wear rate and the MB composite shows the most stable worn surface.

In many composite materials, continuous fibers are usually introduced as a common technique to improve the tribological properties. Particularly, continuous fibers are used to reinforce the polymer matrix in numerous tribological applications. This is due to their excessive wear resistance in parallel or normal to the direction of sliding. When composites reinforced with unidirectional continuous fibers slide on a rough counter face, the tangential and normal forces from the rough surface induce combined shear and tensile stresses in the fiber surface. The magnitude of these stresses depends on the orientation of the unidirectional fiber. These stresses are also functions of the properties of the composite, the geometry of the counter face asperity, and the applied load [39]. Furthermore, the effect of the orientation of the fiber on the tribological properties also depends on the composite material type as well as the tribological system type [40].

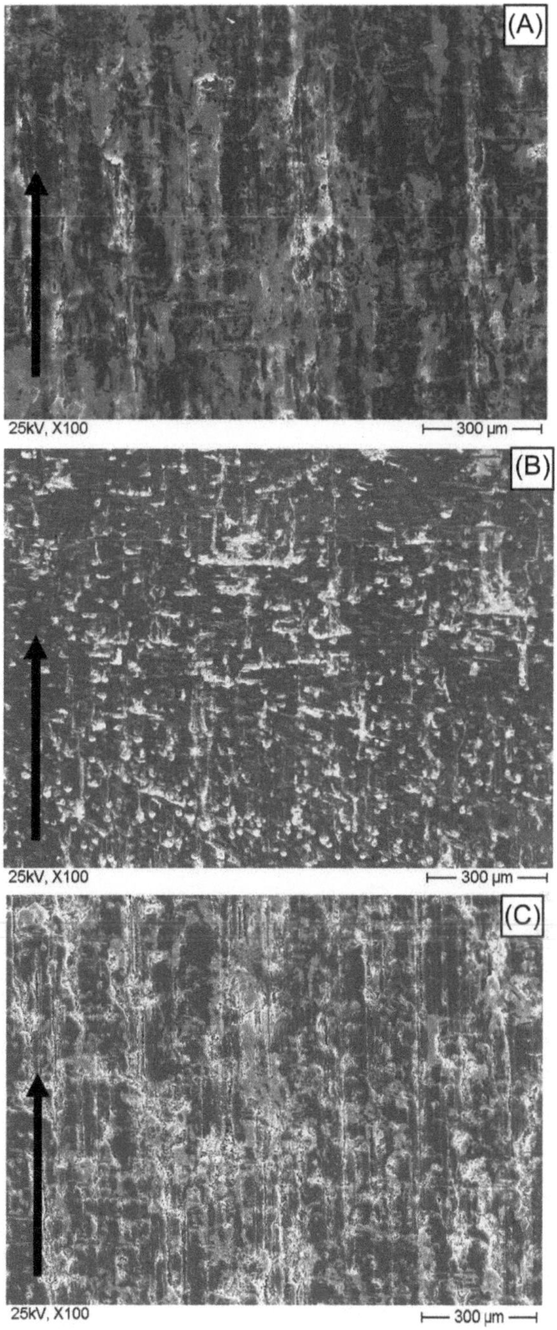

Fig. 6 (A) Fibers perpendicular to the sliding direction at high pressure (MA); (B) fibers perpendicular to the sliding direction at high pressure (MB); and (C) fibers perpendicular to the sliding direction at high pressure (MC). *(From Z. Rasheva, G. Zhang, T. Burkhart, A correlation between the tribological and mechanical properties of short carbon fibers reinforced PEEK materials with different fiber orientations, Tribol. Int. 43 (8) (2010) 1430–1437, https://doi.org/10.1016/j.triboint.2010.01.020, with permission from Elsevier.)*

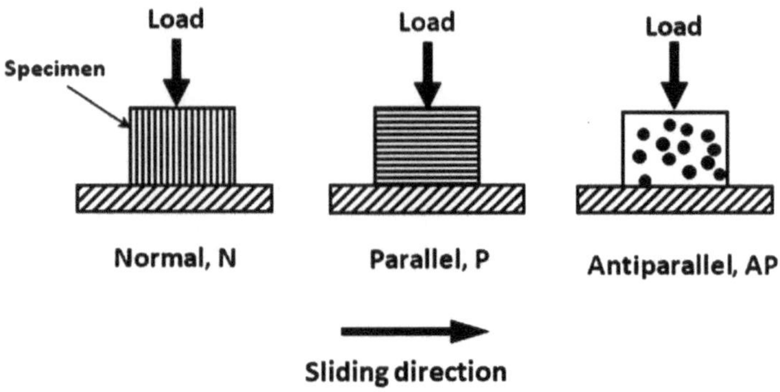

Fig. 7 Systematic illustration of fiber orientations. *(From T.C. Ovaert, Wear of unidirectional polymer matrix composites with fiber orientation in the plane of contact, Tribol. Trans. 40 (2) (1997) 227–234, https://doi.org/10.1080/10402009708983649, with permission from Taylor & Francis.)*

In order to clarify the influence of fiber orientation on the wear performance of fiber-reinforced polymers, carbon fiber polyetherketone (CF-PK) and aramid fiber epoxy (AF-EP) were examined [39]. The three suggested orientations were: normal (N); parallel (P); and antiparallel (AP), as shown in Fig. 7.

It was detected that the sliding direction relative to the fiber has a robust influence on the wear performance of composites. The same observation was detected regarding the fiber orientation relative to the sliding surface. The wear rate is also affected by other factors such as the sliding direction and the type of composite material.

According to microscopy examination of wear mechanisms, the microwear arising in the normal (N) orientation is in the form of cracking, slicing, bending, and debonding of fibers. On the other hand, in the parallel (P) orientation, the failure of the composite takes the form of internal crack propagation, damaging of fiber, debonding of the fiber matrix, and fiber cracking. The essential influential issue is whether an abrasive particle compresses the fiber as it passes or pulls the fiber out. The two mechanisms generate different wear levels, where the first one tends to generate higher material removal.

4.3 Effect of fiber content

Fiber content is a significant consideration for the design with polymer reinforced composites, as it controls the mechanical, thermomechanical, and tribological performance. Therefore, for particular applications, it is important to identify how the polymer composite behavior reacts with the fiber content under given operating circumstances. The contribution of fiber content in the tribological behavior of polymer composite systems has been explored in many studies in the literature [41, 42]. It is broadly established that a better tribological performance of polymer composites can be achieved by the addition of short fibers to the compositions. However, no fixed rule exists to define how the fiber content affects the tribological response of a composite. For example, the sliding wear behavior of polyetheretherketone (PEEK) with different continuous fibers shows that the rate of wear decreases with an increase in the fiber content [43].

The investigation of the erosive wear of glass fiber polypropylene composites at various fiber contents (40–60 wt.%) postulated that the fiber content has a strong effect on the wear rate [41]. In this study, the variation of the erosive wear rate with the fiber content was plotted to the fiber weight fraction (wt.%), as illustrated in Fig. 8. It can be recognized that the addition of brittle GF results in the increase of the erosive wear rate. The figure indicates a linear correlation of the erosive wear rate with a fiber content up to 60 wt.%.

The influence of fiber content (or weight fraction) upon the wear behavior of a short glass fiber (SGF)-reinforced polyurethane (PU) composite established that the higher weight fraction of SGF (45%) in the composites enhanced the resistance to wear compared to the lesser SGF composite (40%), as illustrated in Fig. 9. The abrasive sand action on polyurethane increases when a higher weight percent of SGF is involved in the composite [44, 45].

4.4 Effect of surface treatment and sizing

In many cases, the low interlaminar shear strength of polymer composites can be found from utilizing fibers without surface treatment. Low interlaminar shear strength is often attributed to poor bonding and weak adhesion between the fiber and polymer matrix. It is commonly identified that the surface treatment of fibers increases the surface acidic functional groups and surface area, and therefore increases the mutual adhesion between the polymer matrix and the fiber [46, 47]. For instance, the physical and interfacial properties of carbon fiber (CF) polymer composites are considerably affected by the interfacial characteristics existing between the reinforcing CFs and the polymer matrix. The properties of the CF-matrix interface could be either physically or chemically affected by the surface treatment of the fiber as well as by sizing [47, 48]. The treatment of the fiber surface

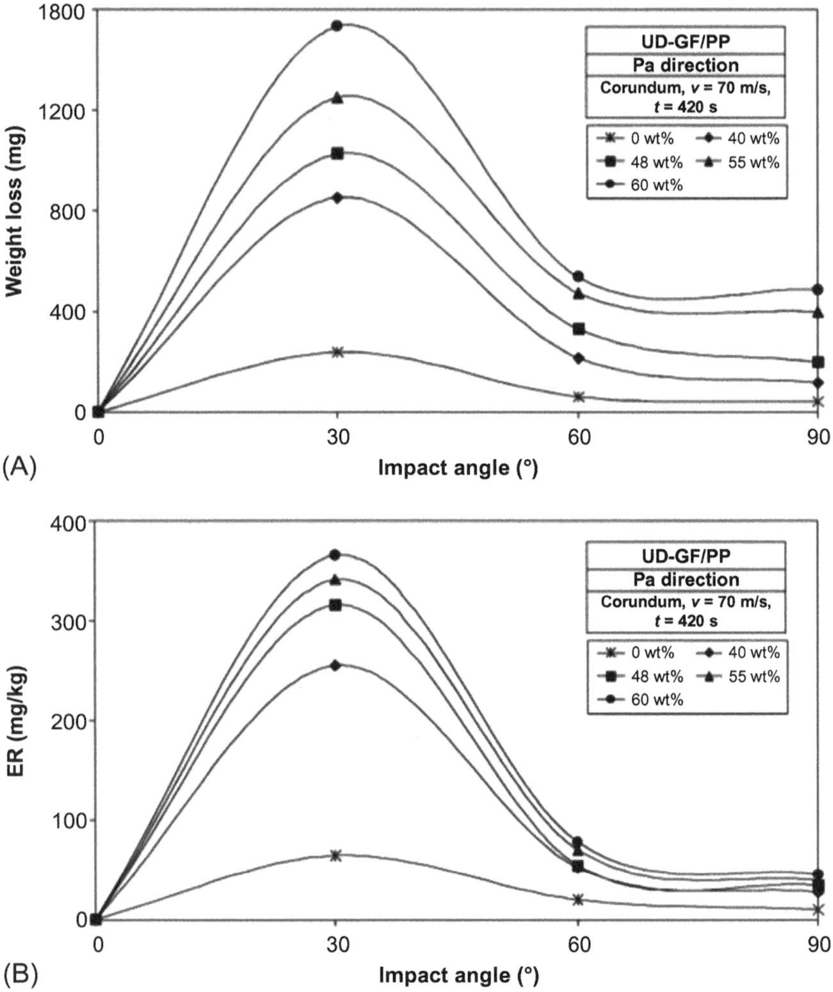

Fig. 8 Weight loss (A) and erosion rate (ER) (B) as a function of impact angle and fiber content of UD-GF/PP composites containing fibers aligned parallel (Pa) to the erosion direction, respectively. *(From N.M. Barkoula, J. Karger-Kocsis, Effects of fibre content and relative fibre-orientation on the solid particle erosion of GF/PP composites, Wear 252 (1) (2002) 80–87, https://doi.org/10.1016/S0043-1648(01)00855-9, with permission from Elsevier.)*

could be implemented by a number of gas processes and oxidizing agents such as electrochemical oxidation, plasma, and thermal and chemical treatments. On the other hand, sizing can be used to coat the surface of fibers with a layer, which is usually either an emulsion or a solution consisting of polymeric components. In addition, gamma-ray radiation has been broadly implemented for the surface alteration of CFs [49].

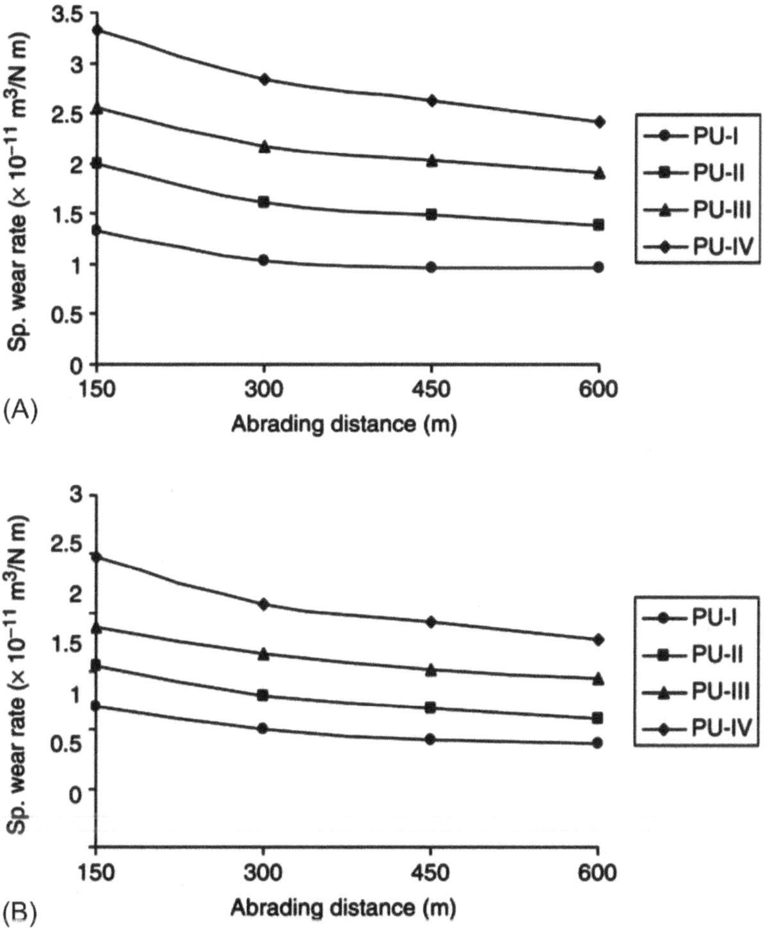

Properties	PU-I	PU-II	PU-III	PU-IV
Short glass fiber (wt%)	10	20	30	40

Fig. 9 Variation of specific wear rate with abrading distance of SGF-PU composites at (A) 22 N and (B) 32 N. *(From B. Suresha, G. Chandramohan, N. Dayananda Jawali, Siddaramaiah, Effect of short glass fiber content on three-body abrasive wear behaviour of polyurethane composites, J. Compos. Mater. 41 (22) (2007) 2701–2713, with permission from Elsevier.)*

Carbon nanofiber-reinforced composites are more appropriate for use under severe operating conditions such as elevated temperatures and/or extraordinary applied pressures. The surface morphology of carbon nanofibers derived from iodine-treated bacterial cellulose produced an enhancement in the wear properties of the composites [50]. However, the treatment of fiber surfaces with oxidizing solutions may result in a simple route to changing fibers with an equally effective surface morphology. Nakamura et al. [48] proved that at fiber concentrations higher than 1.0 wt.%, the wear volumes of aqueous hydrogen peroxide fiber-reinforced composites were considerably lower than those of untreated fiber composites. Scanning microscopy images of the worn surfaces of C/C composites reinforced with 5.0 wt.% of vapor-grown carbon fibers (VGCF) and VGCFs treated with (HNO_3) and aqueous hydrogen peroxide (H_2O_2) are illustrated in Fig. 10. As shown in the figure, the flaking of CNF/C is observed on the rough sliding surface, which suggests that this may have a controlling impact on the abrasive wear mechanism.

4.5 Effect of lubrication

Generally, the life limit and consistency of any tribological system are often governed by the tribological performance of its components under dry and lubrication conditions. Regarding polymer composites, it is believed that various composites reveal higher rates of wear when subjected to water lubrication conditions than those obtained under dry sliding [51, 52]. It is detected that the rate of wear of PEEK sliding against steel under water-lubricated conditions is threefold compared to those under dry sliding [53, 54]. It is generally assumed that the reason for such odd performances could be the reduction of polymer hardness due to water absorption. Also, under such lubricating conditions, water may reduce the transfer of wear debris onto the metallic counter face, and therefore the transfer film formation becomes hindered. Tanaka [55] reported the additional potential influence of water lubrication on the wear of polymeric composites where the wear alters the surface structure of the polymer.

In contrast to the aforementioned, another investigation detected a reduction in the wear rate of polymeric composites sliding in aqueous lubrication [56]. In this study, the authors attributed the reduction in the wear rate to the cooling action of the lubricant. This reason can explain why most of the time a distinguished reduction of friction is detected in water-lubricating conditions [57, 58]. The friction and wear examination of different CF-reinforced polymers sliding against metals in aqueous solutions, water, and organic fluids showed that the coefficient of friction and wear rate of these composites was significantly reduced under lubricating conditions [59].

The impact of hybrid GF and/or CF fillers on the tribological behavior of ultrahigh molecular weight polyethylene (UHMWPE) composites under

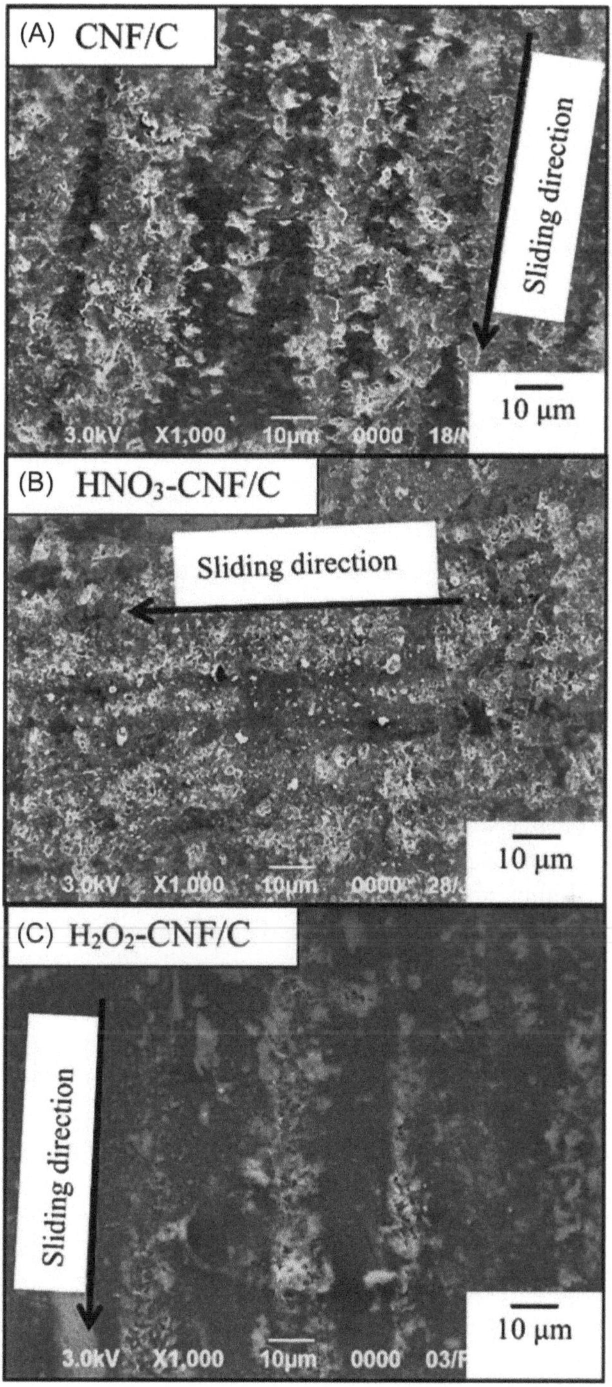

Fig. 10 SEM images of the sliding surfaces of (A) CNF/C, (B) HNO_3-CNF/C, and (C) H_2O_2-CNF/C filled with 5.0 wt.% of VGCFs after wear tests. *(From K. Nakamura, H. Kanno, S. Ishii, Wear properties of carbon composite reinforced by vapor-grown carbon fibers treated with nitric acid and aqueous hydrogen peroxide, Mater. Lett. 209 (2017) 228–230, with permission from Elsevier.)*

dry and water lubrication sliding environments was explored by Wang et al. [60]. Their investigation was performed to offer guidance on the probability of applying such composites in bearing applications and ultimately develop an outstanding water-lubricated bearing material. Ultrahigh molecular weight polyethylene composites were filled with CF (diameter: 7 μm; length/radius: 4; carbon content: 495%; density: 1.75 g/cm^3) and E-glass fibers (diameter: 10 μm; length/radius: 4; density: 2.25 g/cm^3).

Four UHMWPE composite types were prepared:

(1) Pure UHMWPE.
(2) UHMWPE reinforced with 10 wt.% GF (GF/UHMWPE).
(3) UHMWPE reinforced with 10 wt.% CF (CF/UHMWPE).
(4) UHMWPE reinforced with 10 wt.% GF and 10 wt.% CF (GF/CF/UHMWPE).

Friction and wear experiments were conducted using a pin-on-disc tribometer according to the ASTM G99 standard. The lubrication of the sliding surface was performed using distilled water dropped on the surface at about 100 drops/min flow rate. Wear examinations were implemented at an applied load of 1.0–5.0 MPa, a constant sliding speed of 0.2–1.0 m/s, and a disc roughness of $Ra = 0.1$–0.3 μm. The test duration was fixed at 6 h. The results showed the correlation between the coefficient of friction (COF) and the sliding speed in dry and water lubrication conditions at a given counter face roughness. It is concluded that in such lubricating conditions, the friction coefficients of all tested composites decreased as the sliding speed increased, as presented in Fig. 11. The presence of a lubrication film among the metallic counter face and composite leads to the formation of boundary lubrication regimes, which diminish the coefficient of friction and inhibit adhesion wear. With the increase in sliding speed, the adhesive wear becomes weak and consequently a further reduction of COF is obtained.

Under conditions of water lubrication, all composites show a decreasing COF with the increase in contact pressure for counter face roughness ($Ra = 0.3$ μm, 0.1 μm). Because ultrahigh molecular weight polyethylene and its composites are viscoelastic substances, we can expect that their deformations under load will also be viscoelastic, as shown in Fig. 12. Consequently, the variation of the friction coefficient (μ) with the applied load (N) follows the relation $\mu = k * N^{(n-1)}$ [61], where k is a system-dependent constant and n is a constant (where $2/3 < n < 1$). According to this equation, the COF (μ) decreases with an increase in applied load (N) [62]. A similar finding was also suggested by researchers investigating the effect of water lubrication

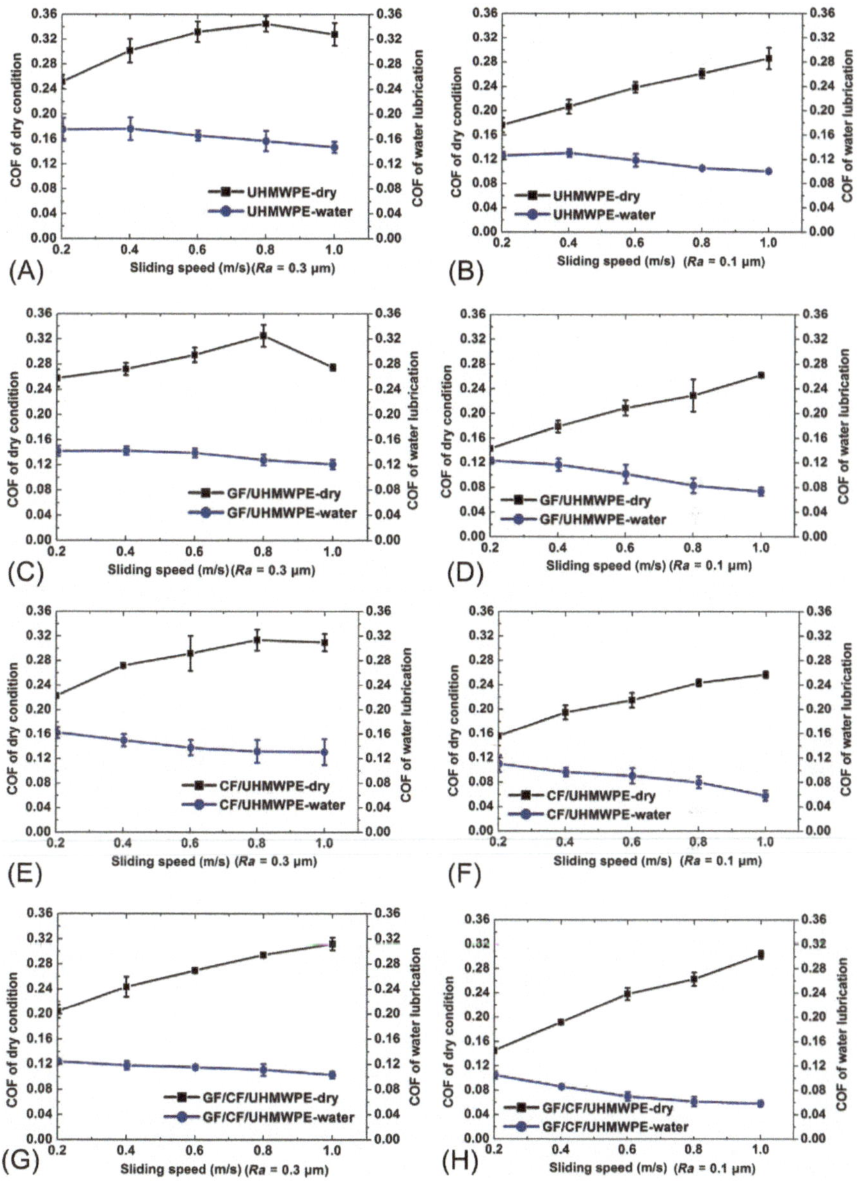

Fig. 11 Friction coefficients and standard deviations of UHMWPE composites under dry and water lubrication conditions at different sliding speed under contact pressures 2.5 MPa, with a counter face roughness *Ra* of 0.3 μm (A, C, E, G) and 0.1 μm (B, D, F, H). *(From Y. Wang, Z. Yin, H. Li, G. Gao, X. Zhang, Friction and wear characteristics of ultrahigh molecular weight polyethylene (UHMWPE) composites containing glass fibers and carbon fibers under dry and water-lubricated conditions, Wear 380–381 (2017) 42–51, with permission from Elsevier.)*

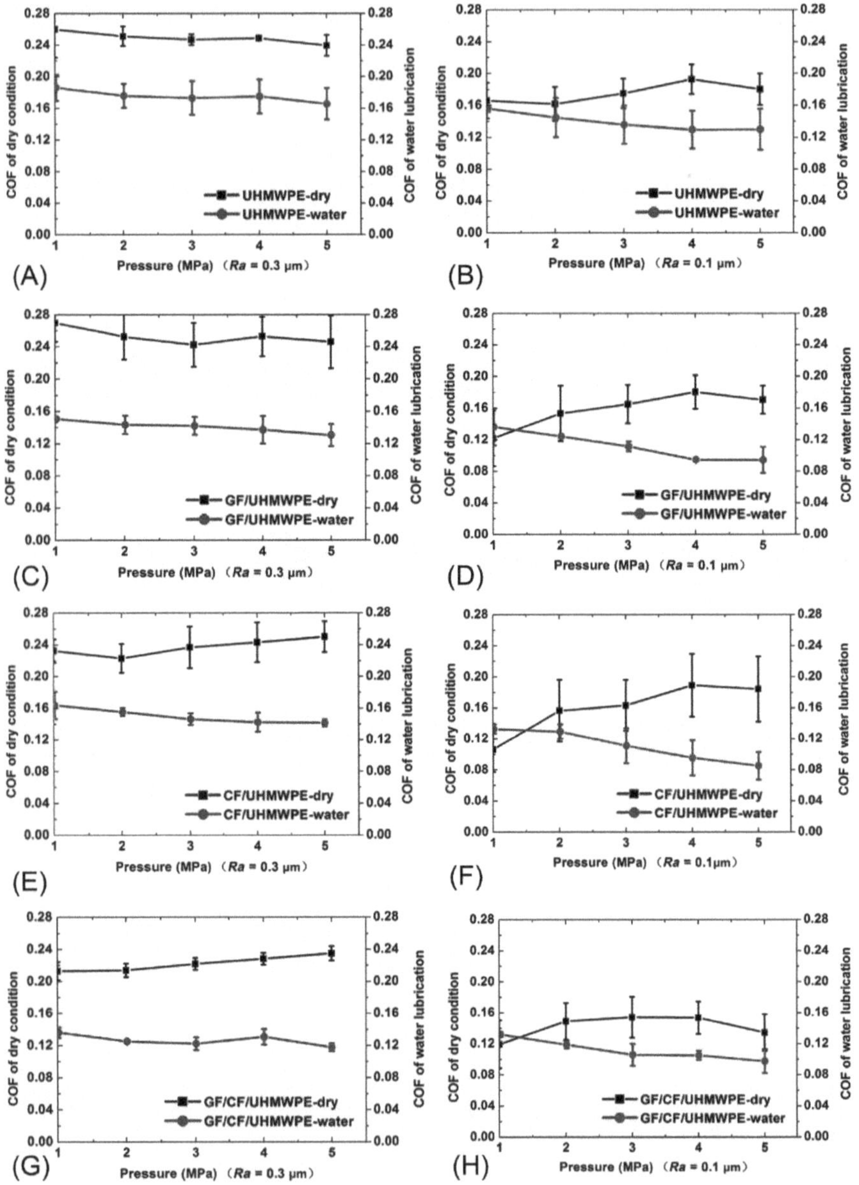

Fig. 12 Friction coefficients and standard deviations of UHMWPE composites under dry and water lubrication conditions at different contact pressures and at sliding speed 0.2 m/s, with a counter face roughness *Ra* of 0.3 μm (A, C, E, G) and 0.1 μm (B, D, F, H). *(From Y. Wang, Z. Yin, H. Li, G. Gao, X. Zhang, Friction and wear characteristics of ultrahigh molecular weight polyethylene (UHMWPE) composites containing glass fibers and carbon fibers under dry and water-lubricated conditions, Wear 380–381 (2017) 42–51, with permission from Elsevier.)*

conditions on the tribological performance of PEEK sliding against silicon nitride [63]. The results indicated that at relatively low loads, the COF of PEEK decreased with an increase in the normal pressure.

In addition to the aforementioned discussion, other investigations have been conducted to study the effect of oil lubrication conditions on the tribological behavior of polymer composites [64–66]. Zhao et al. [64,67] reported the tribological properties of composites reinforced with short carbon fibers (SCF) and short glass fibers (SGF) under oil and water lubrication situations. The tribological behavior and wear mechanism of the SCF and SGF composites sliding with oil lubrication conditions are proposed in Fig. 13.

As established in Fig. 13, at first the fibers crashed into small fragments (forming wear debris) due to the stress concentration and abrasion by counter face asperities. Second, this debris scraped by the short fiber from the counter face to get into the base oil. It may also transfer directly to the worn surface. During sliding, the wear debris is further compacted and fragmented under the pressure of surface asperities. This process occurred on the mating surfaces of SGF and the counter face. As a result, a transfer film containing wear debris is formed on the mating surfaces. Consequently, this film protects the mating surfaces against severe wear by separating the direct contact of friction pairs. Also, it prevents the progressive wear of fibers by filling wear troughs with compacted particles. Fig. 14 gives an overview of the transmission electron microscopy (TEM) micrographs of the cross-section of the transfer (tribo-) film formed on the steel counter face slid by SGF/G/PTFE. The thickness of the transfer film is around 50 nm.

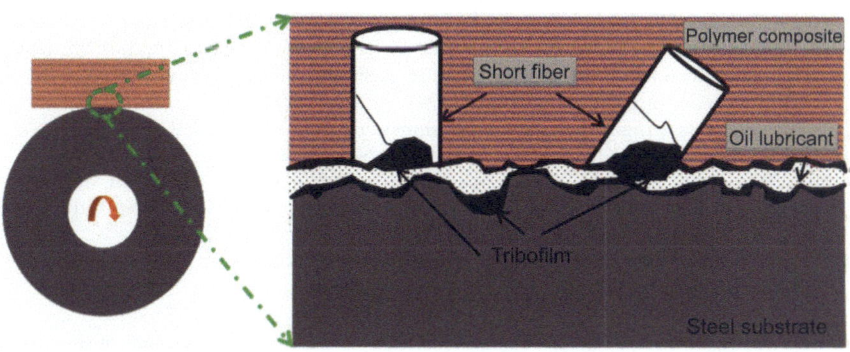

Fig. 13 Schematic presentation of the tribological mechanism of short fiber reinforced polymer composites under oil lubrication conditions. *(From F. Zhao, G. Li, W. Österle, I. Häusler, G. Zhang, T. Wang, Q. Wang, Tribological investigations of glass fiber reinforced epoxy composites under oil lubrication conditions, Tribol. Int. 103 (2016) 208–217, with permission from Elsevier.)*

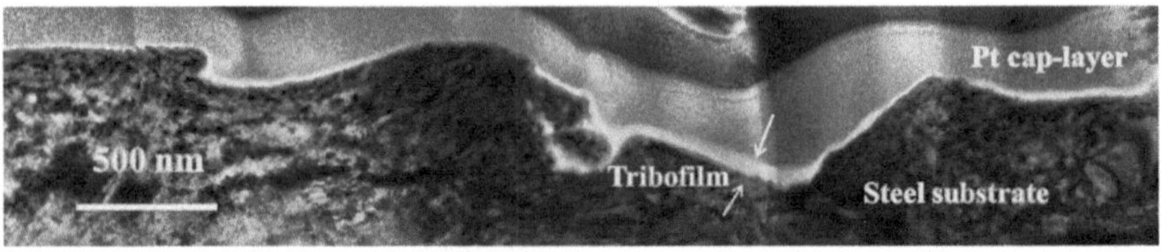

Fig. 14 Cross-section TEM micrographs of tribo-film on counterpart slide by SGF/G/PTFE. *(From F. Zhao, G. Li, W. Österle, I. Häusler, G. Zhang, T. Wang, Q. Wang, Tribological investigations of glass fiber reinforced epoxy composites under oil lubrication conditions, Tribol. Int. 103 (2016) 208–217, with permission from Elsevier.)*

References

[1] B. Vishwanath, A.P. Verma, C.V.S. Kameswara Rao, Effect of matrix content on strength and wear of woven roving glass polymeric composites, Compos. Sci. Technol. 44 (1) (1992) 77–86, https://doi.org/10.1016/0266-3538(92)90027-Z.

[2] A. Abdelbary, Extreme Tribology, CRC Press, USA, 2020.

[3] H. Pihtili, An experimental investigation of wear of glass fibre–epoxy resin and glass fibre–polyester resin composite materials, Eur. Polym. J. 45 (1) (2009) 149–154, https://doi.org/10.1016/j.eurpolymj.2008.10.006.

[4] U. Nirmal, J. Hashim, M.M.H. Megat Ahmad, A review on tribological performance of natural fibre polymeric composites, Tribol. Int. 83 (2015) 77–104, https://doi.org/10.1016/j.triboint.2014.11.003.

[5] A.A. Ticoalu, T. Aravinthan, Cardona, A review of current development in natural fiber composites for structural and infrastructure applications, in: The Southern Region Engineering Conference (SREC '10), Toowoomba, Australia, November, 2010, pp. 113–117.

[6] A. Shalwan, B.F. Yousif, In State of Art: mechanical and tribological behaviour of polymeric composites based on natural fibres, Mater. Des. 48 (2013) 14–24, https://doi.org/10.1016/j.matdes.2012.07.014.

[7] S. Sinha Ray, M. Bousmina, Biodegradable polymers and their layered silicate nanocomposites: in greening the 21st century materials world, Prog. Mater. Sci. 50 (8) (2005) 962–1079, https://doi.org/10.1016/j.pmatsci.2005.05.002.

[8] L. Mohammed, M.N.M. Ansari, G. Pua, M. Jawaid, M.S. Islam, A review on natural fiber reinforced polymer composite and its applications, Int. J. Polym. Sci. 2015 (2015) 15, https://doi.org/10.1155/2015/243947. Art no. 243947.

[9] M.S. Pradeep, P. Ingole Michael, S. Nosonovsky, V. Kailas Michael, R. Lovell, Tribology for Scientists and Engineers, Springer-Verlag, New York, 2013, p. 948.

[10] O. Faruk, A.K. Bledzki, H.-P. Fink, M. Sain, Biocomposites reinforced with natural fibers: 2000–2010, Prog. Polym. Sci. 37 (11) (2012) 1552–1596, https://doi.org/10.1016/j.progpolymsci.2012.04.003.

[11] C. Unterweger, O. Brüggemann, C. Fürst, Synthetic fibers and thermoplastic short-fiber-reinforced polymers: properties and characterization, Polym. Compos. 35 (2) (2014) 227–236, https://doi.org/10.1002/pc.22654.

[12] Z. Zhang, K. Friedrich, K. Velten, Prediction on tribological properties of short fibre composites using artificial neural networks, Wear 252 (7) (2002) 668–675, https://doi.org/10.1016/S0043-1648(02)00023-6.

[13] B. Suresha, K. Shiva Kumar, S. Seetharamu, P. Sampath Kumaran, Friction and dry sliding wear behavior of carbon and glass fabric reinforced vinyl ester composites, Tribol. Int. 43 (3) (2010) 602–609, https://doi.org/10.1016/j.triboint.2009.09.009.

[14] B.F. Yousif, N.S.M. El-Tayeb, The effect of oil palm fibres as reinforcement on tribological performance of polyester composites, surface review letter, Compos. Sci. Technol. 14 (6) (2007) 1–8.
[15] B.F. Yousif, N.S.M. El-Tayeb, Adhesive wear performance of T-OPRP and UT-OPRP composites, Tribol. Lett. 32 (3) (2008) 199–208, https://doi.org/10.1007/s11249-008-9381-7.
[16] B.F. Yousif, N.S.M. El-Tayeb, Wet adhesive wear characteristics of untreated oil palm fibre-reinforced polyester and treated oil palm fibre-reinforced polyester composites using the pin-on-disc and block-on-ring techniques, Proc. Inst. Mech. Eng. Part J: J. Eng. Tribol. 224 (2) (2009) 123–131, https://doi.org/10.1243/13506501JET655.
[17] B.F. Yousif, Frictional and wear performance of polyester composites based on coir fibres, Proc. Inst. Mech. Eng. Part J: J. Eng. Tribol. 223 (1) (2009) 51–59, https://doi.org/10.1243/13506501JET455.
[18] U. Nirmal, J. Hashim, K.O. Low, Adhesive wear and frictional performance of bamboo fibres reinforced epoxy composite, Tribol. Int. 47 (2012) 122–133, https://doi.org/10.1016/j.triboint.2011.10.012.
[19] N.S.M. El-Tayeb, A study on the potential of sugarcane fibers/polyester composite for tribological applications, Wear 265 (1) (2008) 223–235, https://doi.org/10.1016/j.wear.2007.10.006.
[20] N. Singh Gill, B. Yousif, Wear and frictional performance of betelnut fibre-reinforced polyester composite, Proc. Inst. Mech. Eng. Part J: J. Eng. Tribol. 223 (2) (2009) 183–194.
[21] N. Chand, U.K. Dwivedi, Sliding wear and friction characteristics of sisal fibre reinforced polyester composites: effect of silane coupling agent and applied load, Polym. Compos. 29 (3) (2008) 280–284, https://doi.org/10.1002/pc.20368.
[22] K. Sabeel Ahmed, S.S. Khalid, V. Mallinatha, S.J. Amith Kumar, Dry sliding wear behavior of SiC/Al_2O_3 filled jute/epoxy composites, Mater. Des. (1980-2015) 36 (2012) 306–315, https://doi.org/10.1016/j.matdes.2011.11.010.
[23] S.A.R. Hashmi, U.K. Dwivedi, N. Chand, Graphite modified cotton fibre reinforced polyester composites under sliding wear conditions, Wear 262 (11) (2007) 1426–1432, https://doi.org/10.1016/j.wear.2007.01.014.
[24] L.-H. Sun, Z.-G. Yang, X.-H. Li, Tensile and tribological properties of PTFE and nanoparticles modified epoxy-based polyester fabric composites, Mater. Sci. Eng. A 497 (1) (2008) 487–494, https://doi.org/10.1016/j.msea.2008.07.049.
[25] H.-J. Zhang, Z.-Z. Zhang, F. Guo, W. Jiang, W.-M. Liu, Study on the tribological behavior of hybrid PTFE/cotton fabric composites filled with Sb_2O_3 and melamine cyanurate, Tribol. Int. 42 (7) (2009) 1061–1066, https://doi.org/10.1016/j.tribo-int.2009.03.002.
[26] S. Cao, H. Liu, S. Ge, G. Wu, Mechanical and tribological behaviors of UHMWPE composites filled with basalt fibers, J. Reinf. Plast. Compos. 30 (4) (2011) 347–355.
[27] I. Siva, J.T.W. Jappes, B. Suresha, Investigation on mechanical and tribological behavior of naturally woven coconut sheath-reinforced polymer composites, Polym. Compos. 33 (5) (2012) 723–732, https://doi.org/10.1002/pc.22197.
[28] N. Rajini, J.T.W. Jappes, B. Suresha, S. Rajakarunakaran, I. Siva, N. Azhagesan, Effect of organically modified montmorillonite clay on wear behavior of naturally woven coconut sheath/polyester composite, Proc. Inst. Mech. Eng. Part J: J. Eng. Tribol. 228 (5) (2014) 483–497, https://doi.org/10.1177/1350650113515199.
[29] L. Boopathi, P.S. Sampath, K. Mylsamy, Influence of fiber length in the wear behaviour of borassus fruit fiber reinforced epoxy composites, Int. J. Eng. Sci. Technol. 4 (9) (2012) 4119–4129.
[30] H.H. Parikh, P.P. Gohil, Tribology of fiber reinforced polymer matrix composites—a review, J. Reinf. Plast. Compos. 34 (16) (2015) 1340–1346, https://doi.org/10.1177/0731684415591199.

[31] S.S. Mahapatra, V. Chaturvedi, Modelling and analysis of abrasive wear performance of composites using Taguchi approach, Int. J. Eng. Sci. Technol. 1 (1) (2009) 123–135, https://doi.org/10.4314/ijest.v1i1.58072.
[32] S. Kumar, L. Prasad, S. Kumar, V.K. Patel, Physico-mechanical and Taguchi-designed sliding wear properties of Himalayan agave fiber reinforced polyester composite, J. Mater. Res. Technol. 8 (4) (2019) 3662–3671, https://doi.org/10.1016/j.jmrt.2019.06.004.
[33] H. Zhang, Z. Zhang, K. Friedrich, Effect of fiber length on the wear resistance of short carbon fiber reinforced epoxy composites, Compos. Sci. Technol. 67 (2) (2007) 222–230, https://doi.org/10.1016/j.compscitech.2006.08.001.
[34] J. Karger-Kocsis, K. Friedrich, Temperature and strain-rate effects on the fracture toughness of poly(ether ether ketone) and its short glass-fibre reinforced composite, Polymer 27 (11) (1986) 1753–1760, https://doi.org/10.1016/0032-3861(86)90272-7.
[35] D. Hull, T.W. Clyne (Eds.), Fabrication, in: An Introduction to Composite Materials, second ed., Cambridge Solid State Science Series, Cambridge University Press, Cambridge, 1996, pp. 271–294.
[36] C. Lhymn, K.E. Tempelmeyer, P.K. Davis, The abrasive wear of short fibre composites, Composites 16 (2) (1985) 127–136, https://doi.org/10.1016/0010-4361(85)90619-6.
[37] Z. Rasheva, G. Zhang, T. Burkhart, A correlation between the tribological and mechanical properties of short carbon fibers reinforced PEEK materials with different fiber orientations, Tribol. Int. 43 (8) (2010) 1430–1437, https://doi.org/10.1016/j.triboint.2010.01.020.
[38] G. Zhang, Z. Rasheva, A.K. Schlarb, Friction and wear variations of short carbon fiber (SCF)/PTFE/graphite (10 vol.%) filled PEEK: effects of fiber orientation and nominal contact pressure, Wear 268 (7) (2010) 893–899, https://doi.org/10.1016/j.wear.2009.12.001.
[39] T.C. Ovaert, Wear of unidirectional polymer matrix composites with fiber orientation in the plane of contact, Tribol. Trans. 40 (2) (1997) 227–234, https://doi.org/10.1080/10402009708983649.
[40] M. Cirino, K. Friedrich, R.B. Pipes, The effect of fiber orientation on the abrasive wear behavior of polymer composite materials, Wear 121 (2) (1988) 127–141, https://doi.org/10.1016/0043-1648(88)90038-5.
[41] N.M. Barkoula, J. Karger-Kocsis, Effects of fibre content and relative fibre-orientation on the solid particle erosion of GF/PP composites, Wear 252 (1) (2002) 80–87, https://doi.org/10.1016/S0043-1648(01)00855-9.
[42] N. Miyazaki, T. Hamao, Solid particle erosion of thermoplastic resins reinforced by short fibers, J. Compos. Mater. 28 (9) (1994) 871–883, https://doi.org/10.1177/002199839402800905.
[43] M. Cirino, K. Friedrich, R.B. Pipes, Evaluation of polymer composites for sliding and abrasive wear applications, Composites 19 (5) (1988) 383–392, https://doi.org/10.1016/0010-4361(88)90126-7.
[44] B. Suresha, G. Chandramohan, N. Dayananda Jawali, Siddaramaiah, Effect of short glass fiber content on three-body abrasive wear behaviour of polyurethane composites, J. Compos. Mater. 41 (22) (2007) 2701–2713, https://doi.org/10.1177/0021998307078730.
[45] S. Poomali, B. Suresha, J.-H. Lee, Mechanical and three-body abrasive wear behaviour of PMMA/TPU blends, Mater. Sci. Eng. A 492 (1) (2008) 486–490, https://doi.org/10.1016/j.msea.2008.03.018.
[46] S. Tiwari, J. Bijwe, Surface treatment of carbon fibers—a review, Proc. Technol. 14 (2014) 505–512, https://doi.org/10.1016/j.protcy.2014.08.064.
[47] S.-J. Park, Surface treatment and sizing of carbon fibers, in: S.-J. Park (Ed.), Carbon Fibers, Springer Singapore, Singapore, 2018, pp. 105–137.

[48] K. Nakamura, H. Kanno, S. Ishii, Wear properties of carbon composite reinforced by vapor-grown carbon fibers treated with nitric acid and aqueous hydrogen peroxide, Mater. Lett. 209 (2017) 228–230, https://doi.org/10.1016/j.matlet.2017.08.001.
[49] B.-J. Kim, S.-J. Park, A simple method for the preparation of activated carbon fibers coated with graphite nanofibers, J. Colloid Interface Sci. 315 (2) (2007) 791–794, https://doi.org/10.1016/j.jcis.2007.07.013.
[50] K. Nakamura, M. Sato, R. Matsuzaki, T. Takase, Wear properties of carbon composites reinforced by carbon nanofibers derived from iodine-treated bacterial cellulose, TANSO 2016 (274) (2016) 139–144, https://doi.org/10.7209/tanso.2016.139.
[51] A. Abdelbary, Polymer tribology, in: A. Abdelbary (Ed.), Wear of Polymers and Composites, Woodhead Publishing, Oxford, 2014, pp. 1–36 (Chapter 1).
[52] J.K. Lancaster, Lubrication of carbon fibre-reinforced polymers part I—water and aqueous solutions, Wear 20 (3) (1972) 315–333, https://doi.org/10.1016/0043-1648(72)90413-9.
[53] Y. Yamamoto, T. Takashima, Friction and wear of water lubricated PEEK and PPS sliding contacts, Wear 253 (7) (2002) 820–826, https://doi.org/10.1016/S0043-1648(02)00059-5.
[54] Y. Yamamoto, M. Hashimoto, Friction and wear of water lubricated PEEK and PPS sliding contacts: Part 2. Composites with carbon or glass fibre, Wear 257 (1) (2004) 181–189, https://doi.org/10.1016/j.wear.2003.12.004.
[55] K. Tanaka, Effects of various fillers on the friction and wear of PTFE-based composites, in: K. Friedrich (Ed.), Composite Materials Series, vol. 1, Elsevier, 1986, pp. 137–174. (Chapter 5).
[56] J. Jia, J. Chen, H. Zhou, L. Hu, Comparative study on tribological behaviors of polyetheretherketone composite reinforced with carbon fiber and polytetrafluoroethylene under water-lubricated and dry-sliding against stainless steel, Tribol. Lett. 17 (2) (2004) 231–238, https://doi.org/10.1023/B:TRIL.0000032449.32855.3d.
[57] J.P. Davim, N. Marques, A.M. Baptista, Effect of carbon fibre reinforcement in the frictional behaviour of Peek in a water lubricated environment, Wear 251 (1) (2001) 1100–1104, https://doi.org/10.1016/S0043-1648(01)00741-4.
[58] H. Unal, U. Sen, A. Mimaroglu, Dry sliding wear characteristics of some industrial polymers against steel counterface, Tribol. Int. 37 (9) (2004) 727–732, https://doi.org/10.1016/j.triboint.2004.03.002.
[59] S.V. Panin, et al., Wear resistance of composites based on hybrid UHMWPE–PTFE matrix: Mechanical and tribotechnical properties of the matrix, J. Frict. Wear 36 (3) (2015) 249–256, https://doi.org/10.3103/S1068366615030113.
[60] Y. Wang, Z. Yin, H. Li, G. Gao, X. Zhang, Friction and wear characteristics of ultrahigh molecular weight polyethylene (UHMWPE) composites containing glass fibers and carbon fibers under dry and water-lubricated conditions, Wear 380-381 (2017) 42–51, https://doi.org/10.1016/j.wear.2017.03.006.
[61] B.H. Stuart, Surface plasticisation of poly(ether ether ketone) by chloroform, Polym. Test. 16 (1) (1997) 49–57, https://doi.org/10.1016/S0142-9418(96)00026-8.
[62] H. Unal, A. Mimaroglu, U. Kadıoglu, H. Ekiz, Sliding friction and wear behaviour of polytetrafluoroethylene and its composites under dry conditions, Mater. Des. 25 (3) (2004) 239–245, https://doi.org/10.1016/j.matdes.2003.10.009.
[63] Q. Tang, J. Chen, L. Liu, Tribological behaviours of carbon fibre reinforced PEEK sliding on silicon nitride lubricated with water, Wear 269 (7) (2010) 541–546, https://doi.org/10.1016/j.wear.2010.05.009.
[64] F. Zhao, et al., Tribological investigations of glass fiber reinforced epoxy composites under oil lubrication conditions, Tribol. Int. 103 (2016) 208–217, https://doi.org/10.1016/j.triboint.2016.07.002.

[65] A. Arun, K. Kumar Singh, Friction and wear behaviour of glass fibre reinforced polymer composite (GFRP) under dry and oil lubricated environmental conditions, Mater. Today: Proc. 4 (8) (2017) 7285–7292, https://doi.org/10.1016/j.matpr.2017.07.058.
[66] Q.-Q. Shangguan, X.-H. Cheng, On the friction and wear behavior of PTFE composite filled with rare earths treated carbon fibers under oil-lubricated condition, Wear 260 (11) (2006) 1243–1247, https://doi.org/10.1016/j.wear.2005.08.009.
[67] C.P. Gao, et al., Tribological behaviors of epoxy composites under water lubrication conditions. Tribol. Int. 95 (2016) 333–341, https://doi.org/10.1016/j.triboint.2015.11.041.

CHAPTER 5

Tribological characterization of cellulose fiber-reinforced polymer composites

Krishasamy Senthilkumar[a,b], Suchart Siengchin[b], Thiagamani Senthil Muthu Kumar[a], Subramaniam Karthikeyan[a], Muthukumar Chandrasekar[c], Krittirash Yorseng[b], Thitinun Ungtrakul[b], and Nagarajan Rajini[a]

[a]Department of Mechanical Engineering, Kalasalingam Academy of Research and Education, Krishnankoil, Tamil Nadu, India
[b]Department of Mechanical and Process Engineering, The Sirindhorn International Thai-German Graduate School of Engineering (TGGS), King Mongkut's University of Technology North Bangkok, Bangkok, Thailand
[c]School of Aeronautical Sciences, Hindustan Institute of Technology & Science, Chennai, Tamil Nadu, India

1 Introduction

Polymer matrix composites (PMCs) comprising different fillers and/or reinforcements are frequently used in many technical applications, where friction and wear rate are the primary concern. Nowadays, metallic sliding materials are being replaced by PMCs [1]. However, continuous research and development is being done to investigate other areas for the potential use of these materials and to modify their functional characteristics for the increased loading and temperature conditions. For example, a ball joint used in the automotive applications will have to withstand high load and temperature, thus acting like a tribo couple. In this situation, PMCs can be a better choice, since they can operate in higher temperatures, i.e., around 120°C. Hence analyzing the tribological properties such as wear resistance and coefficient of friction (COF) of PMCs is highly valuable.

Poly(lactic) acid (PLA) is a biopolymer that is used in many applications such as medical, food packaging, electronics, 3D printing, etc.. However, its brittle nature, impact strength, and water sensitivity limit its usage in many industrial applications [2–4]. It is suggested that the performance of the polymers can be enhanced in terms of mechanical, thermal, wear properties, etc., by adding them with inorganic compounds. Sreekala and Eger [5] reported that increasing the SiO_2 content in epoxy matrix composites

Tribology of Polymer Composites
https://doi.org/10.1016/B978-0-12-819767-7.00005-0

exhibited improved wear resistance. It was attributed to the increased microhardness of these composites by the gradual addition of SiO_2 with the epoxy matrix. Contrary to this, the coefficient of friction was increased with increasing the SiO_2 content. Gregory et al. [6] observed a monotonic increment in wear resistance by increasing the alumina particles in the polytetrafluoroethylene (PTFE) matrix. Wetzel et al. [7] analyzed the wear resistance of alumina/epoxy composites by varying the volume percentage of alumina from 0.5 to 10. They reported that 2 vol.% of alumina-mixed composites exhibited enhanced wear resistance than pure epoxy. Further, the wear resistance of these composites was found to be increased by the addition of calcium silicate. In another study, Shi et al. [8] fabricated Si_3N_4/epoxy matrix composites and reported higher wear resistance with a reduced coefficient of friction.

Some researchers also reported that the inclusion of inorganic fillers reduced the wear performance of PMCs that could be seen contrary to the reports above. Bahadur and Deli Gong [9] studied the influence of adding three different fillers such as CuO, CuS, and CuF_2 in polyetheretherketone (PEEK). It was reported that the coefficient of friction of the composites increased with the addition of fillers compared to the pure PEEK. However, the wear resistance of all these composites was also found to be increased. In another study, the wear resistance of copper acetate/nylon 11 filled composite was found to be decreased [10]. It was attributed to the poor formation of the transfer film. Bahadur et al. [11] studied the wear performance of CuS/nylon matrix composites. They found that a decrement in wear resistance of these composites was due to the decomposition of CuS during the sliding and the poor adhesion between the transfer film of composites and the glass substrate.

In recent years, bio-based composites are gaining more attention due to their widespread applications such as industrial building, aerospace, construction, sports, etc. [12, 13]. Natural fiber composites are used in the development of automotive body parts such as dashboards, door panels, interior parts, and headliners [14]. Many researchers have attempted to investigate the tribological behavior of natural fiber-based composites [15–22]. Bajpai et al. [23] fabricated three different types of composites, such as PLA/nettle, PLA/*grewia optiva*, and PLA/sisal fiber. It was found that the composites exhibited a reduction in the coefficient of friction of about 10%–44%, and a reduction in the wear rate of about 70% compared to the pure PLA. Chaudhary et al. [24] hybridized three natural fibers such as jute, hemp, and flax in the epoxy matrix. The authors reported that the addition of natural fibers in the epoxy matrix exhibited significant improvement in wear resistance.

Even though there are many reports available on the tribological behavior of natural fiber-based composites, studies on the combination of fiber and/or filler in matrix composites are very minimal and further more investigation is needed for the development of high-performance materials. Hence this work attempted to study the coefficient of friction and specific wear rate of new combination of materials, i.e., Cu_2O/PLA composites and Cu_2O/MF-coated sisal fiber/modified PLA composites. The wear characteristics of Cu_2O/PLA composites were compared with the Cu_2O/MF-coated sisal fiber/modified PLA hybrid composites. The developed composites were subjected to different operating conditions such as normal load (10 N), sliding speed (1 m/s), sliding distance (2000 m), and operating temperatures (35°C and 65°C). The worn surfaces of the developed composites were analyzed using optical micrographs.

2 Experimental details

2.1 Materials

Commercial grade PLA (2003D) having a density of 1.24 g/cm^3, glass transition temperature of 55–60°C, and a melt flow index of 6 g/cm^3 min was obtained from Minnetonka, United States. The sisal fiber was purchased from the local market in North Bangkok, Thailand. Melamine, formaldehyde (40 vol.%), and surfactant sodium dodecyl sulfate were purchased from Sigma Aldrich. Cu_2O was procured from Thermo Fisher Scientific company. The particle size, density, and molar mass of Cu_2O were 350 nm, 6 g/cm^3, and 143.09 g/mol, respectively.

2.2 Melamine-formaldehyde sisal fiber coating

Initially, the sisal fibers were cut to 0.25-nm length using a fiber cutting machine and were thoroughly washed using distilled water. The fibers were then dried in a hot air oven at 60°C overnight. Then 0.3 wt.% of surfactant was mixed with 150 mL of distilled water, in which the dried sisal fibers were dispersed by mixing it for 5 min at 8000 rpm using a high-speed stirrer. Then melamine-formaldehyde (1:3 ratio) was mixed with 50 mL of distilled water by continuous stirring at a temperature of 70°C. This mixture was mixed continuously until a transparent solution was obtained. To this transparent solution, the sisal fiber slurry was mixed and the pH was adjusted to 5 using 10% acetic acid. To obtain polymerization of melamine-formaldehyde over the sisal fiber surface, the mixture was continuously stirred for about 4 h at 70°C. Finally, the melamine-formaldehyde-coated sisal fibers were filtered, washed, and then dried in a hot air oven at 60°C overnight.

2.3 Fabrication of PLA/Cu_2O and MF-coated sisal fiber/modified PLA/Cu_2O hybrid composites

For fabricating PLA/Cu_2O composites, the PLA was melt mixed using an internal mixer at 170°C for 3 min. Then the required amount of Cu_2O was mixed with PLA for about 7 min with a mixing speed of 40 rpm.

In the case of MF-coated sisal fiber/modified PLA/Cu_2O hybrid composites, the required amount of sisal fiber was added after the addition of Cu_2O. The composites were fabricated using a compression molding machine at 170°C and 2500 psi for 10 min. Table 1 gives the details of composite samples with their abbreviations.

2.4 Wear test

The dry sliding wear test was conducted on a pin-on-disc machine (shown in Fig. 1) supplied by DUCOM (TR-20LE-PHM 400-CHM 600). A steel disc of EN31 and hardness of 65HRC was used. The developed composites were tested with the following conditions: sliding speed of 1 m/s, the normal load of 10 N, a sliding distance of 2000 m, and operating temperatures of 35°C and 65°C.

The composite samples were measured before and after the wear test, and the wear loss technique was adopted to analyze the wear results. Wear debris was cleaned after each test with a silicon carbide paper, and the surface of the wear track was washed by acetone.

3 Results and discussion

3.1 Coefficient of friction

Figs. 2–4 show the coefficient of friction of PLA/Cu_2O composites against the sliding distance of 2000 m, and the COF values were measured periodically after every 108 s. The COF of different composites, i.e., unfilled PLA, PLA/Cu_2O (10 pph), PLA/Cu_2O (20 pph), and PLA/Cu_2O (30 pph) were recorded under the normal load of 10 N, while the sample temperatures were maintained as 35°C and 65°C. It is evident from the COF graphs that the PLA/Cu_2O composites exhibited similar behavior in both the temperatures. The COF decreased in the order PLA > PLA/Cu_2O (30 pph) > PLA/Cu_2O (20 pph) > PLA/Cu_2O (10 pph) with increasing sliding distance.

Table 1 Details of composite samples.

Material	Temperature (°C)	Abbreviation	Matrix (pph)	Cu_2O (pph)	Melamine-formaldehyde-coated sisal fiber (pph)
PLA	35	PLA_35	100	–	–
PLA	65	PLA_65	100	–	–
PLA/Cu_2O	35	PLA10@35	100	10	–
PLA/Cu_2O	65	PLA10@65	100	10	–
PLA/Cu_2O	35	PLA20@35	100	20	–
PLA/Cu_2O	65	PLA20@65	100	20	–
PLA/Cu_2O	35	PLA30@35	100	30	–
PLA/Cu_2O	65	PLA30@65	100	30	–
PLA/Cu_2O/MF-coated sisal fiber	35	PLA_Hybrid10@35	100	10	10
PLA/Cu_2O/MF-coated sisal fiber	65	PLA_Hybrid10@65	100	10	10
PLA/Cu_2O/MF-coated sisal fiber	35	PLA_Hybrid20@35	100	20	10
PLA/Cu_2O/MF-coated sisal fiber	65	PLA_Hybrid20@65	100	20	10
PLA/Cu_2O/MF-coated sisal fiber	35	PLA_Hybrid30@35	100	30	10
PLA/Cu_2O/MF-coated sisal fiber	65	PLA_Hybrid30@65	100	30	10

Fig. 1 Pin-on-disc machine.

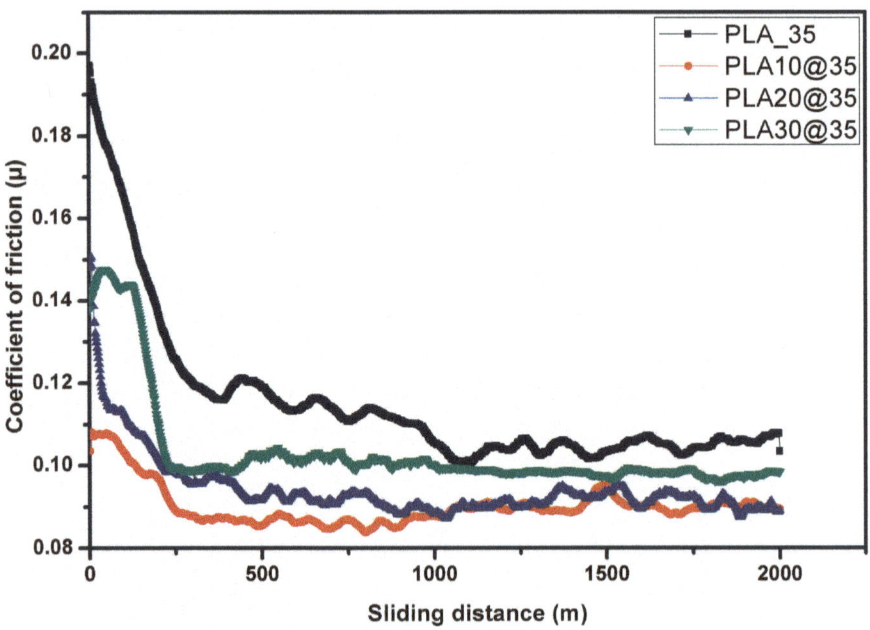

Fig. 2 Coefficient of friction vs. sliding distance for PLA and its composites filled with Cu_2O at 35°C. Test conditions: sliding speed 1 m/s and normal load 10 N.

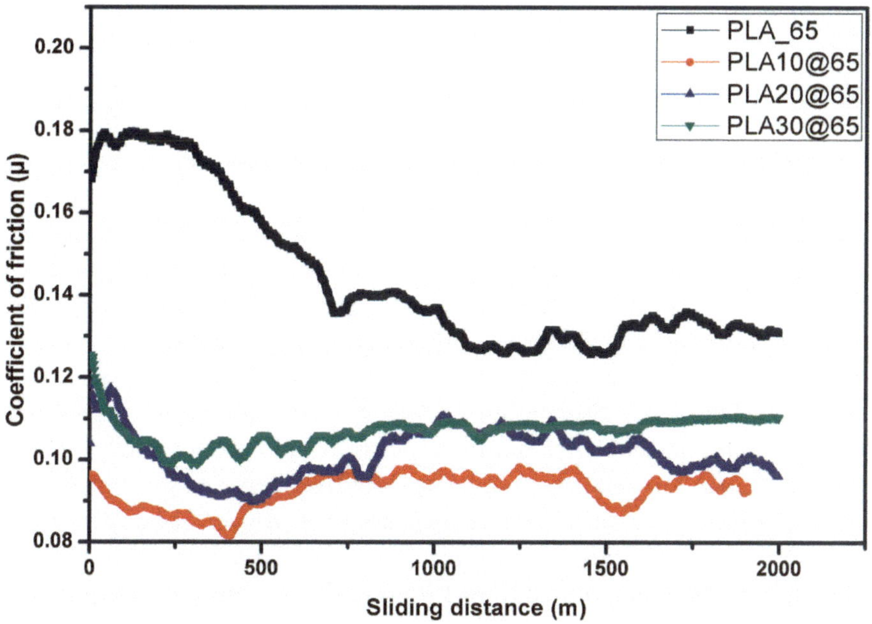

Fig. 3 Coefficient of friction vs. sliding distance for PLA and its composites filled with Cu_2O at 65°C. Test conditions: sliding speed 1 m/s and normal load 10 N.

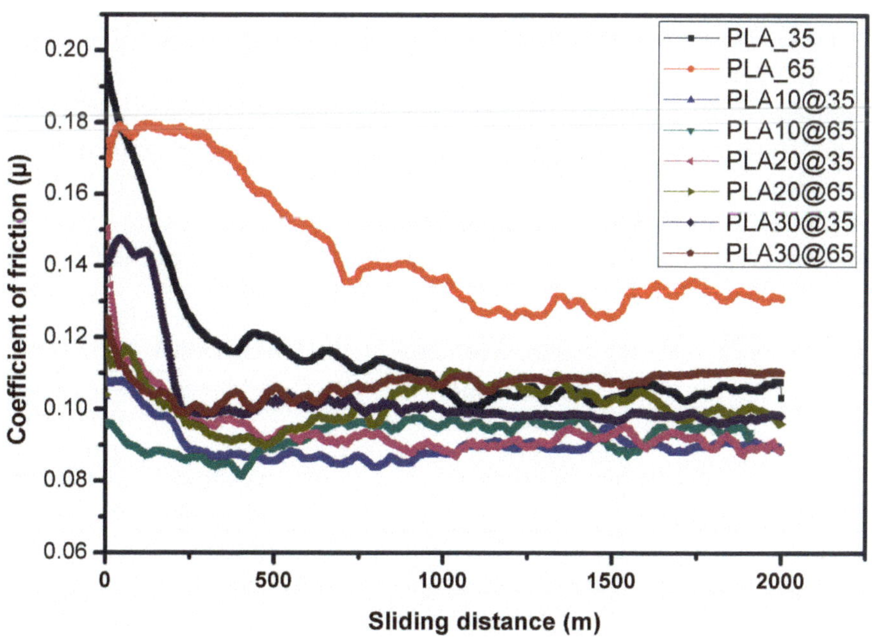

Fig. 4 Coefficient of friction vs. sliding distance for PLA and its composites filled with Cu_2O at 35°C and 65°C. Test conditions: sliding speed 1 m/s and normal load 10 N.

Furthermore, it was observed from Fig. 4 that the COF of 65°C tested composites was higher than the 35°C tested composites. This could be attributed to the weakening of the bonding between the Cu_2O and PLA by increased temperature, i.e., from 35°C to 65°C.

The general performance of COF (increasing trend, decreasing trend, or constant) depends on the formation of polymer film, the interfacial bonding between the fiber and matrix, or nature of the fibers at the interface [18]. During the initial sliding distances (around 200 m, as shown in Figs. 2–4), the COF of PLA/Cu_2O composites were found to be increased. This could be due to the mechanical interlocking of asperities at the interface. Further increasing the sliding distance and repetitive sliding, the temperature could be increased at the contact surface, i.e., between the composite sample and the steel counterface. Hence the PLA matrix could have softened, and the softened matrix can generate an intermediate layer between the sample and the steel counterface. Furthermore, the smooth and glassy layer generated by the PLA matrix prevented the direct contact of the sample with the steel counterface. Therefore, the COF of PLA/Cu_2O composites were found to be decreased (shown in Figs. 2–4).

The effect of Cu_2O filler on the coefficient of friction of PLA matrix composites are presented in Table 2. It is evident that the COF of PLA10@35 was reduced by 17% than that of the PLA 35. Similar studies have been reported on the coefficient of friction of the polyphenylene sulfide filled with CaF_2 composites [25], and a similar kind of trend was reported. In another study, Zhao and Bahadur reported that polyphenylene sulfide/NiS-filled composites exhibited lower COF than the unfilled polyphenylene sulfide [26].

The COF of PLA/Cu_2O composites increased with the increased Cu_2O content, i.e., 10%–30%. Similarly, there was a 17% increase in the COF of

Table 2 Comparison of the coefficient of friction of filled PLA composites.

Composites	Coefficient of friction	Percentage reduction in the coefficient of friction	Composites	Coefficient of friction	Percentage reduction in the coefficient of friction
PLA_35	0.1075	–	PLA_65	0.1290	–
PLA10@35	0.0890	17.2093	PLA10@65	0.0925	28.2945
PLA20@35	0.0935	13.0232	PLA20@65	0.1020	20.9303
PLA30@35	0.0980	8.8872	PLA30@65	0.1085	15.8914

PLA10@65 compared with the PLA30@65 composite. The possible reason behind the increasing trend of COFs could be the particle agglomeration.

The variation of tribological characteristics of PLA/Cu_2O composites studied after the addition of MF-coated sisal fiber (fiber length of 0.25 mm, 10 pph), and their respective COF is plotted in Figs. 5–7. At first glance, the hybrid composites behaved in a similar manner in both the temperatures, i.e., the COF of PLA/Cu_2O/MF-coated sisal fiber hybrid composites exhibited in the order 10 ppm > 20 ppm > 30 ppm > PLA. Besides, the composites tested at 65°C showed higher magnitudes than the 35°C tested composites (shown in Fig. 7).

The behavior of hybrid composites was contrary to the PLA/Cu_2O composites, i.e., with increasing sliding distances, the COF of all the hybrid composites was found to be increased. It could be due to the different characteristics of transfer films formed by these composite materials on the counterface and the contact of MF-coated sisal fibers with the steel counterface. Similar kinds of results were reported while hybridizing the jute/hemp/flax/epoxy matrix hybrid composites. The hybrid composites exhibited higher coefficient of friction than the pure epoxy matrix [24]. In

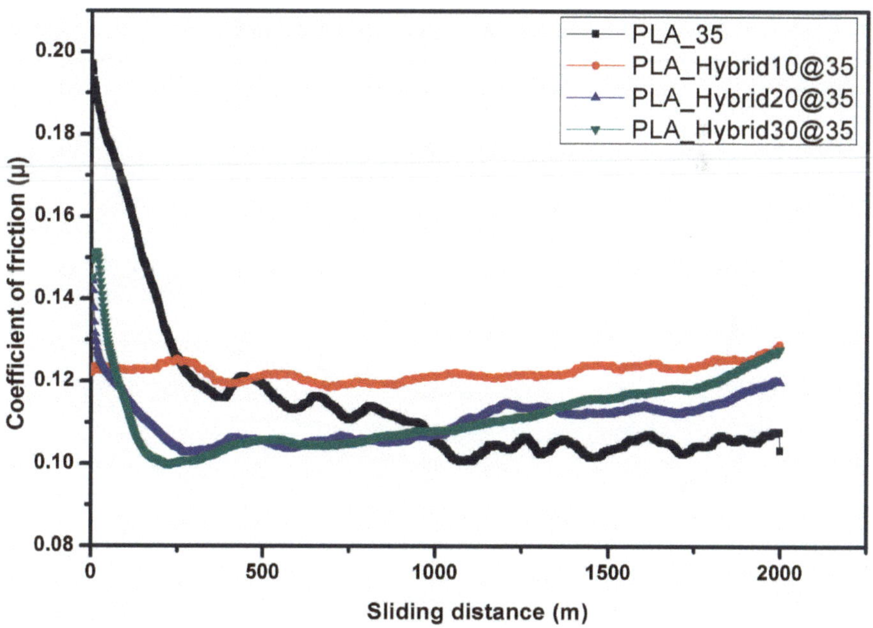

Fig. 5 Coefficient of friction vs. sliding distance for PLA and its composites filled with Cu_2O and MF-coated sisal fiber at 35°C. Test conditions: sliding speed 1 m/s and normal load 10 N.

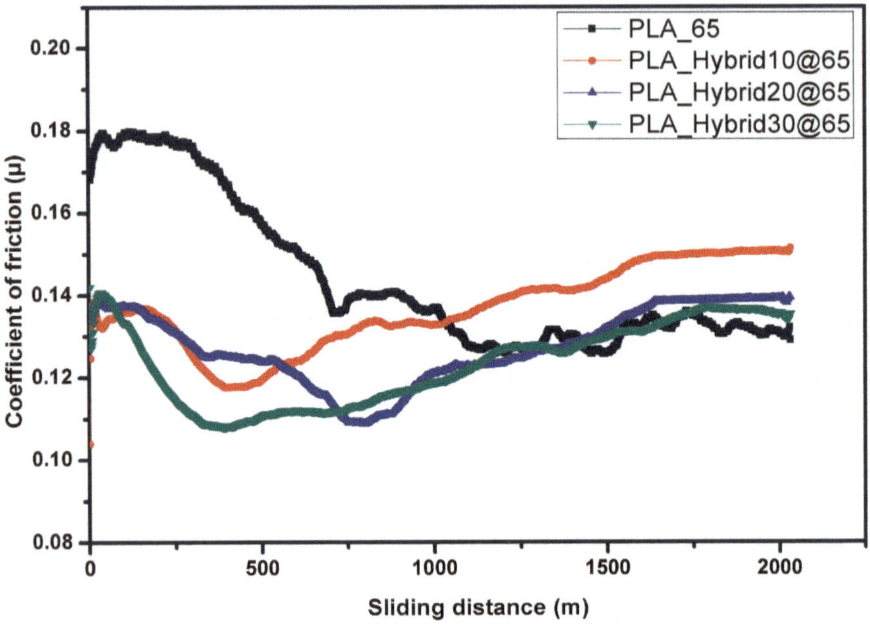

Fig. 6 Coefficient of friction vs. sliding distance for PLA and its composites filled with Cu_2O and MF-coated sisal fiber at 65°C. Test conditions: sliding speed 1 m/s and normal load 10 N.

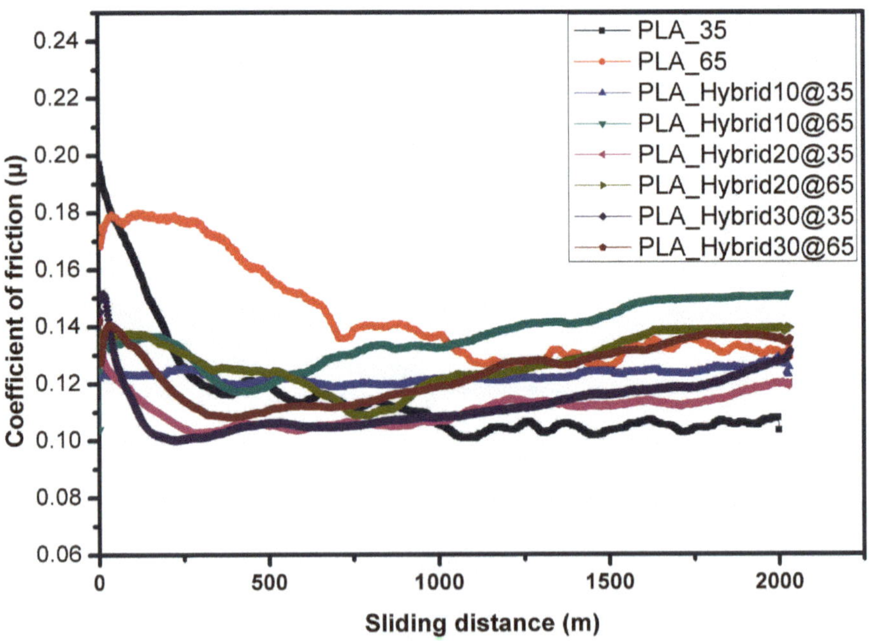

Fig. 7 Coefficient of friction vs. sliding distance for PLA and its composites filled with Cu_2O and MF-coated sisal fiber at 35°C and 65°C, respectively. Test conditions: sliding speed 1 m/s and normal load 10 N.

another study, nylon 11 reinforced with ZnF, ZnS, and PbS fillers exhibited 15%–20% higher COF values compared to that of pure nylon 11 [27].

Table 3 shows the percentage improvement of COF of hybrid composites compared with the PLA/Cu_2O composites. It was observed that the COF of PLA_Hybrid10@65 composites showed a 63% improvement than the PLA10@65 composites. Besides, the steady-state COF of PLA/Cu_2O/MF-coated sisal fiber composites was about 3.5%–20% higher than the unfilled PLA. However, in the case of PLA/Cu_2O composites, the COF values increased by about 9%–28%. Hence the dependence of COF of PLA-filled composites was mainly dependent on the addition of (i) Cu_2O fillers and (ii) MF-coated sisal fibers and not so much on the PLA matrix, whereas the filler and/or fibers were mixed.

3.2 Wear

Fig. 8 shows the wear volume of PLA and its composites as a function of the sliding distance of 2000 m. It was observed that the wear volumes of all the different proportions of Cu_2O-filled composites were found to be higher at 65°C. Likewise, the specific wear rate of PLA/Cu_2O composites increased with increasing temperature (Table 4). For example, the specific wear rate of PLA30@65 composite was found to be 42% higher than that of PLA30@35 composites. It could be due to the absence of transfer film at higher temperature composites.

Table 3 Comparison of the coefficient of friction of PLA/Cu_2O and PLA/Cu_2O/MF-coated sisal fiber hybrid composites.

PLA/Cu_2O composites	COF	PLA/ Cu_2O/MF-coated sisal fiber hybrid composites	COF	COF improvement of hybrid composites compared to PLA/Cu_2O
PLA_35	0.1075	PLA_35	0.1075	–
PLA_65	0.1290	PLA_65	0.1290	–
PLA10@35	0.0890	PLA_Hybrid10@35	0.1255	41%
PLA10@65	0.0925	PLA_Hybrid10@65	0.1510	63%
PLA20@35	0.0935	PLA_Hybrid20@35	0.1200	28%
PLA20@65	0.1020	PLA_Hybrid20@65	0.1335	30%
PLA30@35	0.0980	PLA_Hybrid30@35	0.1295	32%
PLA30@65	0.1085	PLA_Hybrid30@65	0.1350	24%

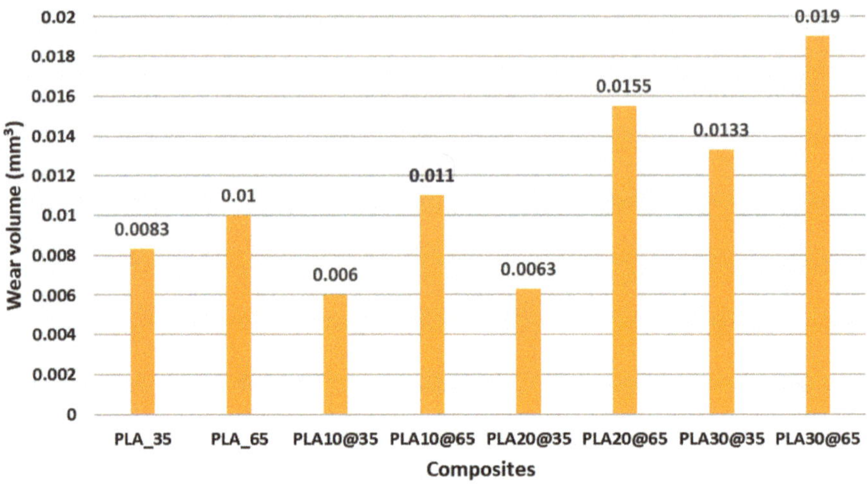

Fig. 8 Variation in wear volume of PLA/Cu_2O composites (test conditions: sliding speed 1 m/s; normal load 10 N; temperature 35°C, 65°C; and sliding distance 2000 m).

Table 4 Comparison of a specific wear rate of PLA/Cu_2O composites.

Composites	Specific wear rate (mm^3/N m)
PLA_35	0.000167
PLA_65	0.000200
PLA10@35	0.000111
PLA10@65	0.000203
PLA20@35	0.000108
PLA20@65	0.000264
PLA30@35	0.000219
PLA30@65	0.000312

In addition, it may be noted from Table 4 that the PLA10@35 composites (10 pph) exhibited higher wear resistance than the PLA20@35 (20 pph) and PLA30@35 (30 pph) composites. This indicates that the transfer film of 10 pph composites could have played a beneficial role. The role of transfer films in filler composites was also reported elsewhere [28–30].

After the wear test, the worn surfaces of PLA-filled composites were examined using optical micrographs, as shown in Figs. 9–14. Fig. 9 shows a clear and smooth surface of PLA10@35 composites compared to PLA10@65 (Fig. 10). It was ascribed to the better bonding between the fillers (10 pph) and the PLA. Thus a higher wear resistance exhibited by the PLA10@35 composites. The wear debris in the case of PLA20@65 and PLA30@65 was found

Fig. 9 Optical micrograph of the clear and smooth surface formed by PLA10@35.

Fig. 10 Optical micrograph of wear debris formed by PLA10@65.

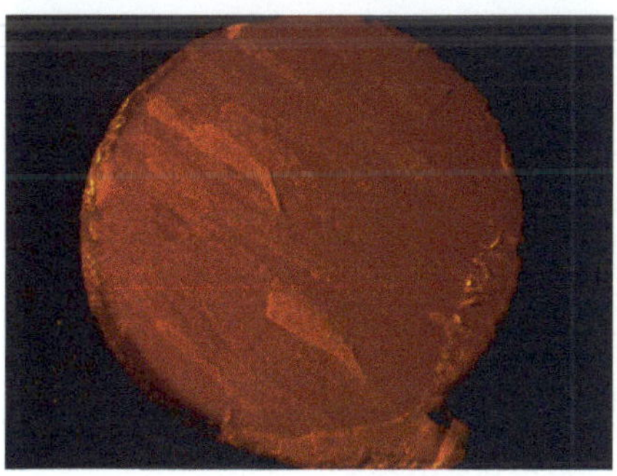

Fig. 11 Optical micrograph of PLA20@35 against 2000 m sliding distance.

to be higher (Figs. 12 and 14) than PLA20@35 and PLA30@35 (shown in Figs. 11 and 13). The increased wear rate of these composites was presumably due to the weak bonding between the transfer films and their substrates.

The wear performance of PLA/Cu_2O composites analyzed after the addition of MF-filled sisal fiber-reinforced hybrid composites. The variation

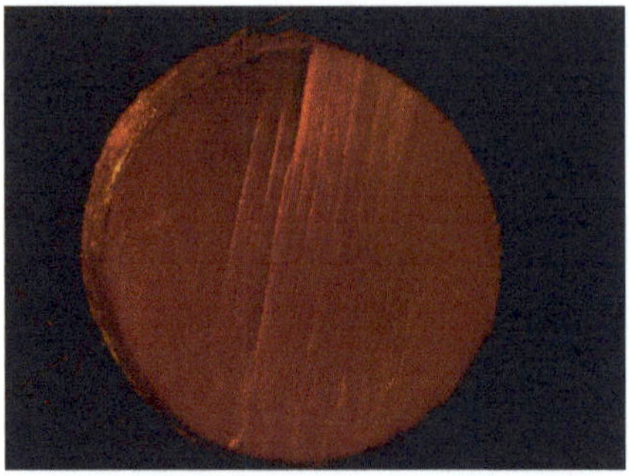

Fig. 12 Optical micrograph of wear debris formed in PLA20@65.

Fig. 13 Wear debris witnessed in an optical micrograph of PLA30@35.

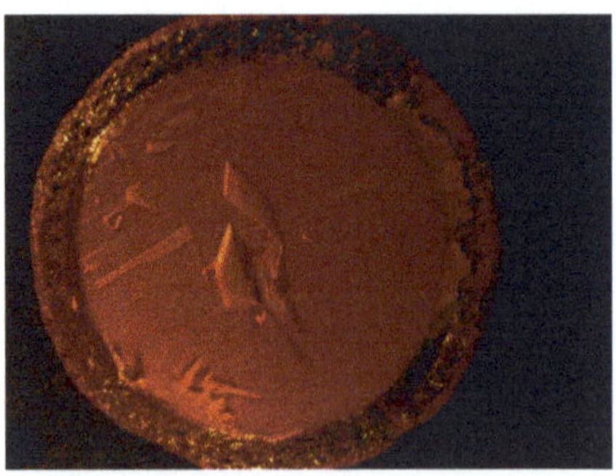

Fig. 14 Optical micrograph of PLA30@65.

of wear volume and specific wear rate of PLA and their hybrid composites are shown in Fig. 15 and Table 5. It can be seen that the wear volume increased after the addition of MF-filled sisal fiber with PLA/Cu_2O composites. For instance, at 35°C, the increasing order of wear volume was PLA_Hybrid30 > PLA_Hybrid20 > PLA_Hybrid10. At 65°C the order was

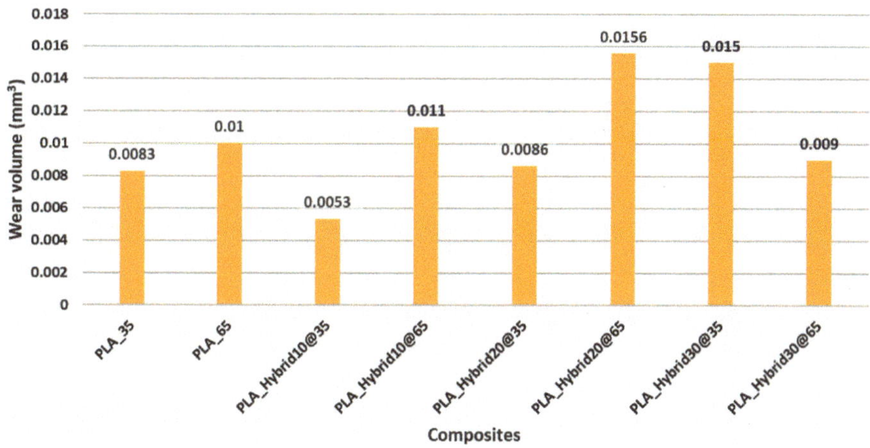

Fig. 15 Variation in wear volume of PLA/Cu_2O/MF-filled sisal fiber composites (test conditions: sliding speed 1 m/s; normal load 10 N; temperature 35°C, 65°C; and sliding distance 2000 m).

Table 5 Comparison of a specific wear rate of PLA/Cu_2O/MF-filled sisal fiber composites.

Composites	Specific wear rate (mm^3/N m)
PLA_35	0.000167
PLA_65	0.000200
PLA_Hybrid10@35	0.000099
PLA_Hybrid10@65	0.000204
PLA_Hybrid20@35	0.000150
PLA_Hybrid20@65	0.000271
PLA_Hybrid30@35	0.000261
PLA_Hybrid30@65	0.000156

followed as PLA_Hybrid20 > PLA_Hybrid10 > PLA_Hybrid30. In addition, it may be noted from Fig. 15 that the PLA_Hybrid10@35 composite was about 56% lower in wear volume than PLA35. However, the addition of MF-filled sisal fiber (20 and 30 ppm) in PLA/Cu_2O composites was not appropriate in terms of reducing wear volume and/or specific wear rate.

In other words, the interfacial bonding between the MF-filled sisal fiber and Cu_2O in the PLA matrix was weak. Moreover, the good interfacial bonding could be the only source of bonding between the transfer films of these hybrid composites and the counterface. Since the bonding was weak in the hybrid composites, the transfer film could be easily peeled off during sliding. It could be ascribed to the reduced wear volume, and the wear resistance of MF-filled sisal fiber-reinforced hybrid composites. For the sake of comparison, the specific wear rates of PLA/Cu_2O and PLA/Cu_2O/MF-filed sisal fiber hybrid composites are shown in Fig. 16. It may be noticed that most of the hybrid composites exhibited a higher specific wear rate than the PLA/Cu_2O composites.

It can be observed from Fig. 17A that there were a lesser number of cracks on the worn surfaces against the sliding distance of 2000 m. This indicated better wear resistance than the PLA_Hybrid20 composites. As compared to Fig. 17A, the worn surface and wear debris could be seen in Fig. 17B. The debris could have contributed to an increased rate of wear volume in PLA_Hybrid20 composites.

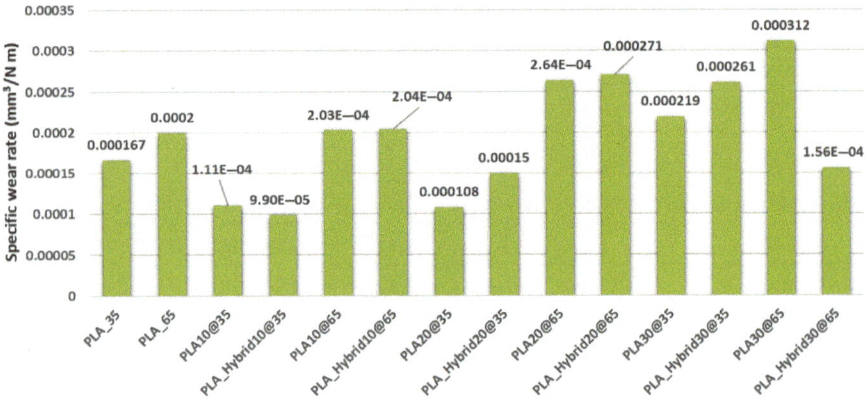

Fig. 16 Variation of specific wear rate of PLA/Cu_2O/MF-filled sisal fiber composites (test conditions: sliding speed 1 m/s; normal load 10 N; temperature 35°C, 65°C; and sliding distance 2000 m).

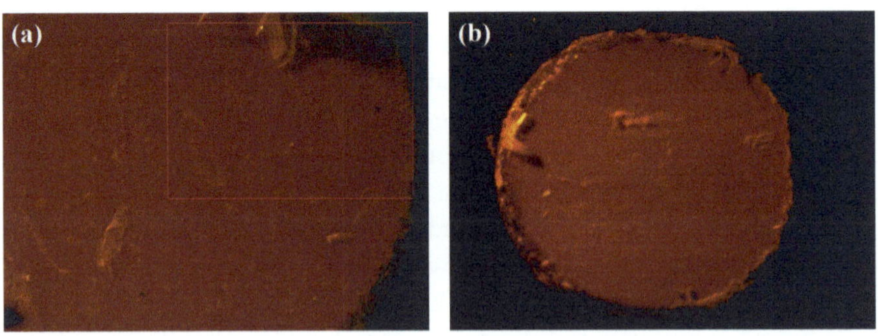

Fig. 17 Optical micrograph of (A) PLA_Hybrid10@35 and (B) PLA_Hybrid20@65 composites.

4 Conclusions

From the sliding tests on PLA composites with Cu_2O and Cu_2O/MF-filled sisal fiber, the following conclusions were drawn:

(1) The coefficient of friction of PLA10@35 composite was low. However, the coefficient of friction was considerably increased with increasing the Cu_2O content. Contrary to the PLA/Cu_2O composites, MF-filled sisal fiber composites exhibited a decreasing trend in COF.

(2) The optical micrograph images of PLA10@35 composites showed the precise and smooth surfaces, but in other cases (i.e., PLA20 and PLA30 composites), wear debris can be noticed. This could be an indication of their increasing rate of wear volume.

(3) Based on the observations from this study, it was suggested that the addition of MF-filled sisal fiber (20 and 30 ppm) in PLA/Cu_2O composites was not appropriate in terms of reducing wear volume and/or specific wear rate.

Acknowledgments

This research was completely supported by King Mongkut's University of Technology North Bangkok (KMUTNB), Thailand, through the Post-Doctoral Program (Grant Nos. KMUTNB-61-Post-03 and KMUTNB-63-KNOW-001).

References

[1] K. Friedrich, Z. Zhang, A.K. Schlarb, Effects of various fillers on the sliding wear of polymer composites, Compos. Sci. Technol. 65 (2005) 2329–2343, https://doi.org/10.1016/j.compscitech.2005.05.028.

[2] J. Tengsuthiwat, K. Yorseng, S. Siengchin, J. Parameswaranpillai, Thermomechanical, water absorption, ultraviolet resistance and laser-assisted electroless plating behavior of Cu_2O and melamine–formaldehyde-coated sisal fiber-modified poly (lactic acid) composites, Polym. Compos. (2019).

[3] J.T. Lee, M.W. Kim, Y.S. Song, T.J. Kang, J.R. Youn, Mechanical properties of denim fabric reinforced poly(lactic acid), Fibers Polym. 11 (2010) 60–66.

[4] P.K. Bajpai, I. Singh, J. Madaan, Development and characterization of PLA-based green composites: a review, J. Thermoplast. Compos. Mater. 27 (2014) 52–81.

[5] M.S. Sreekala, C. Eger, Property improvements of an epoxy resin by nanosilica particle reinforcement, Polym. Compos. (2005) 91–105, https://doi.org/10.1007/0-387-26213-x_6.

[6] W.G. Sawyer, K.D. Freudenberg, P. Bhimaraj, L.S. Schadler, A study on the friction and wear behavior of PTFE filled with alumina nanoparticles, Wear 254 (2003) 573–580, https://doi.org/10.1016/S0043-1648(03)00252-7.

[7] B. Wetzel, F. Haupert, M.Q. Zhang, Epoxy nanocomposites with high mechanical and tribological performance, Compos. Sci. Technol. 63 (2003) 2055–2067, https://doi.org/10.1016/S0266-3538(03)00115-5.

[8] G. Shi, M.Q. Zhang, M.Z. Rong, B. Wetzel, K. Friedrich, Friction and wear of low nanometer Si_3N_4 filled epoxy composites, Wear 254 (2003) 784–796, https://doi.org/10.1016/S0043-1648(03)00190-X.
[9] S. Bahadur, D. Gong, The role of copper compounds as fillers in the transfer and wear behavior of polyetheretherketone, Wear 154 (1992) 151–165.
[10] S. Bahadur, D. Gong, J.W. Anderegg, The role of copper compounds as fillers in transfer film formation and wear of nylon, Wear 154 (1992) 207–223, https://doi.org/10.1016/0043-1648(92)90155-2.
[11] S. Bahadur, D. Gong, J.W. Anderegg, Tribochemical studies by XPS analysis of transfer films of Nylon 11 and its composites containing copper compounds. Wear 165 (1993) 205–212, https://doi.org/10.1016/0043-1648(93)90336-K.
[12] S.M.K. Thiagamani, S. Krishnasamy, S. Siengchin, Challenges of biodegradable polymers: an environmental perspective, Appl. Sci. Eng. Prog, (2019) https://doi.org/10.14416/j.asep.2019.03.002.
[13] A. Atiqah, M. Chandrasekar, T.S.M. Kumar, K. Senthilkumar, M.N.M. Ansari, Characterization and interface of natural and synthetic hybrid composites. in: Encyclopedia of Renewable and Sustainable Materials, vol. 4, Elsevier Ltd, 2020, https://doi.org/10.1016/B978-0-12-803581-8.10805-7.
[14] J. Holbery, D. Houston, Natural-fiber-reinforced polymer composites in automotive applications, JOM 58 (2006) 80–86.
[15] N.S.M. El-Tayeb, B.F. Yousif, T.C. Yap, An investigation on worn surfaces of chopped glass fibre reinforced polyester through SEM observations, Tribol. Int. 41 (2008) 331–340.
[16] U. Nirmal, B.F. Yousif, D. Rilling, P.V. Brevern, Effect of betelnut fibres treatment and contact conditions on adhesive wear and frictional performance of polyester composites, Wear 268 (2010) 1354–1370, https://doi.org/10.1016/j.wear.2010.02.004.
[17] N. Singh Gill, B.F. Yousif, Wear and frictional performance of betelnut fibre-reinforced polyester composite, Proc. Inst. Mech. Eng. Part J: J. Eng. Tribol. 223 (2009) 183–194.
[18] M. El-Tayeb, A study on the potential of sugarcane fibers/polyester composite for tribological applications, Wear 265 (2008) 223–235, https://doi.org/10.1016/j.wear.2007.10.006.
[19] C.W. Chin, B.F. Yousif, Potential of kenaf fibres as reinforcement for tribological applications, Wear 267 (2009) 1550–1557.
[20] S.A.R. Hashmi, U.K. Dwivedi, N. Chand, Graphite modified cotton fibre reinforced polyester composites under sliding wear conditions, Wear 262 (2007) 1426–1432, https://doi.org/10.1016/j.wear.2007.01.014.
[21] X. Xin, C.G. Xu, L.F. Qing, Friction properties of sisal fibre reinforced resin brake composites, Wear 262 (2007) 736–741.
[22] N. Rajini, K. Mayandi, M. Manoj Prabhakar, S. Siengchin, N. Ayrilmis, C. Bennet, et al., Tribological properties of cyperus pangorei fiber-reinforced polyester composites (friction and wear behavior of cyperus pangorei fiber/polyester composites), J. Nat. Fibers (2019) 1–13, https://doi.org/10.1080/15440478.2019.1621232.
[23] P.K. Bajpai, I. Singh, J. Madaan, Tribological behavior of natural fiber reinforced PLA composites, Wear 297 (2013) 829–840, https://doi.org/10.1016/j.wear.2012.10.019.
[24] V. Chaudhary, P.K. Bajpai, S. Maheshwari, An investigation on wear and dynamic mechanical behavior of jute/hemp/flax reinforced composites and its hybrids for tribological applications, Fibers Polym. 19 (2018) 403–415, https://doi.org/10.1007/s12221-018-7759-6.
[25] L. Yu, S. Bahadur, An investigation of the transfer film characteristics and the tribological behaviors of polyphenylene sulfide composites in sliding against tool steel, Wear 214 (1998) 245–251, https://doi.org/10.1016/S0043-1648(97)00234-2.

[26] Q. Zhao, S. Bahadur, The mechanism of filler action and the criterion of filler selection for reducing wear, Wear 225 (1999) 660–668.
[27] S. Bahadur, A. Kapoor, The effect of ZnF2, ZnS and PbS fillers on the tribological behavior of nylon 11, Wear 155 (1992) 49–61, https://doi.org/10.1016/0043-1648(92)90108-K.
[28] S. Bahadur, D. Tabor, The wear of filled polytetrafluoroethylene, Wear 98 (1984) 1–13.
[29] S. Bahadur, D. Tabor, The role of fillers in the friction and wear behavior of high-density polyethylene, in: Abstr. Pap. Am. Chem. Soc., vol. 187, Amer Chemical Soc, Washington, DC, 1984, pp. 48-PMSE.
[30] B.J. Briscoe, D. Tabor, The sliding wear of polymers; a brief review, in: N.P. Suh, K. Sako (Eds.), Fundamentals of Tribology, MIT Press, Cambridge, MA, 1980, pp. 733.

CHAPTER 6

Effect of adding sisal fiber on the sliding wear behavior of the coconut sheath fiber-reinforced composite

Muthukumar Chandrasekar[a], Krishasamy Senthilkumar[b,c,d], Thiagamani Senthil Muthu Kumar[b,c,d], I. Siva[b,c], P.S. Venkatanarayanan[a], Matuphum Phuthotham[d], Nagarajan Rajini[b,c], Suchart Siengchin[d], and M.R. Ishak[e,f]

[a]School of Aeronautical Sciences, Hindustan Institute of Technology & Science, Chennai, Tamil Nadu, India
[b]Department of Mechanical Engineering, Kalasalingam Academy of Research and Education, Krishnankoil, Tamil Nadu, India
[c]Centre for Composite Materials, International Research Centre, Kalasalingam Academy of Research and Education, Krishnankoil, Tamil Nadu, India
[d]Department of Mechanical and Process Engineering, The Sirindhorn International Thai-German Graduate School of Engineering (TGGS), King Mongkut's University of Technology North Bangkok, Bangkok, Thailand
[e]Department of Aerospace Engineering, University of Putra Malaysia, Serdang, Selangor, Malaysia
[f]Aerospace Malaysia Research Centre, Tower Block, University of Putra Malaysia, Serdang, Selangor, Malaysia

1 Introduction

To decrease the dependency on the petroleum-based products and reduce the impact on environment, research focus has been on the use of eco-friendly fibers extracted from natural resources in the composite [1, 2]. Natural fibers are lightweight due to their low density, have higher specific strength, good wear resistance, nontoxic, and more importantly renewable in nature [3, 4]. Due to these attributes, composites with the natural fiber reinforcements have found usage in structural and nonstructural applications [1, 5]. Industrial sectors like automobiles use natural fiber-reinforced composites in different parts of the vehicle like door panels, dashboards and other interior parts. Furthermore, research has been going on to use them in high-performance applications such as aircraft and military applications, space applications, etc. [6, 7]. In all these applications, these composites are easily subjected to adhesive, abrasive, and sliding wear during their life cycle [1].

Sliding wear behavior with the pin-on-disc method is one of the commonly used methods for assessing the wear performance of the composite

Tribology of Polymer Composites
https://doi.org/10.1016/B978-0-12-819767-7.00006-2

under laboratory conditions. Parameters such as sliding velocity, sliding distance, and applied load were varied under wet and dry conditions in the previous studies [7–9]. The wear and frictional characteristics of sisal-, nettle-, and grewia optiva-reinforced polypropylene composites with varying loadings and slide velocities were investigated. It was found that the sisal/polypropylene composites exhibited better wear resistance compared with the other two configurations. They also revealed that the applied normal load had a strong effect on the wear behavior of the composites compared to the sliding velocity [9]. The sliding wear properties of jute fabric-reinforced polypropylene composites were examined, and the results showed that the incorporation of jute fabric into polypropylene composite enhanced the wear resistance up to 45% and the reduction in frictional coefficient up to 65% [10].

Wear behavior of hybrid composites reinforced with two or more fibers has been well documented in the literature. Published studies have either synthetic reinforcement or a combination of synthetic and natural fibers. However, studies with two different natural fibers in various stacking sequences are limited and hence provide lot of room for research in this area. Thus, in this work, sliding wear behavior of hybrid composite reinforced with the sisal and coconut sheath fibers in the polyester matrix was investigated.

2 Materials and methods

2.1 Materials

Sisal fibers were purchased from Shiva Exports, Tirunelveli, Tamil Nadu, India, and naturally woven coconut sheath was extracted from the rural region in Madurai, Tamil Nadu, India. The fiber properties used in this study are given in Table 1. The matrix isophthalic unsaturated polyester resin, methyl ethyl ketone peroxide (MEKP) as a catalyst, and cobalt naphthenate as hardener were procured from Vasavibala Resins Pvt. Ltd., Chennai, India.

The procured sisal fibers were chopped to a length of 3 mm, while the coconut sheath fibers were used as reinforcement in their natural woven state. The fiber surface was modified by treating sisal and coconut sheath fiber with alkali and trichloro vinyl silane. The natural fibers were separately immersed in containers with 4% NaOH solution for 1 h subsequently followed by immersion in 0.3% silane solution for 1 h. The treated fibers were then dried in sunlight and were used as reinforcement in the composite.

Table 1 Fiber specifications [11].

Fiber	Coconut sheath	Sisal
Density (g/cm^3)	1.375	1.450
Cellulose (%)	68	65–70
Hemicellulose (%)	22	12
Lignin (%)	20.6	9.9
Moisture content (%)	8.79	10
Tensile strength (MPa)	88.63	68
Young's modulus (GPa)	4.4	3.8

2.2 Composite preparation

Sisal/coconut sheath fibers at various layering arrangements as shown in Table 2 were fabricated through the compression moulding process. The fiber layers were stacked into the mold of a dimension $300 \times 12 \times 3\,mm^3$ and precompressed at 15 MPa for 1 h. The resin mixture was prepared by mixing polyester resin with 1.5 wt.% of the MEKP and cobalt naphthenate. The resin mixture was then applied onto the stacked fiber layers and was allowed to settle for few minutes. The mold setup with the resin/fiber mixture was compressed in a hydraulic machine under the pressure of 17 MPa for 24 h at room temperature.

2.3 Density and void content

Void contents for the composites were calculated according to ASTMD2734. Experimental density of three specimens was measured using Densometer, and their average value was reported. Theoretical density was calculated as per Eq. (1).

$$T = 100 / (R/D + r/d) \tag{1}$$

Table 2 Various composite configurations with sisal and coconut sheath fibers fabricated for this study.

Composite	Individual fiber (wt.%)		Fiber/resin (wt.%)
	Coconut sheath	Sisal	
CCC	100	0	50/50
SSS	0	100	50/50
CSC	70	30	50/50
SCS	30	70	50/50

where T is the theoretical density (g/cm^3), R is the weight% of resin in the composite, D is the density of the resin (g/cm^3), r is the fiber weight% in the composite, and d is the density of the fiber (g/cm^3).

The void content can be evaluated using Eq. (2):

$$V = 100\left(T - M\right) / T \tag{2}$$

where V is the void content (%), T is the theoretical density (g/cm^3), and M is the measured density (g/cm^3).

2.4 Wear characterization

The sliding wear test was conducted with the "pin-on-disc" machine supplied from DUCOM (TR-20LE-PHM 400-CHM 600). A composite specimen of thickness 3mm and diameter 20mm was abraded against a stainless steel disc EN31 of hardness 65HRC. Wear test was carried out at sliding distances 600, 1800, 2400 and 3000mm, respectively. The disc velocity and applied load was maintained to 1 m/s and 30N, respectively. Specific wear rate was calculated as per Eq. (3).

$$k = \frac{m}{\rho LF} \tag{3}$$

where k is the specific wear rate in mm^3/Nm, ρ is the density of composite specimen in g/cm^3, L is the sliding distance in mm, and F is the applied load in N.

3 Results and discussion

3.1 Density and void content

Voids or porosity are micro defects that occur as a result of trapped air molecules in the resin during fabrication of the composite in hot press molding. In terms of measurement, porosity in the composite is the difference between measured density and theoretical density [12]. Table 3 presents the void content for the composites employed in this study.

Composite with the coconut sheath fibers (CCC) had greater extent of voids with 8.23% as opposed to the least void content of 1.9% for the composite with sisal fiber. Higher void content in the case of CCC could be explained by the fiber characteristic and naturally woven pattern of cocout sheath that comprises of core fiber and outer fiber varying in thickness within their naturally woven arrangement. The varying fiber thickness in

Table 3 Density and voids in the composite configurations employed in this study.

Composite	Experimental density (g/cm^3)	Theoretical density (g/cm^3)	Voids (%)
CCC	1.125	1.226	8.23
SSS	1.231	1.255	1.90
CSC	1.143	1.236	7.51
SCS	1.211	1.245	2.76

the coconut sheath fibers was documented by Naveen et al. in their study [13]. Hence this pysical characteristic of the coconut sheath fibers led to micropores on the composite surface as shown in Fig. 1A and B. The presence of voids as shown in the optical micrograph (Fig. 2A) was responsible for higher voids in case of CCC, while the least amount of voids in case of SSS is due to the lack of visible voids and good fiber distribution within the matrix as shown in Fig. 2B. Hybrid composite specimens such as CSC and

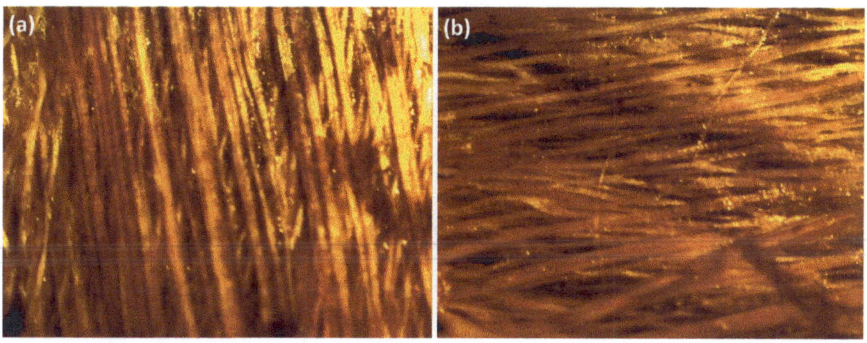

Fig. 1 Optical micrographs. (A) CSC and (B) CCC.

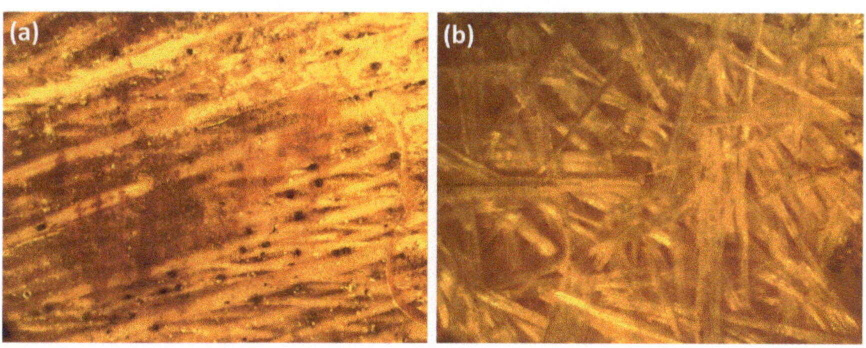

Fig. 2 Optical micrographs of composite. (A) CCC and (B) SSS.

SCS with sisal and core in the outer layer and core arrangement exhibited intermediate void content of 7.51% and 2.76% between the CCC and SSS configuration.

Fig. 3 shows weight loss of the composite configurations against the sliding distance at an applied load of 30 N and a sliding velocity of 1 m/s. At a sliding distance of 600 mm, composite with coconut sheath fibers (CCC) showed the highest mass loss of 2.7 mg, while the composite with sisal fibers (SSS) showed the least mass loss of 0.9 mg. Optical micrographs of CCC shown in the figure indicated surface discontinuities in the form of micro-voids. During the initial contact of the polymer surface, deformation takes place due to the rubbing action of the abrasive grit with the surface. The abrasive grit particles could easily penetrate into the surface discontinuities and form a mechanical interlocking. Hence the deformation becomes more pronounced leading to a higher rate of material removal.

Mass loss at the initial sliding distance gives an indication that the surface has began to worn out from the sliding against the abrasive grit. This is because at a load as high as 30 N, large frictional heat generated between the polymer surface and abrasive grit softens the polyester matrix and forms

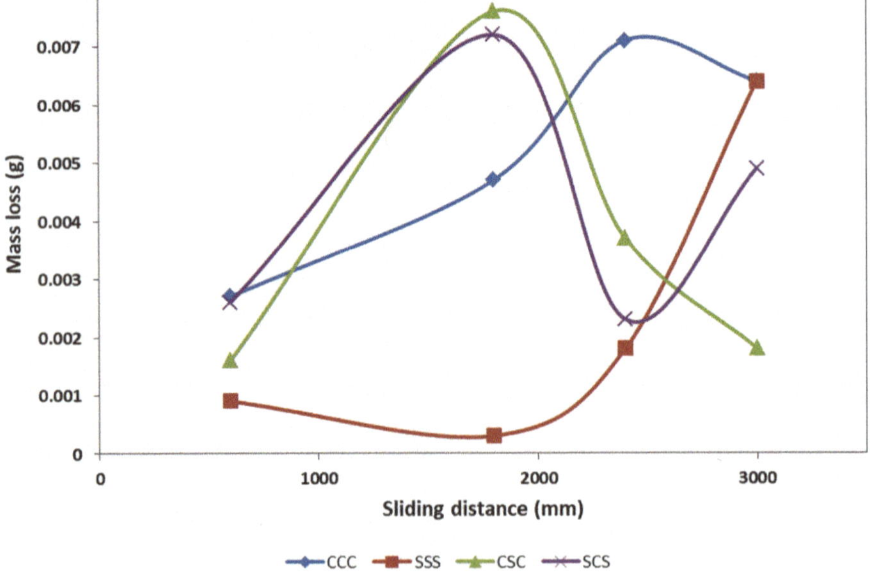

Fig. 3 Effect of hybridization on the mass loss of the sisal/coconut sheath-reinforced polyester composite at an applied load of 30 N and velocity of 1 m/s.

a film. Hence the polymer could be easily removed from the composite during the sliding process.

As the sliding distance was changed to 1800 mm, more volume of material was removed from the hybrid composite specimens. Nearly 7 and 7.5 mg have been removed as compared to the composite with individual fibers that had a mass loss of 4.5 and 0.5 mg. As the sliding distance was changed from 600 to 1800 mm, the composite specimens were subjected to fresh abrasive grit and it was easier for the abrasives to penetrate the matrix surface causing more material to be removed. With the further increase in sliding distance to 2400 and 3000 mm, the hybrid composite specimens showed lower mass loss than the control specimens at 2400 and 3000 mm, respectively. One possible reason for such trend could be due to uneven material removal from the polymer surface leading to irregular patches (Fig. 4A) spreading across the polymer surface exposed to wear. Similar observations were made in the previous studies on natural fiber-reinforced composite [14, 15]. Among the composite configurations, maximum material removal of 6.4 mg was observed for CCC and SSS followed by SCS and CSC at a sliding distance of 3000 mm. Fig. 4A shows the optical micrograph of SCS. Wear deformation in the form of wear tracks created by the continuous rubbing action of the abrasive grit on the polyester matrix is visible in Fig. 4B. In this study, wear deformation occurred only in the polyester matrix. However, as reported in a recent study, continuous sliding against the abrasive grit and increasing the sliding distance not only affected the matrix but also the natural fibers in the composite. Failure mechanisms such as fiber splitting, loose fibers, fiber pull-out, and fiber stripping were observed in the study [16].

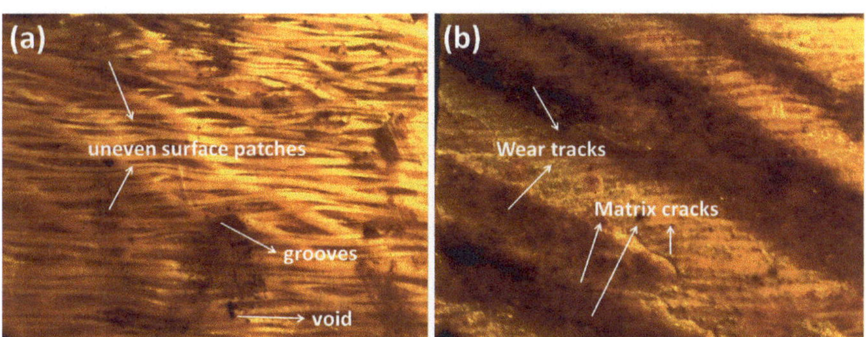

Fig. 4 Wear deformation in the composite. (A) Uneven surface patches and grooves in CCC and (B) wear tracks and Matrix cracks in SCS.

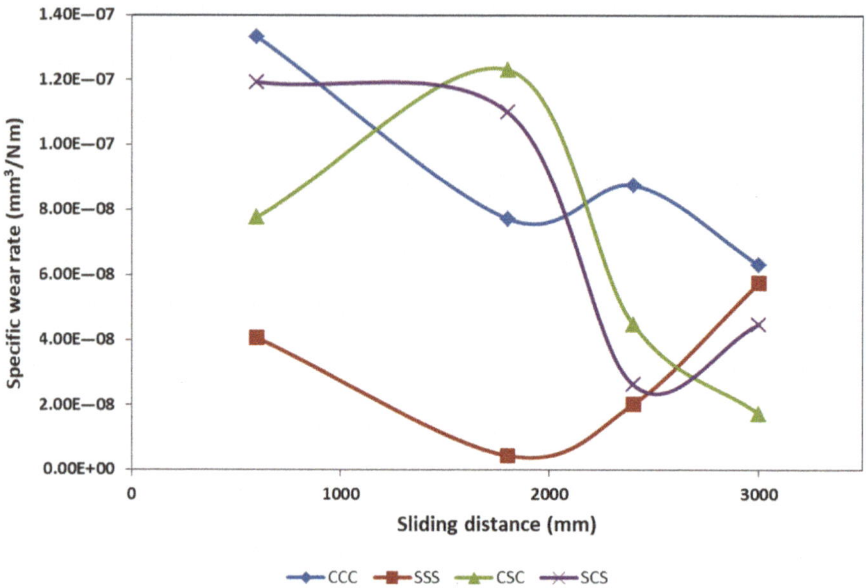

Fig. 5 Effect of hybridization on the specific wear rate of the sisal/coconut sheath-reinforced polyester composite at an applied load of 30 N and a velocity of 1 m/s.

The specific wear rate of the composite specimens at various sliding distances is given in Fig. 5. Specific wear rate is directly proportional to the weight loss and inversely proportional to density of the sample, sliding distance, and applied load as per Eq. (3).

At the start of sliding at 600 mm, maximum specific wear rate was observed. According to Deo et al. [16], since the abrasive paper was fresh, abrasive grits were more effective in causing wear deformation. As the sliding distance was further increased, CCC, CSC, and SCS showed a typical declining trend while the specific wear rate increased with the sliding distance for SSS. The reasons for decrease in specific wear rate during the subsequent runs and increase in the sliding distance are possibly due to (a) ineffectiveness of the abrasive grits due to the change in their surface texture from rough to smooth and (b) reduced depth of penetration as the wear debris from the previous runs fills the space between the polymer surface and abrasive paper [17]. In case of the hybrid composite specimens, specific wear rate was comparatively lower than the composite with individual fibers at all sliding distances except 1800 mm. The sudden increase in specific wear rate at 1800 mm is attributed to the higher magnitude of weight loss over the composite with individual fibers.

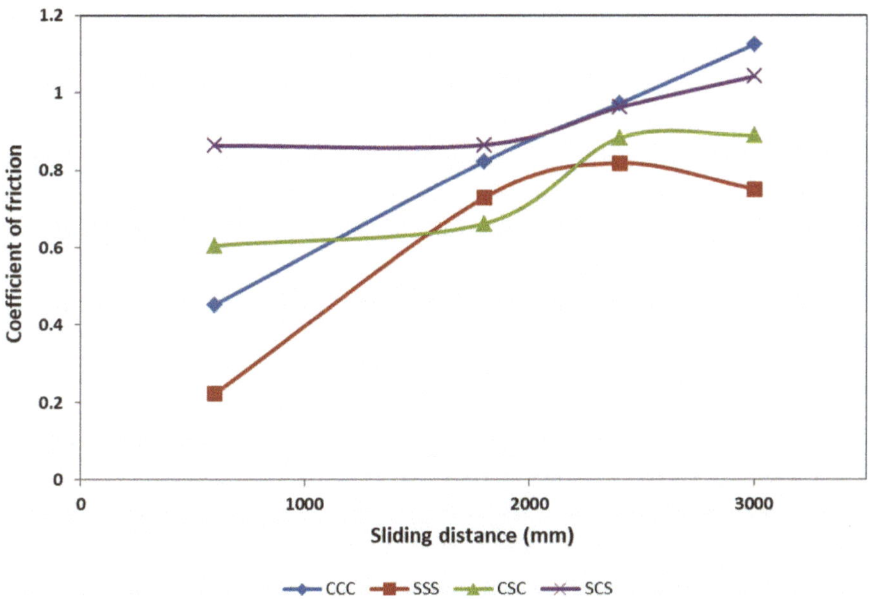

Fig. 6 Effect of hybridization on coefficient of friction of the sisal/coconut sheath-reinforced polyester composite at an applied load of 30 N and a velocity of 1 m/s.

Coefficient of friction (COF) shown in Fig. 6 was initially higher for hybrid composite specimens than the control specimens with individual fibers. It could be noticed that COF value kept increasing as the sliding distance was changed and maximum COF was observed at a sliding distance of 3000 mm. Initially, the surface temperature of the polyester matrix increases due to the frictional heat generated from the rubbing action of abrasive grit on the polymer surface. As the sliding continues, more surface area of the polymer comes into contact exposing new surface to the abrasive grit. This could cause a rise in the frictional heat, thereby leading to larger coefficient of friction [12].

4 Conclusion

Hybrid composite with sisal and coconut fibers showed better wear resistance than their counterparts with the individual fibers. Hybrid composites with sisal in the core/coconut sheath fibers in the outer layer and vice versa were found to have lower mass loss, decreased specific wear rate, and smaller value of coefficient of friction against the increasing sliding distance. The optical micrographs showed that hybrid composite specimens had less porosity or voids as opposed to the significantly higher void content in composite with the coconut sheath fibers. This physical attribute is believed to have influenced the wear behavior of the hybrid composites with coconut sheath fiber as the outer layer. The poor wear resistance of composite with coconut sheath fiber was also evident from the optical micrographs that showed the presence of grooves, wear deformation, and uneven surface patches on the polyester matrix. Other than the grooves and surface patches, wear tracks and matrix cracks could also be noticed on the surface of the specimens subjected to wear. However, damage to fiber reinforcement in the form of fiber splitting, fiber pull-out, and fiber stripping was not observed.

References

[1] V. Chaudhary, P.K. Bajpai, S. Maheshwari, An investigation on wear and dynamic mechanical behavior of jute/hemp/flax reinforced composites and its hybrids for tribological applications, Fibers Polym. 19 (2) (2018) 403–415.
[2] T. Senthil Muthu Kumar, N. Rajini, K. Obi Reddy, A. Varada Rajulu, S. Siengchin, N. Ayrilmis, All-cellulose composite films with cellulose matrix and Napier grass cellulose fibril fillers, Int. J. Biol. Macromol. 112 (2018) 1310–1315.
[3] M. Chandrasekar, M.R. Ishak, S.M. Sapuan, Z. Leman, M. Jawaid, A review on the characterisation of natural fibres and their composites after alkali treatment and water absorption, Plast. Rubber Compos. 46 (3) (2017).
[4] M. Chandrasekar, et al., Flax and sugar palm reinforced epoxy composites: effect of hybridization on physical, mechanical, morphological and dynamic mechanical properties, Mater. Res. Express 6 (10) (2019) 105331.
[5] S. Krishnasamy, et al., Effect of fibre loading and $Ca(OH)_2$ treatment on thermal, mechanical, and physical properties of pineapple leaf fibre/polyester reinforced composites, Mater. Res. Express 6 (8) (2019) 085545.
[6] S.M.K. Thiagamani, et al., Investigation into mechanical, absorption and swelling behaviour of hemp/sisal fibre reinforced bioepoxy hybrid composites: effects of stacking sequences, Int. J. Biol. Macromol. 140 (2019) 637–646.
[7] A. Atiqah, M. Chandrasekar, T. Senthil Muthu Kumar, K. Senthilkumar, M.N.M. Ansari, Characterization and Interface of Natural and Synthetic Hybrid Composites, Elsevier Ltd., 2019.
[8] P. Sampathkumaran, S. Seetharamu, A. Murali, R.K. Kumar, Dry sliding wear behavior of glass-epoxy composite, J. Reinf. Plast. Compos. 18 (1) (1999) 55–62.
[9] P.K. Bajpai, I. Singh, J. Madaan, Frictional and adhesive wear performance of natural fibre reinforced polypropylene composites, Proc. Inst. Mech. Eng. Part J: J. Eng. Tribol. 227 (4) (2013) 385–392.
[10] T.B. Yallew, P. Kumar, I. Singh, Sliding wear properties of jute fabric reinforced polypropylene composites, Procedia Eng. 97 (2014) 402–411.
[11] K.S. Kumar, I. Siva, N. Rajini, P. Jeyaraj, J.T.W. Jappes, Tensile, impact, and vibration properties of coconut sheath/sisal hybrid composites: effect of stacking sequence, J. Reinf. Plast. Compos. 33 (19) (2014) 1802–1812.
[12] M. Aslan, M. Tufan, T. Küçükömeroğlu, Tribological and mechanical performance of sisal-filled waste carbon and glass fibre hybrid composites, Compos. Part B Eng. 140 (2018) 241–249.
[13] J. Naveen, M. Jawaid, E.S. Zainudin, M.T.H. Sultan, R. Yahaya, Mechanical and moisture diffusion behaviour of hybrid Kevlar/Cocos nucifera sheath reinforced epoxy composites, J. Mater. Res. Technol. 8 (1) (2019) 1308–1318.

[14] K. Mylsamy, I. Rajendran, Influence of fibre length on the wear behaviour of chopped agave americana fibre reinforced epoxy composites, Tribol. Lett. 44 (1) (2011) 75.
[15] N.S.M. El-Tayeb, Abrasive wear performance of untreated SCF reinforced polymer composite, J. Mater. Process. Technol. 206 (1–3) (2008) 305–314.
[16] C. Deo, S.K. Acharya, Effects of load and sliding velocity on abrasive wear of Lantana camara fibre-reinforced polymer matrix composite, Proc. Inst. Mech. Eng. Part J: J. Eng. Tribol. 224 (5) (2010) 491–496.
[17] K.K. Basumatary, N. Mohanta, S.K. Acharya, Effect of fiber loading on abrasive wear behaviour of Ipomoea carnea reinforced epoxy composite, Int. J. Plast. Technol. 18 (1) (2014) 64–74.

CHAPTER 7

Influence of coconut and graphite fillers on the wear and friction behavior of epoxy composites

T.P. Sathishkumar[a], PON. Maheshkumar[b], and S. Ramakrishnan[c]
[a]Department of Mechanical Engineering, Kongu Engineering College, Perundurai, Tamil Nadu, India
[b]Department of Mechanical Engineering, Nandha College of Technology, Perundurai, Tamil Nadu, India
[c]Faculty of Mechanical Engineering, School of Building and Mechanical Sciences, Kongu Engineering College, Perundurai, Tamil Nadu, India

1 Introduction

Fiber-reinforced polymer composites are widely used in various parts of tribological applications. The incorporation of various fibers with varying fiber volume or weight fractions in polymer composites induces different wear rates and coefficients of friction at different applied loads, sliding velocities, and sliding distances. Sathishkumar et al. [1] investigated the synergy of a cashew nut shell filler (CNS) on the tribological behavior of a natural fiber-reinforced epoxy composite. The adhesive wear and friction performance were conducted in a pin-on-disk machine and under dry sliding conditions. The CNS content of 5%, 10%, 15%, 20%, 25%, 30%, 35%, and 40% weight ratio was added to epoxy and mixed with a hardener solution. The higher specific wear rate and friction coefficient were found for composites containing 5%, 10%, 15%, and 20% CNS filler content due to the sufficient accumulation of the CNS filler in the composites. With the uniform distribution of filler in the composites, the lowest wear rate and coefficient of friction were found for composites containing 30% raw CNS filler. The coefficient of friction (μ) was around 0.076 at 30% of CNS. Shalwan and Yousif [2] concluded that the addition of graphite filler in date palm fiber reinforced epoxy composites reduced the friction and improved the wear characteristic due to enhanced surface hardness. Shalwan and Yousif [2] investigated the effect of date palm fiber and graphite filler on the friction and wear behaviors of epoxy composites. The microstructure failure of tensile and worn surface samples are discussed by a scanning electron microscope. The addition of graphite was highly suitable for preparing

Tribology of Polymer Composites
https://doi.org/10.1016/B978-0-12-819767-7.00007-4

the natural fiber-reinforced polymer composite. This can help to reduce the friction coefficient and in turn reduce the wear rate of the epoxy composites. However, the high content of graphite deteriorates the mechanical properties. Yousif et al. [3] investigated the sliding wear and frictional behavior of kenaf fiber-reinforced polymer composites. The sliding surface used was a polished stainless-steel counter face in a block on disk. The applied loads, sliding distances, and sliding velocities were varied, such as 30–100 N, 0–5 km, and 1.1–3.9 m/s. The experiment was analyzed with three fiber orientations, parallel, antiparallel, and normal, with respect to the sliding direction. The results revealed that the presence of kenaf fibers enhanced the tribological performance of the kenaf fiber-reinforced epoxy composite. The specific wear rate of the composite in all three orientations was less against a normal applied load and sliding velocity. The composite exhibited better wear performance in normal compared to the other two composites.

Nirmal et al. [4] investigated the adhesive wear performance of epoxy composite with a reinforcing of bamboo fiber. It was shown that the AP-O was superior compared to the R-O and P-O composites. At a low sliding velocity, the frictional performance for AP-O was found to be greater by 44% relative to the high sliding velocity. The AP-O bamboo fiber composites gave lower interface temperatures along the sliding direction. The R-O, P-O, and AP-O composites experienced high wear performance at lower sliding velocities (i.e., 1.7 m/s) of 28.8%, 39.7%, and 31.3% compared to the higher sliding velocity. The AP-O had superior wear and frictional performance compared to the R-O and P-O composites, and it was a good alternative for the sliding component. Nirmal et al. [4] investigated the three-body abrasive wear and frictional performance of a treated betelnut fiber-reinforced epoxy composite. The three-body abrasion was carried out using different abrasive particle sizes (500, 714, and 1430 μm) and sliding velocities (0.026–0.115 m/s) at a constant applied load (5 N). The betelnut fiber composite experienced higher weight losses at higher sliding velocities. Debnath et al. [5] investigated the dry sliding wear behavior of glass fiber-reinforced epoxy composites with the effect of natural fillers. Two natural fillers, rice husk and wheat husk, were used in the glass fiber-reinforced epoxy composite. The friction and wear performance of the developed composites was examined with different sliding conditions. The wear performance of the developed composites was carried out on a pin-on-disk wear test machine. The maximum weight loss for the rice husk filled glass fiber-reinforced epoxy composites was 3.37×10^{-3} g, whereas the maximum weight loss for wheat husk filled glass fiber-reinforced epoxy composites was 3.89×10^{-3} g.

Chaudhary et al. [6] investigated the morphological behaviors of jute/hemp/flux-reinforced hybrid composites. The hand lay-up method was adopted to fabricate laminated composites with the fiber orientation of bidirectional woven mats. The hardness of the kenaf/phenol formaldehyde composite increased linearly with an increase in fiber loading up to 43% of the kenaf fiber.

Similarly, the fiber loading increased the surface hardness of the hybrid composites. The epoxy hybrid composites of jute/hemp/flus reinforcements have the highest tensile strength of 58.59 MPa as compared to other composites. Flax-epoxy and hemp-flax-epoxy composites have 46.21 and 44.17 MPa tensile strengths. The neat resins have 34.69 MPa and 0.6 GPa flexural strength and modulus. The maximum flexural strength and modulus of the jute/hemp/epoxy hybrid composite was 86.6 MPa and 0.78 GPa. The flexural strength of the jute/hemp/epoxy hybrid composite was comparable to the hemp/epoxy composite. Deepak et al. [7] analyzed the mechanical and wear behaviors of jute/banana fiber-reinforced epoxy hybrid composites with an epoxy modifier of molybdenum disulfide. By increasing the wear properties of the hybrid composites, the molybdenum disulfide was mixed with the epoxy resin at various weight fractions to prepare the composites. It was found that the abrasion loss of the composites decreased with an increase in W_f of the molybdenum disulfide. At the same time, an enhancement in wear resistance was noted. Shakuntala et al. [8] attempted to increase the tribological properties of the epoxy matrix composite by reinforcing the wood apple shell particles as the filler material. Epoxy matrix composites reinforced with various filler content (5–20 wt%) were prepared by a simple hand lay-up technique. The optimum filler content for effective load transfer and better fiber-matrix bonding was determined as 15 wt%. From the obtained wear results, it was observed that the wear erosion rate varied with the impingement angle at the filler content. Due to the semi-ductile nature of composites, the maximum erosion was observed at an impingement angle of 45–60 degrees. Thongsang et al. [9] investigated the tribological behaviors of NR vulcanizates reinforced with fly ash and precipitated silica fillers. The optimum wear resistance was obtained by the introduction of a 75% PSi fraction into the FASi/PSi-filled NR compounds. Chang et al. [9a] investigated the tribological properties of monodispersed nanoparticles in epoxy composite-filled carbon fibers and graphite lubricants. The effect of the tribological performance on the initial roughness was tested. The nanoparticle results were compared with submicron particle results. The tribological behaviors of the three composites were compared according to various pv factors and the roughness value of the counter face. The addition of nanoparticles was one of the best ways to reduce the coefficient of friction and the rate of wear.

Chand et al. [10] investigated the tribological properties of polyvinylchloride composites filled with rice husks modified by maleic anhydride. The surface treatment and rice husk reinforcement highly influenced the tribological and mechanical properties in the polyvinylchloride composite. At a lower filler concentration of rice husk (10 wt%), the polyvinylchloride composites had maximum mechanical strength and tribological properties. An SEM image of a worn-out surface showed the wear mechanisms of the surface-modified rice husk in the polyvinylchloride composites. Xian and Zhang [11] investigated graphite flakes of 5–20 vol% on the sliding wear of polyetherimide composites. The tribological properties were investigated against a steel counter face by varying temperatures. The addition of 5–20 vol% of graphite filler in composites improved the tribological behaviors at room temperature and at elevated temperatures up to 220°C. A higher filler content (≥ 15 vol%) led to a very low coefficient of friction (μ), that is, around 0.05, when temperatures exceeded 150°C. Meanwhile, the specific wear rate was also reduced. The worn-out image taken from the microscopy suggested the formation of friction films on both the counterpart and specimen. The layer was formed on the surface at an elevated temperature of 150°C due to the higher graphite content. Matejka et al. [12] investigated a semimetallic brake with effects of silicon carbide on tribological behaviors. The SiC content was varied by 0%, 3.4%, 5.6%, 9%, and 14.6% volume fraction. The experiment was conducted against cast iron disk. The friction coefficient of the semimetallic increased by increasing the SiC. The SiC of 14.6% of the volume fraction in the material showed an increasing coefficient of friction. The rate of wear of the composite increased by increasing the SiC volume fraction and the sliding temperature. The SEM was used to analyze the formation of the friction layer on the surface. The wear film I and II have showed on the composites surface with and without SiC filler.

The graphite and natural filler were blended with a polymer resin at different weight or volume fractions and the natural and synthetic fiber-reinforced polymer composites were developed. By the incorporation of various fillers in the polymer composites, the rate of wear and the friction coefficient were significantly reduced. The lower sliding velocity showed less wear characteristics at a lower applied load. The sliding distance increased the rate of wear. The present work is aimed at investigating the wear rate and friction coefficient by incorporating two fillers, coconut and graphite, in the epoxy resin with different concentrations. The effect of wear parameters such as sliding distance (1, 2, and 3 km), sliding velocities (1, 2, and 3 m/s), and normal loads (10, 20, and 30 N) on the dry sliding wear characteristics by varying the coconut and graphite filler concentration is investigated using the Taguchi technique.

2 Materials

In this work, the dry coconut filler (Fig. 1) is used as a natural filler to prepare the epoxy composites with graphite filler. The coconut shells were dried under sunlight for 5 days after removing the white flesh. This process is used to remove the water content in the shells. The dried coconut shells were crushed between two rotating rollers to make small pieces. These pieces were fed into a flour mill chopper to prepare the coconut fillers. Then the filler was collected in a steel vessel and kept in a hot air oven to remove the moisture further. The laboratory-type graphite filler (Fig. 2) was purchased from the National Scientific Laboratory, Erode, Tamil Nadu, India.

Fig. 1 Dry coconut filler.

Fig. 2 Graphite filler.

3 Preparation of composites

The coconut and graphite fillers were mixed by three different weight contents, such as 5g of coconut and 5g of graphite filler (5+5), 10g of coconut and 5g of graphite filler (10+5), and 5g of coconut filler and 10g of graphite filler (5+10). A weight balancing machine was used to measure the weight of both fillers. The weighted fillers were mixed in epoxy resin one by one, then mixed thoroughly by a mechanical stirrer for about 15min. Then the epoxy and hardener were mixed at a ratio of 1:10 in the resin, and mixed by a mechanical stirrer for about 5min. The composite mixture was filled in a hand-type injection molding machine and then injected into a mold (Fig. 3) to prepare the composite pins. After pouring, both sides of the mold were tightly clamped by a C-clamp using steel plates. After solidification, the cured composite pins in a mold were taken by a small hydraulic ejection pin mechanism. These pins were postcured in a hot air oven at 45°C about 10min before the tribological study.

4 Wear and co-efficient of friction test

A pin-on-disk (POD) Tribo testing machine was used to measure the wear and friction coefficient on the influence of the coconut filler- and the graphite filler-reinforced epoxy composites. The experiments were conducted according to the ASTM G99 standard. The experiments were conducted by varying the speeds at 1000, 1500, and 2000rpm, varying the normal applied loads at 10, 20, and 30N, and a sliding distance of 1200m. The pin dimensions are 10mm diameter and 25mm length. Before conducting the experiments, the stainless-steel counter surface (disk) was polished using SiC abrasive paper, grade 1200.

Fig. 3 Die for composite pin fabrication.

5 Results and discussion

5.1 Effect of parameters on dry sliding wear rate

Fig. 4 shows the dry sliding wear rate of coconut and graphite filler-mixed epoxy composites by varying velocities of 1, 1.5, and 2 m/s and varying applied loads of 10, 20, and 30 N at a constant sliding distance of 1200 m. Fig. 4A shows the rate of wear of composites against a normal load at 10 N. At a 5 + 5 filler content, the rate of wear was found low at 1 m/s velocity. The maximum wear rate was found at 1.5 m/s velocity. At a 5 + 10 filler content, the low wear rate was found at 1.5 m/s velocity and the maximum wear was found at 1 m/s velocity. At a 10 + 5 filler content, the low wear rate was found at 1 m/s velocity and the maximum wear rate was found at 2 m/s velocity. The percentage differences between 1–1.5 and 2 m/s velocities at a 5 + 5 composite are 30.8% and 25%. The percentage differences between 1–1.5 and 2 m/s velocities at a 5 + 10 composite are − 94.9% and − 40%. The percentage differences between 1–1.5 and 2 m/s velocities at a 5 + 10 composite are 27.26% and − 31%. Overall, the low wear rate was found at 1.5 m/s velocity and the composites containing 5 g of coconut filler and 10 g of graphite filler. This variation is based on the uniform distribution of fillers in the epoxy composites. The higher content of the graphite filler in the composites leads to a lower wear rate because the graphite is a hard surface that leads to a reduction in the rate of wear.

Fig. 4B shows the rate of wear of epoxy composites against a normal load at 20 N. At a 5 + 5 filler content, the rate of wear was found low at 1 m/s velocity. The maximum wear rate was found at 1.5 m/s velocity. At a 5 + 10 filler content, the low wear rate was found at 1.5 m/s velocity and the maximum wear was found at 1 m/s velocity. At a 10 + 5 filler content, the low wear rate was found at 2 m/s velocity and the maximum wear rate was found at 1.5 m/s velocity. The percentage differences between 1–1.5 and 2 m/s velocities at a 5 + 5 composite were 72.7% and 66.67%. The percentage differences between 1–1.5 and 2 m/s velocities at a 5 + 10 composite were − 73.3% and − 50%. The percentage differences between 1–1.5 and 2 m/s velocities at a 5 + 10 composite were 76.5% and − 55%. Overall, the low wear rate was found at 1 m/s velocity and the composites containing 5 g of coconut filler with 5 g of graphite filler. This variation is based on the applied load. The uniform filler content in the composites leads to a lower wear rate because the material removal is taken uniformly and also the coconut and graphite fillers are made more coherent with epoxy.

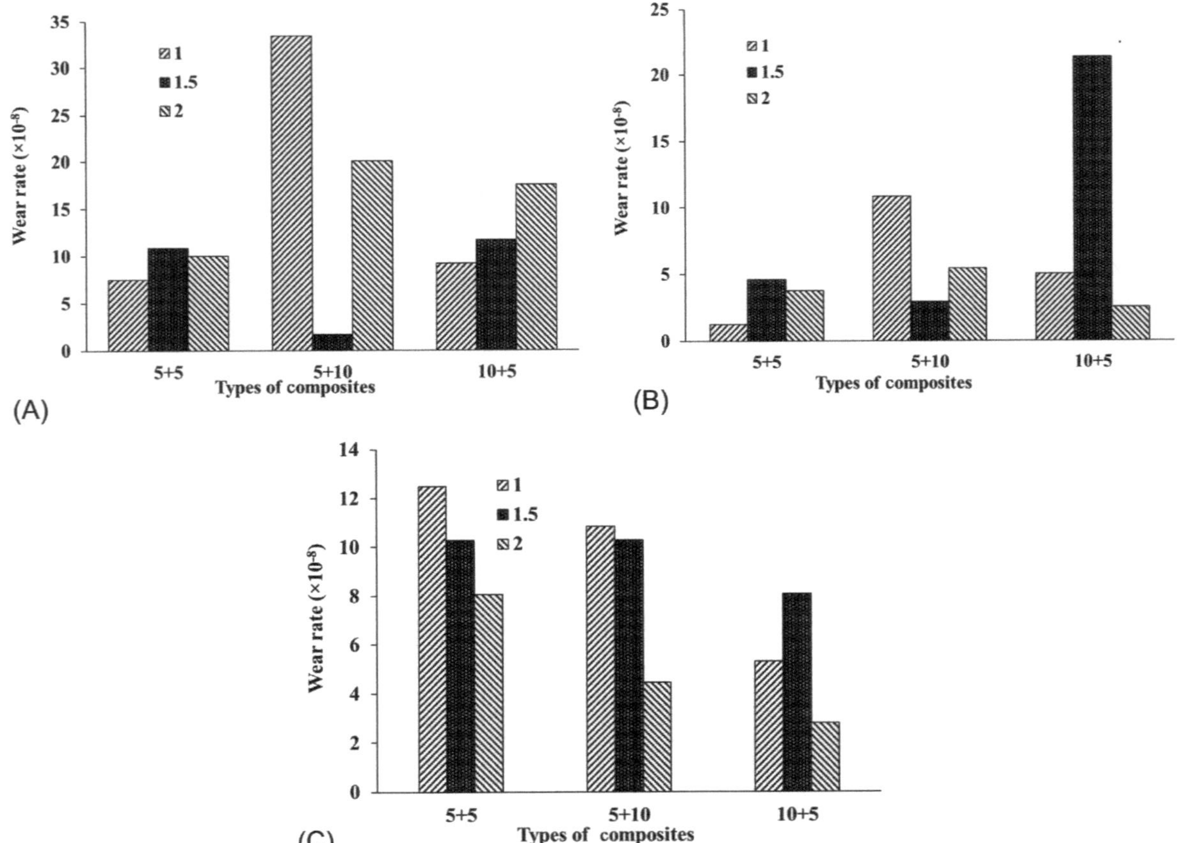

Fig. 4 Wear rate (m^3/mm) against sliding velocity (m/s); (A) normal applied load at 10 N; (B) normal applied load at 20 N; and (C) normal applied load at 30 N.

Fig. 4C shows the rate of wear of composites against a normal load at 30 N. At a 5 + 5 filler content, the rate of wear was found low at 2 m/s velocity. The maximum wear rate was found at 1 m/s velocity. At a 5 + 10 filler content, the low wear rate was found at 2 m/s velocity and the maximum wear was found at 1 m/s velocity. At a 10 + 5 filler content, the low wear rate was found at 2 m/s velocity and the maximum wear rate was found at 1.5 m/s velocity. The percentage differences between 1–1.5 and 2 m/s velocities at a 5 + 5 composite were −17.7% and −35.5%. The percentage differences between 1–1.5 and 2 m/s velocities at a 5 + 10 composite were −5% and −56.8%. The percentage differences between 1–1.5 and 2 m/s velocities at a 5 + 10 composite were 34.5% and −74.7%. Overall, the low wear rate was found at 2 m/s velocity and the composites containing 10 g of coconut filler with 5 g of graphite filler. This variation was based on the applied load and filler contents. The higher coconut filler content in the composites leads to lower wear rate because the natural filler could make a better bonding with epoxy resin. However, among the three loads, the lower rate of wear was found for composites containing 10 g of coconut and 5 g of graphite filler at a 30 N applied load.

5.2 Effect of parameters on the dry sliding coefficient of friction

Fig. 5 shows the dry sliding friction coefficient of coconut and graphite filler-reinforced epoxy composites by varying the velocity at 1, 1.5, and 2 m/s and varying the applied loads of 10, 20, and 30 N at a constant sliding distance of 1200 m. Fig. 5A shows the friction coefficient of the composites against a normal applied load at 10 N. At a 5 + 5 filler content, the coefficient of friction was found low at 1 m/s velocity. The maximum coefficient of friction was found at 1.5 m/s velocity. At a 5 + 10 filler content, the low coefficient of friction was found at 1.5 m/s velocity and the maximum coefficient of friction was found at 2 m/s velocity. At 10 + 5 filler content, the low coefficient of friction was found at 1 m/s velocity and the maximum coefficient of friction was found at 2 m/s velocity. The percentage differences between 1–1.5 and 2 m/s velocities at a 5 + 5 composite were 78.9% and 72.5%. The percentage differences between 1–1.5 and 2 m/s velocities at a 5 + 10 composite were −68.6% and 16.4%. The percentage differences between 1–1.5 and 2 m/s velocities at a 5 + 10 composite were 39.4% and 42.7%. Overall, the low coefficient of friction was found at 1 m/s velocity and the composites containing 5 g of coconut filler with 5 g of graphite filler. This variation is based on the uniform distribution and contribution on the Tribo test of fillers in epoxy composites.

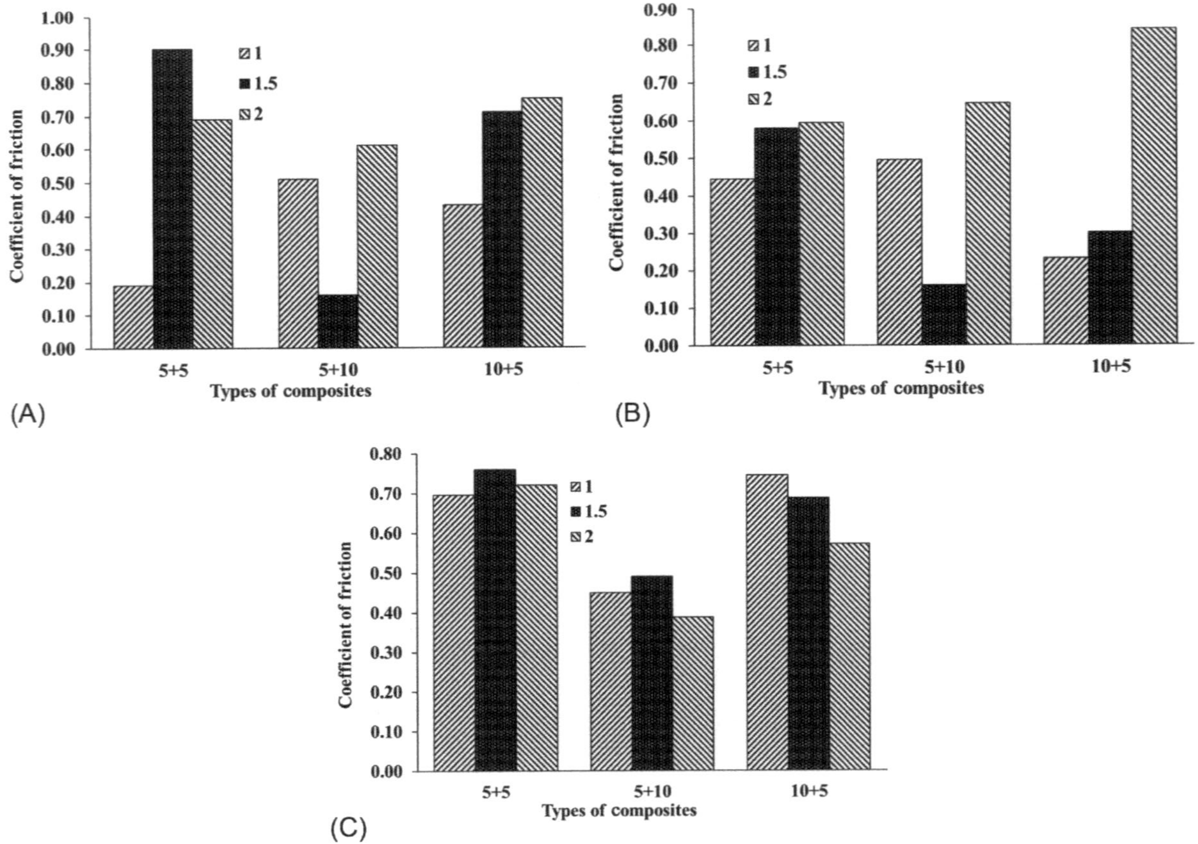

Fig. 5 Coefficient of friction (μ) versus types of composites at sliding velocities (m/s); (A) normal applied load at 10N; (B) normal applied load at 20N; and (C) normal applied load at 30N.

Fig. 5B shows the coefficient of friction versus types of composites at a normal load at 20 N. At a 5 + 5 filler content, the friction coefficient was found low at 1 m/s velocity. The maximum coefficient of friction was found at 1.5 m/s velocity. At a 5 + 10 filler content, the low coefficient of friction was found at 1.5 m/s velocity and the maximum coefficient of friction was found at 2 m/s velocity. At a 10 + 5 filler content, the low coefficient of friction was found at 1 m/s velocity and the maximum coefficient of friction was found at 2 m/s velocity. The percentage differences between 1–1.5 and 2 m/s velocities at a 5 + 5 composite were 22.4% and 25%. The percentage differences between 1–1.5 and 2 m/s velocities at a 5 + 10 were is − 68.6% and 23.1%. The percentage differences between 1–1.5 and 2 m/s velocities at a 5 + 10 composite were 23.33.4% and 72.6%. Overall, the low coefficient of friction was found at 2 m/s velocity and the composites containing 5 g of coconut filler with 10 g of graphite filler. The graphite filler plays an important role on the coefficient of friction at a 20 N applied load. Fig. 5C shows the coefficient of friction of the composites against a normal load at 30 N. At a 5 + 5 filler content, the coefficient of friction was found low at 1 m/s velocity. The maximum coefficient of friction was found at 1 m/s velocity. At a 5 + 10 filler content, the low friction coefficient was found at 2 m/s velocity and the maximum coefficient of friction was found at 1.5 m/s velocity. At a 10 + 5 filler content, the low friction coefficient was found at 2 m/s velocity and the maximum coefficient of friction was found at 1 m/s velocity. Increasing the sliding velocity reduced the coefficient of friction. The percentage differences between 1–1.5 and 2 m/s velocities at a 5 + 5 composite were 8% and 2.7%. The percentage differences between 1–1.5 and 2 m/s velocities at a 5 + 10 composite were 8.2% and 13.33%. The percentage differences between 1–1.5 and 2 m/s velocities at a 5 + 10 composite were − 6.7% and − 29.7%. Overall, the low coefficient of friction was found at 2 m/s velocity and the composites containing 10 g of coconut filler with 5 g of graphite filler. At a higher sliding speed (2 m/s), the coconut filler plays an important role on the friction coefficient at a 20 N applied load.

5.3 Morphology study

A study of the worn-out surface of the coconut and graphite filler-reinforced epoxy composites was carried out using SEM (JEOL—JSM 6380). The morphologies of the wear tracks of the graphite and coconut fillers are shown in Figs. 6–8. It is clear that Fig. 6 shows the wear track of a composite pin with 5 g of coconut filler and 10 g of graphite filler at 1.5 m/s and

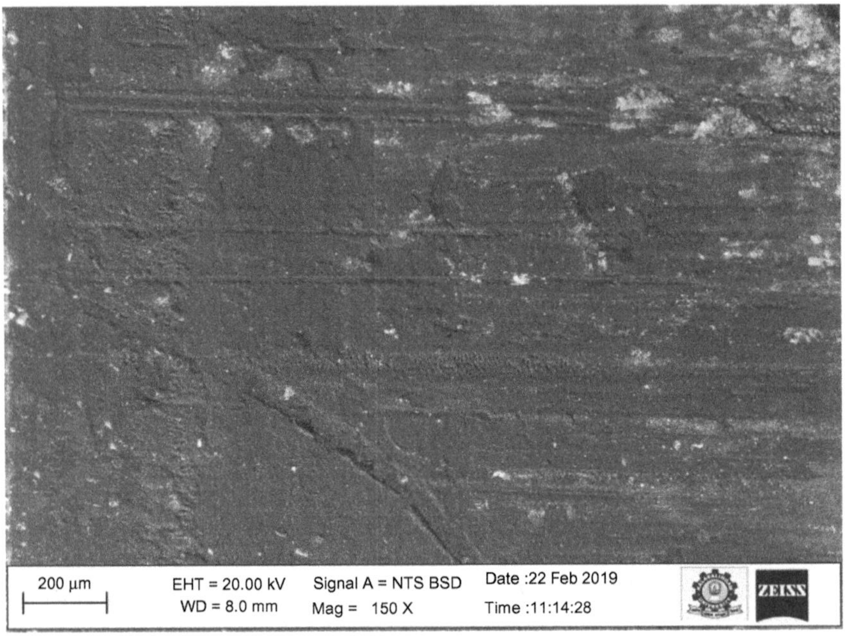

Fig. 6 SEM image of a worn-out surface of 5+10 filler content at 1.5 m/s and a 10 N applied load.

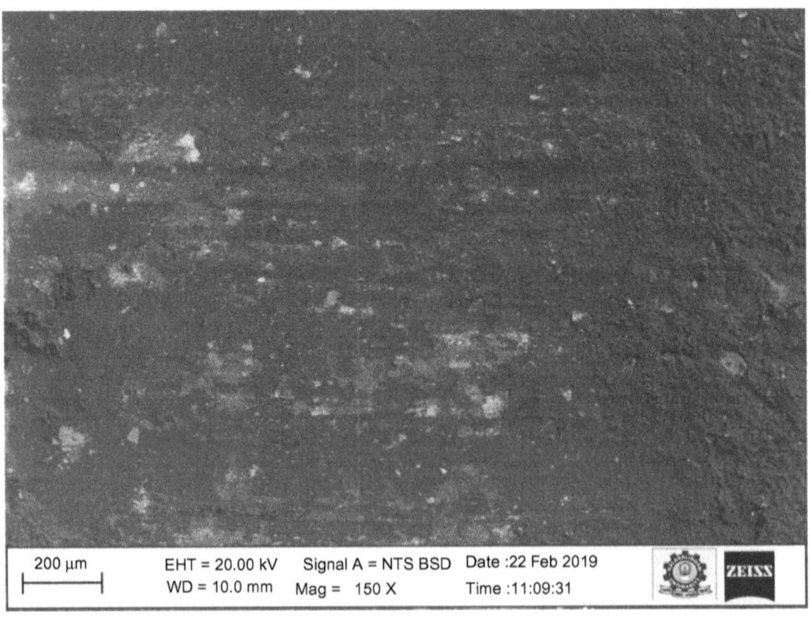

Fig. 7 SEM image of a worn-out surface of 10+5 filler content at 2 m/s and 20 N applied load.

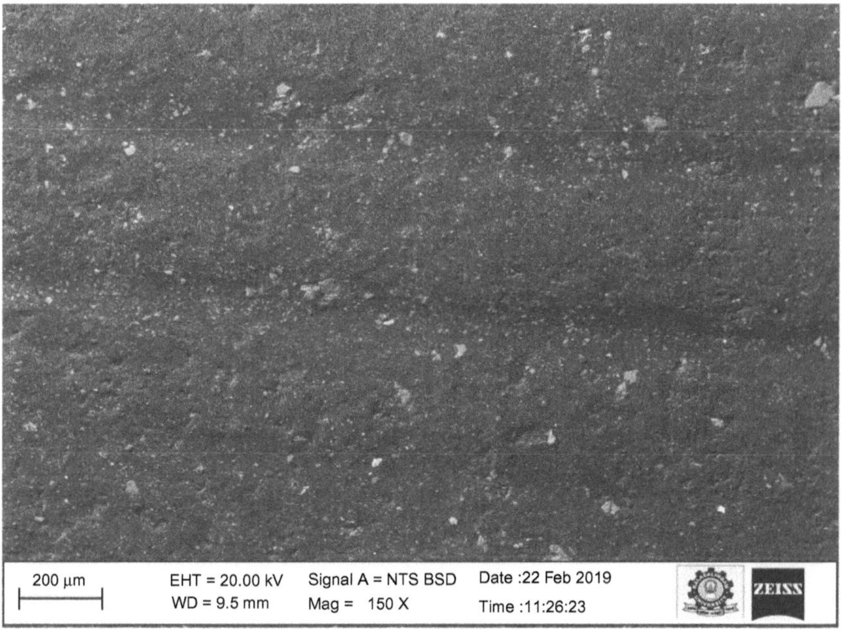

Fig. 8 SEM image of a worn-out surface of 10 + 5 filler content at 2 m/s and 30 N applied load.

a 10 N applied load. This surface shows higher plastic deformation. By this deformation, the wear track shows many cluster lines. Fig. 7 shows the wear track of a composite pin with 10 g of coconut filler and 5 g of graphite filler at 2 m/s and a 20 N applied load. This surface shows the clear wear track and the absence of a cluster line due to more content of coconut filler in the composites. During experiments, the coconut fillers came out from the and pin surface and weren't deposited on the counter surface. Fig. 8 shows the wear track of a composite pin with 10 g of coconut filler and 5 g of graphite filler at 2 m/s and a 30 N applied load. This surface shows up and down wear track which shows less surface wear and coefficient of friction. Increasing the applied load was decreased the wear rate and coefficients due to absence of thin film formation on the pin and also counter part.

6 Conclusion

The wear and frication behaviors of coconut and graphite filler-reinforced epoxy composites were examined by using the POD method. The composite pins were fabricated by a simple injection molding technique. Before the Tribo test, the pin surface was polished by SiC abrasive paper. From the extensive study, at 10 and 20 N applied load, the rate of wear was increased by increasing the sliding velocities. But at a 30 N applied load, the wear rate was decreased by increasing the sliding velocity. The minimum rate of wear was achieved for composites containing 10 g of coconut filler and 5 g of graphite filler in epoxy composites at a 30 N normal applied load and 2 m/s sliding velocity. The coefficient of friction was varied by varying the applied loads and sliding velocities. Increasing the applied load and sliding velocity was increasing the friction coefficient. The maximum friction coefficient was found for a composite containing 5 g of coconut and graphite filler at 1 m/s sliding velocity at a 10 N applied load, and 10 g of coconut and 5 g of graphite filler at a 2 m/s sliding velocity at a 20 N applied load. The minimum friction coefficient was found for a composite containing 5 g of coconut and 10 g of graphite filler at a 1 m/s sliding velocity, at 10 and 20 N applied loads. Overall, the composites containing 5 g of coconut filler and 10 g of graphite filler have better results at a 20 N normal applied load and 2 m/s sliding velocity.

References

[1] T.P. Sathishkumar, S. Arun Kumar, P. Navaneethakrishnan, I. Siva, N. Rajini, Synergy of cashew nut shell filler on tribological behaviors of natural-fiber-reinforced epoxy composite, Sci. Eng. Compos. Mater. 25 (4) (2018) 761–772.

[2] A. Shalwan, B.F. Yousif, Influence of date palm fibre and graphite filler on mechanical and wear characteristics of epoxy composites, Mater. Des. 59 (2014) 264–273.

[3] B.F. Yousif, U. Nirmal, K.J. Wong, Three-body abrasion on wear and frictional performance of treated betelnut fibre reinforced epoxy (T-BFRE) composite, Mater. Des. 31 (9) (2010) 4514–4521.

[4] U. Nirmal, J. Hashim, K.O. Low, Adhesive wear and frictional performance of bamboo fibres reinforced epoxy composite, Tribol. Int. 47 (2012) 122–133.

[5] K. Debnath, V. Dhawan, I. Singh, A. Dvivedi, Effect of natural fillers on wear behavior of glass-fiber-reinforced epoxy composites, in: Proceedings of the International Conference on Research and Innovations in Mechanical Engineering, 2014, pp. 441–450.

[6] V. Chaudhary, P.K. Bajpai, S. Maheshwari, Studies on mechanical and morphological characterization of developed jute/hemp/flax reinforced hybrid composites for structural applications, J. Nat. Fibers 18 (2018) 80–87.

[7] P. Deepak, H. Sivaraman, R. Vimal, S. Badrinarayanan, R.V. Kumar, Study of wear properties of jute/banana fibres reinforced molybdenum disulphide modified epoxy composites, Mater. Today Proc. 4 (2) (2017) 2910–2919.

[8] O. Shakuntala, G. Raghavendra, A. Samir Kumar, Effect of filler loading on mechanical and tribological properties of wood apple shell reinforced epoxy composite, Adv. Mater. Sci. Eng. 2014 (2014) 9.

[9] S. Thongsang, W. Vorakhan, E. Wimolmala, N. Sombatsompop, Dynamic mechanical analysis and tribological properties of NR vulcanizates with fly ash/precipitated silica hybrid filler, Tribol. Int. 53 (2012) 134–141. (a) L. Chang, Z. Zhang, C. Breidt, K. Friedrich, Tribological properties of epoxy nanocomposites: I. Enhancement of the wear resistance by nano-TiO_2 particles, Wear 258 (1–4) (2005) 141–148.

[10] N. Chand, P. Sharma, M. Fahim, Tribology of maleic anhydride modified rice-husk filled polyvinylchloride, Wear 269 (11–12) (2010) 847–853.
[11] G. Xian, Z. Zhang, Sliding wear of polyetherimide matrix composites: II. Influence of graphite flake, Wear 258 (5–6) (2005) 783–788.
[12] V. Matejka, Y. Lu, Y. Fan, J. Kukutschová (Lešková), Effects of silicon carbide in semi-metallic brake materials on friction performance and friction layer formation, Wear 265 (7) (2008) 1121–1128.

CHAPTER 8

Tribological behavior of nanoclay-filled basalt fiber-reinforced polymer composites

A. Jumahat[a], A.A.A. Talib[a], H. Sharudin[b], and N.W.M. Zulfikli[c]
[a]Faculty of Mechanical Engineering, Universiti Teknologi MARA, Shah Alam, Selangor, Malaysia
[b]Faculty of Mechanical Engineering, Universiti Teknologi MARA Cawangan Pulau Pinang, Permatang Pauh, Pulau Pinang, Malaysia
[c]Department of Mechanical Engineering, Faculty of Engineering, University of Malaya, Kuala Lumpur, Malaysia

1 Introduction

Polymers are evolving rapidly as structural and tribo-materials in various applications and industries because of their outstanding properties such as high strength, light weight, excellent thermal stability, and wear and solvent resistance [1]. Polymers are also used extensively as sliding components, cams, brakes, bearings, gears, pipes, chute liners, rotors, and blades due to their self-lubricating properties, lower friction, and better wear resistance [2–4]. The fiber-reinforced polymer (FRP) composite is one of the most rapidly growing classes of materials, as it has a good combination of high specific strength, specific modulus, and wear resistance [5, 6]. The reinforcement of fiber into polymer also produced better mechanical and tribo-engineering materials.

With today's increased environmental awareness, natural fibers are widely emerging such as kenaf, flax, hemp, jute, and coir. Although they hold advantages over traditional fibers, they also face several drawbacks such as very high thermal sensitivity, incompatibility due to their hydrophilic nature, and the scattering of mechanical properties [7]. Therefore, mineral natural fibers such as basalt can serve as an alternative solution. Basalt is produced from basic volcanic rock by melt technology without any other additives in a single production process [8]. Basalt fiber has recently gained increasing attention as a possible replacement for traditional glass fibers or carbon fibers, as it has excellent mechanical properties, high abrasion resistance, a relatively

Tribology of Polymer Composites
https://doi.org/10.1016/B978-0-12-819767-7.00008-6

low price, and a nature that is environmentally friendly, naturally emerging, and safe for recycling [7, 9, 10].

Besides fibers, incorporating special fillers (micro- to nanosized particles) into a polymer matrix is also known to improve a polymer's hardness, stiffness, and strength as well as its friction and wear behavior [2, 3, 11]. The nanoscale particles such as nanoclay [2], carbon nanotubes [12], nanosilica [13], nanoZrO_2 [14], and nanoTiO_2 [15] are able to reside at the interface and lead to strong interfacial interaction, given that they are well-dispersed [2]. They are favored over microscale particles because they are only needed at small amounts ($\leq 5\%$) [16], making them an important member in wear-resistance and friction-reducing materials. Nanoclay platelets particularly provide significant improvements to the mechanical properties of a polymer matrix due to their high intercalation chemistry, high aspect ratio, high surface area, ease of availability, and low cost [16, 17]. Well-dispersed (exfoliation) nanoclay platelets in the polymer matrix will lead to significant improvements in the stiffness, strength, dimensional stability, and thermomechanical behavior of the polymer [18, 19]. The latest development of advanced polymer composites provides a combined effect of nanoparticle fillers and fibers where a composite can efficiently operate under much more severe circumstances than a composite with nanoparticles or fibers alone, in which the nanoparticle acts as a secondary reinforcement that either enhances the adhesion between the reinforcing fibers and matrix or protects the fibers by the "rolling effect" [3, 20, 21].

The tribological performance of the composite material is closely related to the properties of the polymer matrix, the properties of reinforcement, the reinforcement loading, the relative movement between the contact materials, and the fiber orientation and its operating parameters such as sliding velocity, normal load, temperature, and contact geometry [17, 22]. Several studies have been conducted on the mentioned advanced polymer materials in different sets of tribological environments that mimic their real-time applications. Zhang et al. [23] conducted a study on the effect of particle size on the wear properties of a graphite-modified CFRP composite, where a submicron particle of TiO_2/ZnS was compared with nano-SiO_2. The study concluded that nanofiller (SiO_2) leads to a significant reduction in wear and friction compared to the submicron filler (TiO_2/ZnS) when combined with graphite-modified CFRP composites. This conclusion was in agreement with the findings from Wang et al. [24] and Chang et al. [25, 26], where the incorporation of both nanofiller and traditional filler gave better tribological results. Besides that, these researchers also studied the effect of varying parameters on the friction and wear properties of the composites. Similar results were found by Wang et al. [24] and Chang et al. [25, 26], whereby the friction coefficient and wear rate reduced when the load, velocity, and PV factor increased. However, Zhang et al. [23] and Chang and Friedrich [26] found that the friction coefficient reduced but the wear rate increased when the PV factor was increased.

On the other hand, Chang et al. [25] compared two different types of apparatus—pin on disk (POD) and block on ring (BOR)—where the composite sample was in the pin and block shapes, respectively. The wear rate result from BOR was lower than those tested using POD when the sample tested was filled with a nanoscale filler, which is totally contrary for a sample without a nanoscale filler. This was due to the different contact surfaces between the two types of configurations where the BOR consists of arc surface contact instead of linear contact. The results therefore inferred that a composite with nanofiller has a good mechanical load-carrying capacity compared to one without nanofiller.

Based on a literature review, it is believed that nanoclay is one of the most rapid growing nanofillers in assisting a polymer matrix to improve its mechanical and tribological properties. Positive results from previous research conducted on a nanoclay-modified polymer composite revealed that this type of intercalated silicate is an efficient wear-reducing material. However, the role of nanoclay in improving the tribological properties of a fiber-reinforced polymer composite is still inadequate, especially when utilized with natural fiber instead of synthetic fibers. Therefore, in the present work, the role of nanoclay particles on the friction and wear behavior of epoxy composites reinforced with basalt fiber was studied. The influence of nanoclay fraction, PV (pressure × velocity) factors, motions, and counter faces on nanoclay-filled BFRP composites was of special interest. The wear mechanisms of worn surfaces after sliding were also investigated.

2 Experimental details

2.1 Materials

Epoxy resin and hardener (Miracast 1517; density: 1.13 g/cm^3) were supplied by Miracon (M) Sdn. Bhd. The Miracast 1517A epoxy resin is a low-viscosity laminating resin that can produce a high-performance composite structure. Miracast 1517B is an amine-curing agent that provides curing at room temperature. The mixing ratio of Miracast 1517A and 1517B was set at 100:30. Basalt fibers (BF) in a unidirectional roving form (density: 2.65 g/cm^3, diameter: 9–15 μm, tensile strength: 1.96 GPa, tensile modulus: 94.77 GPa, elongation: 1.84%) were supplied by the Haining Anjie Composite Material Company. The nanoclay (Nanomer I.28) (particle size: 8–10 μm, density: 1.9 g/cm^3) was supplied by Sigma Aldrich. The surface of the clay was modified with 25–30 wt% trimethyl stearyl ammonium.

The nanoclay-filled epoxy matrix was prepared by first mixing the preweighted quantities of epoxy with the desired nanoclay content (1.0, 3.0, and 5.0 wt%) using manual mixing and three roll mill mixing techniques. The three roll mill mixing technique used the concept of shear force to uniformly disperse the nanoclay in the matrix, and it is believed to be one of the most effective methods to disperse clay platelets [27]. The milling machine was set at 12.7 m/s speed and 60°C to reduce the epoxy's viscosity during mixing. Then the modified resin was mixed with hardener and left to degas in an oven for 15 min to remove entrapped air bubbles before being poured onto the fiber-wound frame. The frame was fabricated using aluminum plates with dimensions of 350 mm × 240 mm, designed to wind 15 vol% basalt fiber. The mixture was distributed evenly onto the frame using a roller. The frame was then compressed with 10 N weight to ensure a flat surface and thickness before it was left to cure for 24 h at room temperature. A pair of waxed thin films was placed on the top and bottom of the composites to ensure a smooth surface upon curing. The composite surface roughness, Ra, was measured at 0.10–0.20 μm using a Mitutoyo Surftest SV-500 machine.

2.2 Friction and wear tests

A dry friction wear test in a unidirectional motion was performed using a TR-20LE model pin-on-disk tribometer in accordance with the ASTM G99-95a standard. During the test, a stationary pin was pressed against a rotating disk. The composite samples were cut into disk shapes with a diameter of 760 mm and a thickness of 40 mm while the pin used was of 10 mm diameter (GCr15, HRC62 ± 2, Ra: 0.1 μm). The schematic diagram of the pin-on-disk test is shown in Fig. 1A. The operating parameters were set at low and high PV factors of 0.23 MPa m/s and 0.93 MPa m/s, respectively, resulting from two varied loads and velocities and sliding for 2000 m, as summarized in Table 1.

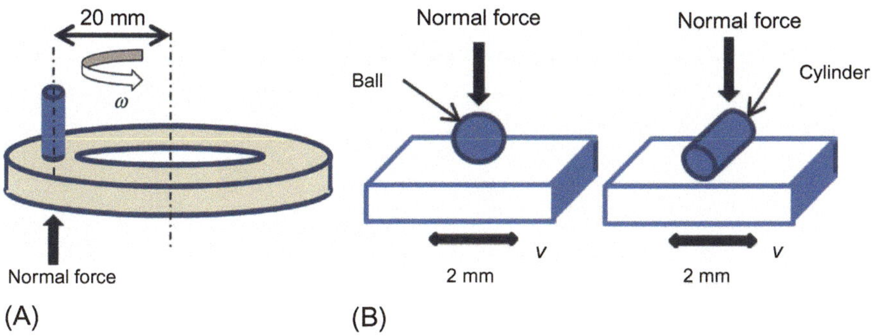

Fig. 1 Schematic diagram of (A) pin on disk, and (B) high-frequency reciprocating rig.

Table 1 Operating parameters of friction tests.

Parameter	**Experimental condition**	
Type of motion	**Unidirectional**	**Reciprocating**
Applied load	30, 70 N	70 N
Velocity	0.63, 1.05 m/s	0.03 m/s
PV factor	0.23, 0.93 MPa m/s	6.18, 0.11 MPa m/s
Sliding distance	2000 m	120 m
Counterface	Pin	Ball, cylinder
Contact area	78.5 mm^2	0.34, 19.2 mm^2

A dry friction wear test in reciprocating motion was performed using a TR-281M8 model high-frequency reciprocating rig in accordance with the ASTM G133-95 standard. The tribometer operates in linear repeated back and forth movements creating reciprocating motion, as shown in Fig. 1B, where the moving counter face was pressed against a stationary flat. The composite samples were cut into a flat square shape with dimensions of 15 mm × 15 mm and a thickness of 4 mm. There were two types of counter faces used, a stainless steel ball (diameter 6 mm, HRC60) and a cylinder (diameter 6 mm, thickness 9.2 mm, HRC60), creating ball-on-flat (B-O-F) and cylinder-on-flat (C-O-F) configurations, respectively. The operating parameters for this test were set at 70 N load, 0.03 m/s velocity, and 120 m distance. The configuration of B-O-F and C-O-F were set to create a difference in contact area, therefore a high and low PV factor, as summarized in Table 1.

The specific wear rate (*Ks*) of the sample was calculated using Eq. 1 [28] while the friction coefficient was calculated by dividing the friction force with the applied load, as shown in Eq. 2, read from the software attached to the tribometer. The initial and final weights were determined in a high-precision digital balance (0.1 mg accuracy). At least five replicate tests were performed to calculate the average value.

$$Ks = \frac{\Delta V}{L \times F_N} = \frac{\Delta m}{L \times F_N \times \rho} \tag{1}$$

where Δm is the mass loss (g) of the samples, ΔV is the volume loss (mm^3) of the samples, is the density of the sample (g/mm^3), L is the distance traveled (m), and F_N is the normal load (N).

$$\mu = \frac{F}{F_N} \tag{2}$$

where μ is the friction coefficient, F is the resistance force (N), and F_N is the normal load (N).

2.3 Characterization

The density of the composites was measured using a density balance based on the Archimedes principle in distilled water. The composite samples were prepared and calculated in accordance with ASTM D792. A Rockwell hardness test type R (HRR) was conducted using the Instron A654 R Rockwell Hardness Tester following the ASTM D785-08 standard. The degree of dispersion of the nanoclay in the epoxy resin was observed using transmission electron microscopy (TEM). The TEM images showed that the nanoclay platelets dispersed uniformly, presenting an intercalated and exfoliated structure of epoxy-clay nanocomposites. The gap between the two adjacent platelets of more than 2.4 nm but less than 8 nm showed an intercalated structure while the gap of more than 9 nm showed an exfoliated structure. The morphologies of the worn surfaces were examined using optical microscopy and a Hitachi TM3000 Tabletop scanning electron microscope (SEM). In order to increase the resolution of the SEM observations, the samples were plated with a platinum coating to render them electrically conductive. The surface roughness was measured using a Mitutoyo Surftest SV-500 machine.

3 Results and discussions

3.1 Characterization of composites

The densities and hardnesses of 0.0, 1.0, 3.0, and 5.0 wt% NC BFRP composites are shown in Table 2. The densities of 1.0, 3.0, and 5.0 wt% NC

Table 2 Properties of composites.

Composites	Density (g/cm^3)	Hardness (HRR)
0.0 wt% NC BFRP/Pure BFRP	1.324 ± 0.009	119.50 ± 0.41
1.0 wt% NC BFRP	1.240 ± 0.008	121.58 ± 0.70
3.0 wt% NC BFRP	1.242 ± 0.014	122.15 ± 0.44
5.0 wt% NC BFRP	1.246 ± 0.006	121.63 ± 0.34

BFRP composites were much lower compared to the 0.0 wt% NC BFRP composite [11]. The density decreased by 6.30%, 6.20%, and 5.80% for 1.0 wt%, 3.0 wt%, and 5.0 wt% NC content, respectively, compared to its pure state. However, the density was increased as the nanoclay content increased from 1.0 wt% to 5.0 wt% [6, 12, 29, 30], although, the difference was marginal. This might be due to the poor interfacial compatibility between the matrix, the filler, and the fiber in the composites as well as the porosity entrapped in the composites.

On the other hand, the hardness of the nanoclay-filled BFRP composites increased until the NC content reached 3.0 wt%. Beyond that, the hardness decreased. The highest hardness was achieved by those samples containing 3.0 wt% NC with a recorded improvement of 2.22%. Meanwhile, the 1.0 wt% NC and 5.0 wt% NC BFRP composites showed improvement by 1.74% and 1.78%, respectively, compared to the 0.0 wt% NC BFRP composite. As the nanoclay content increased, the load-carrying capacity of the composites also improved due to the reduced interparticle distance in the epoxy matrix [31] and the uniform distribution of the nanoclay dispersion [32], as observed in Fig. 2. This well-dispersed nanoclay eventually leads to an improvement of the interface strength between the epoxy matrix and the nanoclay filler [6]. Any further increase in nanoclay content will definitely decrease the hardness value of the composite. Higher nanoclay content leads to higher viscosity of the epoxy matrix during the fabrication process, which possibly increases the porosity of the composites because the air entrapped during mixing and curing has difficulty escaping from the viscous matrix [32]. Besides that, the higher nanoclay content also caused poor dispersion and higher agglomeration [2] of the nanoclay platelets, where it increased the bonding limitation between the epoxy matrix and the nanoclay filler [19]. Therefore, the nanoclay filler could not be utilized fully in improving the properties of the epoxy system and the FRP composite systems.

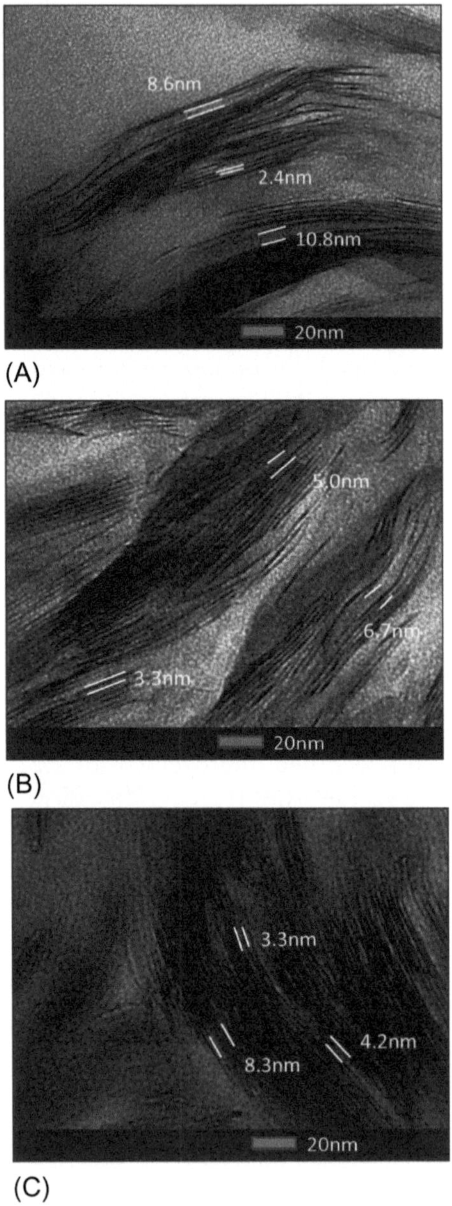

Fig. 2 TEM micrographs showing intercalated structure of (A) 1.0wt%, (B) 3.0wt%, and (C) 5.0wt% clay nanoplatelets in epoxy, prepared using a three roll mill mixing technique.

3.2 Friction and wear results

The friction coefficient (μ) for the composites under different conditions is summarized in Table 3. Frictional friction is largely influenced by the normal load as well as the physical, mechanical, and geometrical properties of the contact pairs [22]. Under unidirectional motion, the μ for low PV exhibited a higher value than the μ at high PV. When sliding at high PV, the μ was lower because there was not enough time to produce more adhesive points due to the decrease in surface contact time. This lack of frictional force component by adhesion has led to the easier and faster formation of integrated transfer film that eventually reduces the surface softening by frictional heat [24]. Besides that, the high load also forces big particle-shaped debris to be crushed and sheared into thinner flakes that act as lubricants, therefore contributing to lower two-body abrasive wear that results in low μ [33]. As nanoclay is incorporated into pure composites, the changes in μ at low and high PV were not consistent. At low PV, the μ value decreased about 6.26% while at high PV, the μ value increased slightly by 1.60% compared to its pure state. However, at both low and high PVs, the μ values were increased as the nanoclay content increased. This might be due to the higher possibility of the high nanoclay content to create disruption instead of aiding the formation of the transfer film, leading to three-body abrasive wear that dominates the wear mechanism of the BFRP composite, which leads to an increase of the friction coefficient [34]. The increase/decrease of the μ value of the nanoclay-filled composite compared to the pure BFRP composite might also be due to the slide/roll state of the nanoclay when experiencing low/high PV factor [1].

Under reciprocating motion, the μ for the B-O-F configuration was lower than the μ obtained from the C-O-F configuration. The reason behind the smaller μ value is due to the smaller contact area of the B-O-F configuration compared to the C-O-F configuration. The smaller contact area makes the pressure at the worn surface higher, thus lowering the result

Table 3 Friction coefficient (μ) of composites under different conditions.

	Unidirectional sliding		Reciprocating sliding	
Composites	**0.23 MPa m/s**	**0.93 MPa m/s**	**B-O-F**	**C-O-F**
0.0 wt% NC BFRP/Pure BFRP	0.703	0.499	0.202	0.426
1.0 wt% NC BFRP	0.659	0.507	0.212	0.406
3.0 wt% NC BFRP	0.691	0.513	0.227	0.399
5.0 wt% NC BFRP	0.695	0.521	0.249	0.388

of μ and also *Ks* [13, 25]. Besides that, the μ values are different in the set/range from one configuration to another due to the nature of the μ value, which is not an intrinsic property but strongly depends on the tribological system the material is involved with [22]. When nanoclay was added in the BFRP composites, the trend of μ obtained from those configurations was different. In the B-O-F configuration, the μ increased steadily as the nanoclay content increased, exhibiting a negative impact of nanoclay incorporation toward the friction behavior of the BFRP composites. On the other hand, for those under the C-O-F configuration, the μ decreased gradually but with a lesser slope as the nanoclay content increased. The BFRP composite with 5.0 wt% NC content exhibited the lowest μ value. Therefore, in each configuration, the 5.0 wt% NC BFRP composites demonstrated the worst and best wear resistances, with 24.50% deterioration and 8.92% improvement for B-O-F and C-O-F, corresponding to the 0.0 wt% NC BFRP composite, respectively. The different trend of μ might be due to the different placement and entrapment of wear debris during sliding because the steel counter face is in different sizes and shapes. The presence of nanoclay may either be involved in the polishing effect [34] or become third-body abrasive [2, 6] during sliding.

Fig. 3 shows the specific wear rate (*Ks*) of composites sliding under unidirectional motion at low and high PV factors of 0.23 and 0.93 MPa m/s for a 2000 m distance. The *Ks* for the FRP composites shows a very distinct value with respect to different PVs, with *Ks* at a high PV increased by one order of magnitude for all composites compared to a low PV factor. The trend is consistent with the general relationship of *Ks* with load and velocity, where as the load and/or velocity increased, the *Ks* should increase as well [35]. The effect of the thermal softening of the epoxy polymer at a higher load and velocity due to increased frictional heat might lead to a higher *Ks* as the thicker soft surface will be detached from the composite [3]. With nanoclay inclusion in the BFRP composite, the *Ks* decreased at both low and high PV factors. The addition of 5.0 wt% NC improved the *Ks* of the pure BFRP composite by 35.14% and 22.46% at low and high PVs, respectively. It can be seen that the nanoclay addition improved the *Ks* more prominently at the higher PV factor [25, 26]. The nanoclay's uniform dispersion and microstructural homogeneity improved the wear of the resulting BFRP composites [14], where it acts as a barrier and prevents the large-scale fragmentation of the polymer [3, 6]. Nanoclay bears some of the load because it has a high surface energy that strengthens the interface bonding between the particle and epoxy matrix, thus improving the mechanical strength and the wear resistance of the FRP composite [32].

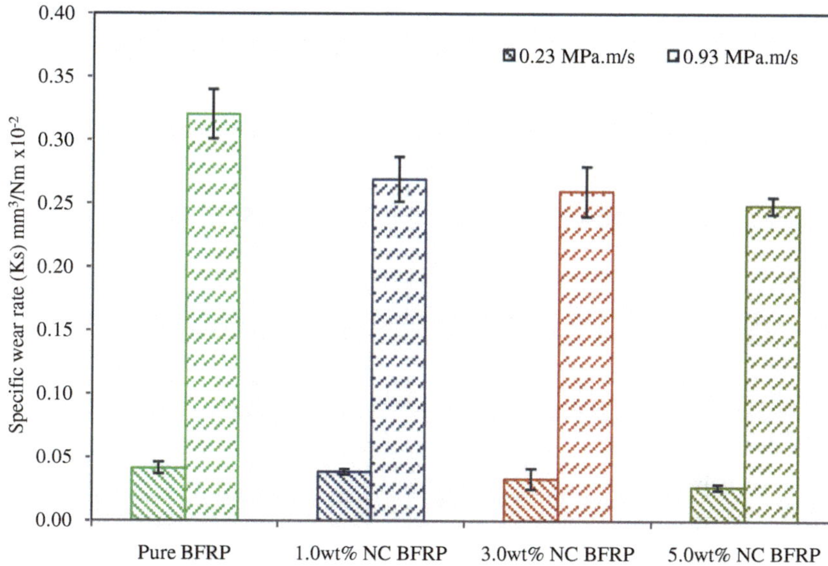

Fig. 3 Specific wear rate of composites sliding under unidirectional motion at PV factor 0.23 and 0.93 MPa m/s for 2000 m distance.

Fig. 4 shows the *Ks* of the composites sliding under reciprocating motion against ball (ball-on-flat) and cylinder (cylinder-on-flat) counter faces at 70 N load and 0.03 m/s velocity for a 120 m distance. In general, the *Ks* obtained from the B-O-F configuration was lower than the *Ks* obtained from the C-O-F configuration. The B-O-F also has a smaller range of *Ks* compared to COF, which indicates a more stable wear mechanism. The higher pressure exerted by the steel ball counter face onto composites at the B-O-F configuration also plays an important role toward lowering the *Ks* values [13, 25]. The smaller surface contact of the steel ball counter face provided only a smaller contact area compared to that obtained from the C-O-F configuration.

As nanoclay is added to the BFRP composites, the samples subjected to both configurations showed decreasing trends of *Ks*. The *Ks* decreased gradually with increasing nanoclay content from 1.0 wt% to 5.0 wt%. For those obtained from the B-O-F configuration, the decrement of *Ks* is small while for those subjected to the C-O-F configuration, the decrement is quite large. It was mentioned before that for those obtained from the B-O-F configuration, the *Ks* values have a smaller range of wear improvement. However, it is also possible to mention that under the C-O-F configuration, nanoclay plays a more prominent role than when the samples were subjected to the B-O-F configuration. The *Ks* of a 5.0 wt% NC BFRP composite under

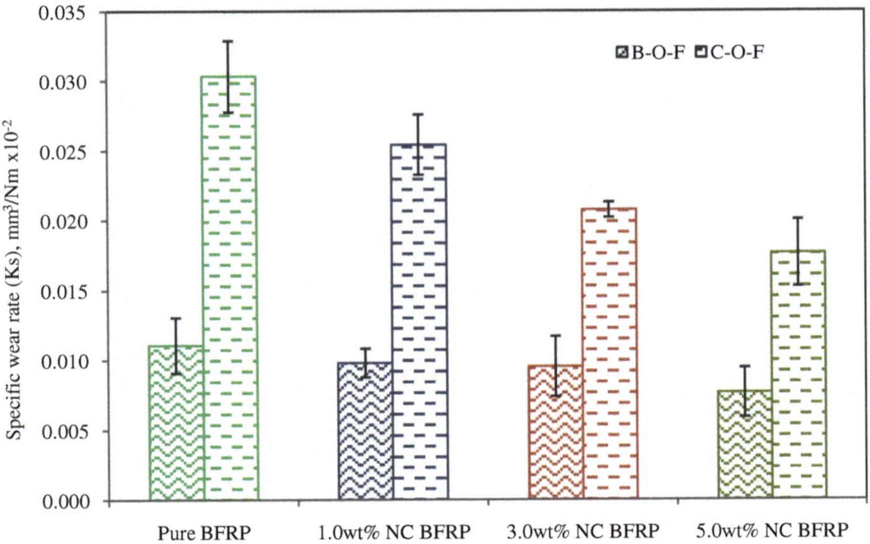

Fig. 4 Specific wear rate of composites sliding under reciprocating motion of B-O-F and C-O-F configurations for 120 m distance.

C-O-F configuration showed a higher improvement of 41.58% than those under the B-O-F configuration, which recorded a 30.90% improvement. Nevertheless, nanoclay incorporation has effectively acted as a barrier and as additional reinforcement to the BFRP composite to improve its wear resistance with reducing its *Ks* value [32]. Its uniform dispersion and structural homogeneity have improved the surface integrity of composites as verified by the TEM micrographs in Fig. 2 and the hardness value in Table 2, which therefore improved its wear properties [14].

3.3 Worn surface analysis

The surface roughness, Ra, of worn surfaces of 0.0, 1.0, 3.0, and 5.0 wt% NC BFRP composites after sliding under both motions is presented in Table 4. Under unidirectional sliding, the Ra of the 0.0 wt% NC BFRP composites at high PV was quite high compared to the Ra at low PV. This might be due to its higher surface damage, as its *Ks* was also high. As nanoclay was added, the Ra became reduced, matching their low *Ks* value. However, at low PV, the Ra results showed no significant changes when nanoclay was added, perhaps due to much less surface damage because their *Ks* was low. Under reciprocating sliding, the B-O-F configuration showed a lower Ra value compared to the C-O-F configuration at the end of the

Table 4 Surface roughness (μm) of composites after sliding tests.

	Unidirectional sliding		Reciprocating sliding	
Composites	**0.23 MPa m/s**	**0.93 MPa m/s**	**B-O-F**	**C-O-F**
0.0 wt% NC BFRP/Pure BFRP	3.36	4.69	1.97	2.81
1.0 wt% NC BFRP	3.46	3.14	1.94	2.73
3.0 wt% NC BFRP	3.27	3.10	1.88	2.59
5.0 wt% NC BFRP	3.49	3.19	1.85	2.75

sliding test for all composites. This might be due to their lower *Ks* and μ values. Nanoclay incorporation seemed to assist in lowering the Ra value at the end of the tests, consistent with their lower *Ks* value.

The SEM examination of worn surfaces of 0.0, 1.0, 3.0, and 5.0 wt% NC BFRP composites sliding under unidirectional motion at low and high PVs is given in Figs. 5 and 6. Fig. 5A–D present the features of worn surface and transfer film of 0.0 wt% NC BFRP composites under unidirectional sliding motion when conducted under low and high PVs, respectively. The arrow lines show the sliding direction. At a low PV, the worn surface showed less matrix damage with less fiber exposure, as shown in Fig. 5A. The surface was relatively rough and plowed where adhesive wear dominates. At a high PV, the worn surface showed more severe matrix damage with mild fiber exposure and fiber breakage, as shown in Fig. 5B. The surface was relatively rough with brittle fragments and multiple microcracks visible at various regions, indicating wear by brittle fracture and abrasive wear. Some fibers were exposed, indicating the gradual wear of the matrix where the fibers support the most load after the matrix has been removed, with some fibers fractured and detached. The detached fiber was introduced back to the surface, creating three-body abrasives. The corresponding transfer films for low and high PV are shown in Fig. 5C and D, respectively. At a high PV, the transfer film was more uniform and integrated compared to the transfer film for a low PV, resulting in a lower μ. The transfer film for a low PV is less integrated and quite lumpy, and some have light scuffing. These signs resulted in a higher μ for the composites.

Fig. 6A–D present the features of worn surface and transfer film of 5.0 wt% NC BFRP composites under a unidirectional sliding motion at low and high PVs, respectively. Under both PV factors, the matrix wear seems to be improving. The microcracks and plowing at the matrix and the fiber wear/fracture were found to be limited, perhaps due to the presence of the nanoclay that improved the load-carrying capacity of the composite,

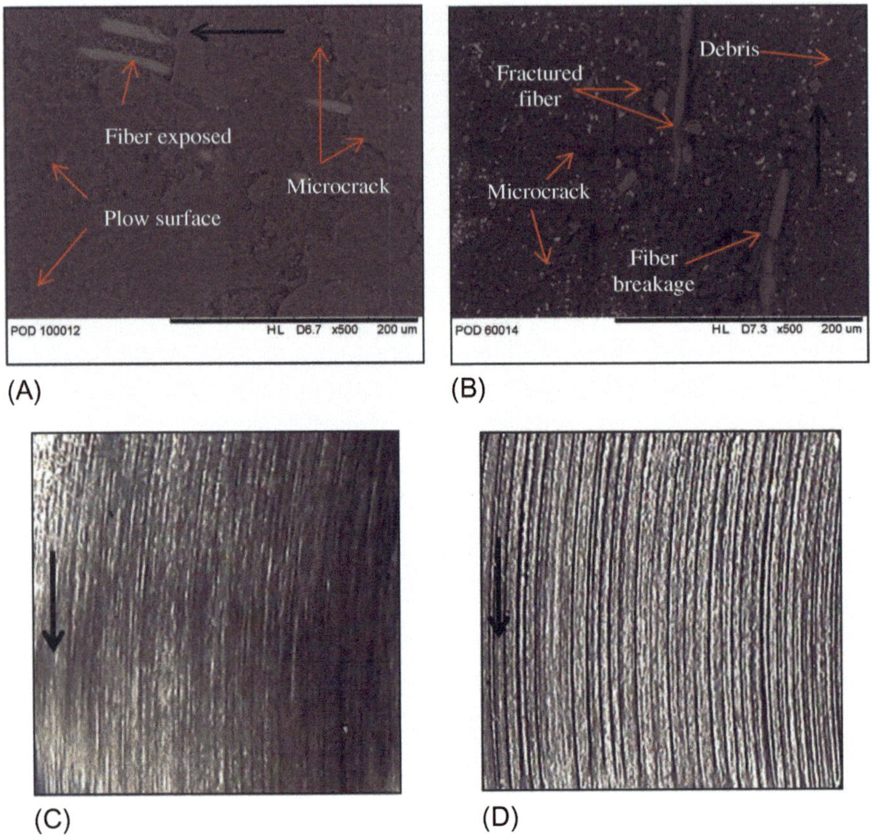

Fig. 5 SEM micrographs of worn surface of 0.0 wt% NC BFRP composites sliding under unidirectional motion at (A) low PV, (B) high PV with their corresponding transfer layer, (C) low PV, and (D) high PV.

thus improving the adhesion between the fiber and matrix. At low PV, the worn surface was relatively smoother with no fiber exposure, as shown in Fig. 6A. There was only a groove caused by brittle fragmentation due to repeated shear loading. At a high PV, the surface also seemed smoother with much less fiber exposure, as shown in Fig. 6B. The fiber showed no signs of debonding, but there was fiber breakage. In Fig. 6C–D, the transfer films were more thin and uniform indicating strong adhesive strength between the film layer and counter face [32]. The presence of nanoclay is believed to generate a reaction product where it enhanced the bonding between the transfer film and the counter face and became localized at the nanoscale cracks [20]. The transfer film seemed more thin and uniform, indicating

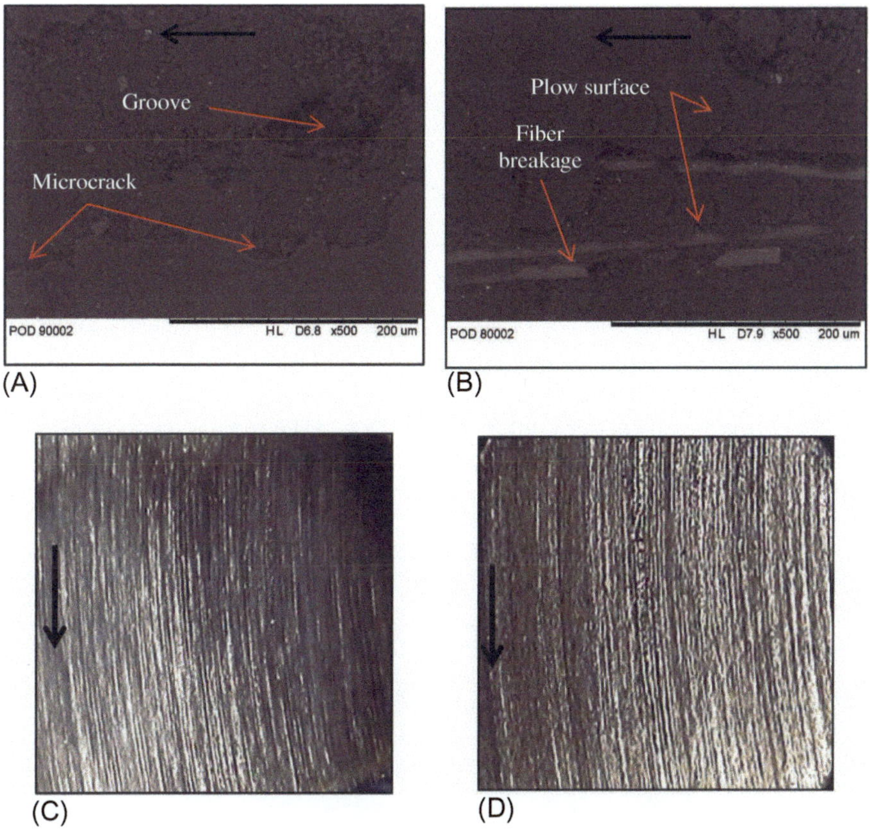

Fig. 6 SEM micrographs of worn surface of 5.0 wt% NC BFRP composites sliding under unidirectional motion at (A) low PV, (B) high PV with their corresponding transfer layer, (C) low PV, and (D) high PV.

strong adhesive strength between them [32]. However, the μ did show a slight increase as the nanoclay content increased for both test parameters in the range of 2.0%–6.0%; this might be due to the agglomeration of the nanoclay content that disrupted the formation of the continuous transfer film at the nanoscale [34].

Figs. 7 and 8 indicate the worn surface of 0.0 wt% NC BFRP and 5.0 wt% NC BFRP composites sliding under the reciprocating motion of B-O-F and C-O-F configurations at a load of 70 N. Based on the micrographs obtained, the overall observation can be made that the surface removal and fiber exposure were severely affected. One important observation to take from these micrographs is the condition of the exposed fibers. The fibers originally were all oriented in one direction, which was

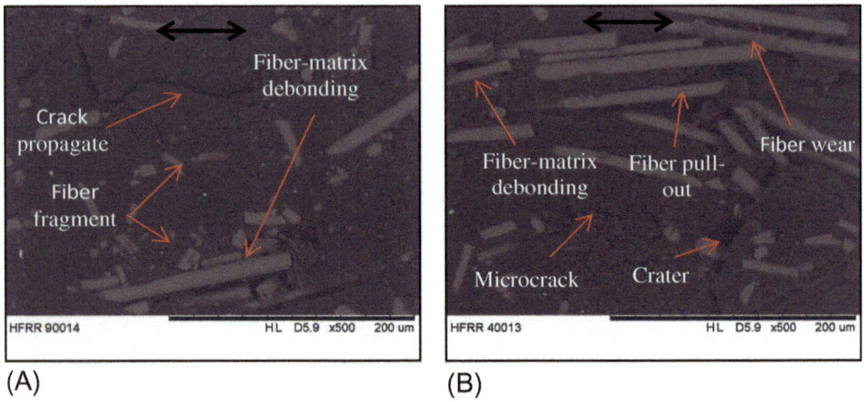

Fig. 7 SEM micrograph of worn surface of 0.0 wt% NC BFRP composites under reciprocating motion obtained from (A) B-O-F, and (B) C-O-F configurations.

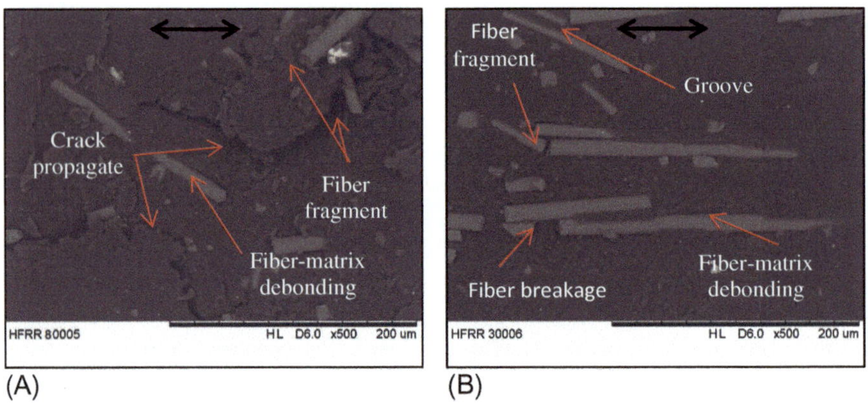

Fig. 8 SEM micrograph of worn surface of 5.0 wt% NC BFRP composites under reciprocating motion obtained from (A) B-O-F, and (B) C-O-F configurations.

parallel to the sliding direction of the counter face. At the end of the test, however, the fibers seemed to be displaced from their original orientation and rearranged randomly without specific direction. This might be due to the type of motion of the test, that is, back and forth at a high-frequency reciprocating motion. The reciprocating motion was believed to trap the removed material between the surface and the counter face. These are commonly called back-transferred materials, especially for those subjected to a B-O-F configuration where the pressure exerted at the dented wear track was deeper and smaller in size. Therefore, the fibers that were fractured and pulled out break into even smaller pieces and are embedded again into the matrix because it is soft due to the high temperature rise, resulting in scattered fibers as seen at the worn surface.

The worn surfaces of the 0.0wt% NC BFRP composites shown in Fig. 7 basically exhibit matrix wear, fiber fracture, and interfacial debonding visible at various places of the composite. Fig. 7A portrays the 0.0wt% NC BFRP composite that was subjected to the B-O-F configuration. It is shown that the matrix surface was relatively smooth at certain regions with scattered fiber exposure and fiber fragments at various regions. Crack propagation can be seen at the top region, indicating surface fatigue, with signs of fiber debonding and fiber pullout nearby. This shows that the shear loading is bigger than the interfacial strength between the fiber and matrix. On the other hand, for the sample subjected to the C-O-F configuration (Fig. 7B), the worn surface was relatively rough with a lot of fiber exposed as well as fiber fragments. Microcracks are seen propagating to form small craters that will be removed after repeated loading. There is fiber wear visible at the end of the exposed fibers together with fiber-matrix debonding. The backscattered fibers and wear debris embedded at certain regions were due to the reciprocating motion of the counter face. The main wear mechanism has changed from adhesive wear to abrasive wear.

In Fig. 8A, the micrograph captured on the 5.0wt% NC BFRP composite that was subjected to B-O-F configuration did not show much difference from its pure composite. The matrix surface was full with microcracks that propagated along the micrograph at various regions, resulting in a high μ although the fiber exposure was limited. Small amounts of fiber fracture and fiber fragments were seen with signs of fiber-matrix debonding. For that obtained from the C-O-F configuration as shown in Fig. 8B, the matrix surface was relatively smooth with shallow grooves visible at certain regions. This might be due to surface fatigue wear. The fiber exposed was greatly limited with most fibers still protected by the matrix. Fiber fragments are visible with fiber-matrix debonding, but with no sign of fiber wear.

4 Conclusion

The friction and wear behaviors of nanoclay-filled BFRP composites were successfully evaluated under unidirectional and reciprocating sliding motions. Based on the results obtained, these conclusions can be made:

- Under unidirectional and reciprocating sliding motions, the incorporation of nanoclay in the BFRP composite showed enhancement in terms of *Ks* at low and high PV factors, and when obtained under B-O-F and C-O-F configurations, respectively.

- A 5.0 wt% nanoclay content exhibited *Ks* improvements by 35.07% and 22.46% corresponding to a 0.0 wt% NC BFRP composite when subjected to low PV and high PV under unidirectional sliding, respectively.
- Improved *Ks* by 31.53% and 41% was also recorded for a 5.0 wt% NC BFRP composite with the motion obtained against B-O-F and C-O-F configurations under reciprocating sliding, respectively.
- Under unidirectional sliding, the μ improved by 6.26% at a low PV factor for 1.0 wt% nanoclay content, compared to its 0.0 wt% NC BFRP composite. Under reciprocating sliding, the 5.0 wt% NC BFRP composite exhibited a μ improvement of 8.92% when subjected to the C-O-F configuration.
- The surface morphology analyses discovered different types of wear mechanisms observed on the worn surfaces of the composites with combinations of abrasive wear, adhesive wear, plowing, surface fatigue, and brittle fracture. These were depicted from signs of matrix surface damage, microcracking, microplowing, grooves, fiber exposure, fiber fracture, fiber-matrix debonding, fiber peel-off, and fiber wear.

References

[1] G. Zhao, I. Hussainova, M. Antonov, Q. Wang, T. Wang, Friction and wear of fiber reinforced polyimide composites, Wear 301 (1–2) (2013) 122–129.
[2] N. Rashmi, M. Renukappa, B. Suresha, R.M. Devarajaiah, K.N. Shivakumar, Dry sliding wear behaviour of organo-modified montmorillonite filled epoxy nanocomposites using Taguchi's techniques, Mater. Des. 32 (8–9) (2011) 4528–4536.
[3] B. Suresha, K. Siddaramaiah, S. Seetharamu, P.S. Kumaran, Investigations on the influence of graphite filler on dry sliding wear and abrasive wear behaviour of carbon fabric reinforced epoxy composites, Wear 267 (9–10) (2009) 1405–1414.
[4] H. Rajashekaraiah, S. Mohan, P.K. Pallathadka, S. Bhimappa, Dynamic mechanical analysis and three-body abrasive wear behaviour of thermoplastic copolyester elastomer composites, Adv. Tribol. (2014).
[5] P. Hrabě, M. Müller, Three-body abrasive wear of polymer matrix composites filled with *Jatropha curcas* L, in: The 20th International Conference: Machine Modeling and Simulations, vol. 136, 2016, pp. 169–174.
[6] S.R. Chauhan, S. Thakur, Effects of particle size, particle loading and sliding distance on the friction and wear properties of cenosphere particulate filled vinylester composites, Mater. Des. 51 (2013) 398–408.
[7] S. Carmisciano, I.M. De Rosa, F. Sarasini, A. Tamburrano, M. Valente, Basalt woven fiber reinforced vinylester composites: flexural and electrical properties, Mater. Des. 32 (1) (2011) 337–342.
[8] C.A. Chairman, S.P.K. Babu, Mechanical and abrasive wear behavior of glass and basalt fabric-reinforced epoxy composites, J. Appl. Polym. Sci. 130 (1) (2013) 120–130.

[9] F. Fazillah, M. Fadzli, B. Abdollah, A. Kalam, M. Hassan, H. Amiruddin, Tribological characteristics comparison for oil palm fi bre/epoxy and kenaf fi bre/epoxy composites under dry sliding conditions, Tribol. Int. 101 (2016) 247–254.
[10] A. Greco, A. Maffezzoli, G. Casciaro, F. Caretto, Mechanical properties of basalt fibers and their adhesion to polypropylene matrices, Compos. B Eng. 67 (2014) 233–238.
[11] V. Manakari, et al., Dry sliding wear of epoxy/cenosphere syntactic foams, Tribol. Int. 92 (2015) 425–438.
[12] V. Kostopoulos, A. Baltopoulos, P. Karapappas, A. Vavouliotis, A. Paipetis, Nano-enhanced impact properties of carbon fibre reinforced composites, in: European Conference on Composite Materials, 2008.
[13] W. Österle, A.I. Dmitriev, B. Wetzel, G. Zhang, I. Häusler, B.C. Jim, The role of carbon fibers and silica nanoparticles on friction and wear reduction of an advanced polymer matrix composite, Mater. Des. 93 (2016) 474–484.
[14] R.V. Kurahatti, A.O. Surendranathan, A.V. Ramesh Kumar, C.S. Wadageri, V. Auradi, S.A. Kori, Dry sliding wear behaviour of epoxyreinforced with nanoZrO_2 particles, in: International Conference on Advances in Manufacturing and Materials Engineering, vol. 5, 2014, pp. 274–280.
[15] C.R. Mahesha, Shivarudraiah, C.R. Chandra, R. Suprabha, Effect of nano TiO_2/clay on the erosive wear behavior of basalt-epoxy hybrid composites at elevated temperatures, Appl. Mech. Mater. 813–814 (2015) 40–45.
[16] A.A. Azeez, K.Y. Rhee, S.J. Park, D. Hui, Epoxy clay nanocomposites—processing, properties and applications: a review, Compos. B Eng. 45 (1) (2013) 308–320.
[17] B. Suresha, B. Ravishankar, L. Sukanya, Dynamic mechanical analysis and three-body abrasive wear behavior of epoxy nanocomposites, J. Reinf. Plast. Compos. (2012) 1–11.
[18] A. Jumahat, C. Soutis, J. Mahmud, N. Ahmad, Compressive properties of nanoclay/epoxy nanocomposites, in: International Symposium on Robotics and Intelligent Sensors, vol. 41, 2012, pp. 1607–1613.
[19] M. Esteves, A. Ramalho, J.A.M. Ferreira, J.P. Nobre, Tribological and mechanical behaviour of epoxy/nanoclay composites, Tribol. Lett. 52 (1) (2013) 1–10.
[20] Q.B. Guo, M.Z. Rong, G.L. Jia, K.T. Lau, M.Q. Zhang, Sliding wear performance of nano-SiO_2/short carbon fiber/epoxy hybrid composites, Wear 266 (7–8) (2009) 658–665.
[21] K. Kanny, T.P. Mohan, Resin infusion analysis of nanoclay filled glass fiber laminates, Compos. B Eng. 58 (2014) 328–334.
[22] A.P. Harsha, R. Wasche, M. Hartelt, Tribological studies on polyetherketone composite under reciprocating sliding condition against steel cylinder, Proc. Inst. Mech. Eng. Part J 229 (7) (2015) 795–806.
[23] G. Zhang, R. Sebastian, T. Burkhart, K. Friedrich, Role of monodispersed nanoparticles on the tribological behavior of conventional epoxy composites filled with carbon fibers and graphite lubricants, Wear 292–293 (2012) 176–187.
[24] Q. Wang, X. Zhang, X. Pei, Study on the friction and wear behavior of basalt fabric composites filled with graphite and nano-SiO_2, Mater. Des. 31 (2010) 1403–1409.
[25] L. Chang, Z. Zhang, C. Breidt, K. Friedrich, Tribological properties of epoxy nanocomposites I. Enhancement of the wear resistance by nano-TiO_2 particles, Wear 258 (1–4) (2005) 141–148.
[26] L. Chang, K. Friedrich, Enhancement effect of nanoparticles on the sliding wear of short fiber-reinforced polymer composites: a critical discussion of wear mechanisms, Tribol. Int. 43 (12) (2010) 2355–2364.
[27] A. Jumahat, C. Soutis, F.R. Jones, A. Hodzic, Compressive behaviour of nanoclay modified aerospace grade epoxy polymer, Plast. Rubber Compos. 41 (6) (2012) 225–232.
[28] J.F. Archard, Wear Theory and Mechanisms, 1980.
[29] A. Jumahat, C. Soutis, F.R. Jones, A. Hodzic, Effect of silica nanoparticles on compressive properties of an epoxy polymer, J. Mater. Sci. 45 (21) (2010) 5973–5983.

[30] S. Han, D.D.L. Chung, Strengthening and stiffening carbon fiber epoxy composites by halloysite nanotubes, carbon nanotubes and silicon carbide whiskers, Appl. Clay Sci. 83–84 (2013) 375–382.
[31] C.K. Lam, K.T. Lau, Tribological behavior of nanoclay/epoxy composites, Mater. Lett. 61 (18) (2007) 3863–3866.
[32] P. Jawahar, R. Gnanamoorthy, M. Balasubramanian, Tribological behaviour of clay—thermoset polyester nanocomposites, Wear 261 (7–8) (2006) 835–840.
[33] X. Zhang, X. Pei, Q. Wang, Friction and wear properties of polyimide matrix composites reinforced with short basalt fibers, J. Appl. Polym. Sci. 111 (2009) 2980–2985.
[34] L. Chang, Z. Zhang, Tribological properties of epoxy nanocomposites part II. A combinative effect of short carbon fiber with nano-TiO_2, Wear 260 (7–8) (2006) 869–878.
[35] S.K. Basu, S.N. Sengupta, B.B. Ahuja, Fundamentals of Tribology, second ed., Prentice-Hall India, New Delhi, 2006.

CHAPTER 9

Slurry jet erosion wear of particulate polymer composites and metal coating

Peerawatt Nunthavarawong[a], Chatchapol Chungchoo[b], and Panadda Niranatlumpong[c]
[a]Tribo-Systems for Industrial Tools and Machinery Research Center, Department of Teacher Training in Mechanical Engineering, Faculty of Technical Education, King Mongkut's University of Technology North Bangkok, Bangkok, Thailand
[b]Department of Mechanical Engineering, Faculty of Engineering, Kasetsart University, Bangkok, Thailand
[c]National Metal and Materials Technology Center, Pathum Thani, Thailand

1 Introduction

Rapid tooling plays an important role in several industries, especially in the production of light-usage or the short-run manufacturing of large parts for the automotive, space and aircraft, and shipbuilding industries. For the above-mentioned applications, the mold making is often difficult, expensive, or cannot be produced in a short lead time using a conventional machining process [1–3]. The rapid tools used in plastic injection mold applications are typically made by an aluminum-filled epoxy composite (AL-EP) via a vacuum casting technique [2, 4]. Nevertheless, this type of mold material does not possess the lifetime as that found in a steel mold. This is often due to the low mechanical strength and the low wear resistance of the mold material, resulting in defects on the mold surface [2, 5–7].

The observations on the failure and wear of the epoxy-based materials for injection molding were investigated in previous works. The surface failure and wear on certain regions of mold surfaces were predominantly obtained where the melted plastics pass through. The chipping failures were found on the sharp curves of mold cavities during 14 cycles of injected parts [8] while the mold failure due to plastic deformation, in particular on the AL-EP mold surface, was observed through 200 cycles, whereby thermal stress concentrations occurred in the cavity of the mold [9]. Unlike the failure of a steel mold, other examinations on the failure of the AL-EP material have mentioned hairline cracks in the mold cavity due to the effects of the injection pressure and the operation temperature; this work reported that the tool life was around 400 cycles [4]. In terms of wearing in plastic

Tribology of Polymer Composites
https://doi.org/10.1016/B978-0-12-819767-7.00009-8

injection molding, the wear mechanisms are often referred to as three main mechanisms: abrasion, erosion, and adhesion. Most studies on the worn mold surfaces from the injection of fiber-filled polymer parts have found the wear mechanism to be the abrasive wear. Especially on the surface regions such as on the core and cavity walls, the runner, and the gate, the erosive wear has been obtained [10–12]. Pitting due to the effects of the high-injection velocity of the melted polymer during material processing was also often found in a hardened steel mold. Furthermore, dry sliding wear was found in the ejection system of a steel mold. The metal-to-metal phenomena occurred in the reciprocating sliding system between the ejector pins and their sliding holes under dry conditions. In the case of a copper-alloy mold material, the shallow crater wear appeared in the critical areas; previous authors remarked that the erosive wear is a main reason for reducing mold lifespans. Some damaged areas on the core surfaces were also determined to be abrasive wear [11]. In an investigation on the worn surface of the AL-EP mold material after material processing, the erosion pitting occurred due to the aluminum filler detachment [8, 9]. This finding was confirmed by an abrasion tribotest [12]. In the solid particle erosion tests, the effects of the angles of impact were dominant. The high wear rate of epoxy-based composites occurred during impingement angles of 60–90 degrees [13].

An alternative in extending the lifetime of the AL-EP material is to use a coating. The thermal-sprayed coating technique covers many commercial methods such as frame spraying and arc spraying that can be used to reduce the corrosion and wear of the component surfaces [14, 15]. Electric arc spraying offers an advantage over other spraying techniques, particularly for rapid tool making, due to its possible low temperature in operation [16–19]. Zinc coating is widely used in the arc spraying technique, due to its high deposition rate and efficiency [18, 20, 21]. The melting point of zinc is low enough to be suitable for the coating fabrication on soft materials. This process could be used to increase the lifetime of the soft tooling by constructing a metal-coated polymer composite [1, 16–18, 20]. Zn coating not only reduces the wear of the AL-EP surface, but it can also protect the mold surface from atmospheric corrosion. Several parameters decreased the Zn material by erosion such as impact velocity, impact angles, erodent shape and size, erosive medium, etc. In one study of an erosion test, Zn-based materials were tested by a slurry test under corrosive media (H_2SO_4 and NaCl) conditions [22]. The increasing volume of the suspended erodent particle in the slurry test induced erosion and abrasion; this is more prominent than the corrosion effect [22, 23]. Some reports included a jet erosion test where it was found that the erosion rate depended on the velocity at impact, and a high wear rate was observed at impingement angles of 30–50 degrees [23, 24].

To date, a study of the slurry erosion of the aluminum-filled epoxy composite and the Zn-sprayed coating to describe the erosive wear properties does not appear to have been reported. Therefore, this study was carried out to gain a greater understanding of the wear mechanism, which is necessary for an improvement in the lifetime of rapid mold materials. In this work, the influence of the mold material made by the Zn-alloy coating and its AL-EP backing material with different surface conditions was investigated on the surface and slurry erosion wear properties with emphasis on slurry jet erosion testing apparatus, surface roughness parameters, wear volume, wear mechanisms, and erosion wear index.

2 Experimental procedure

2.1 Material preparation

The epoxy resin used as a base material in the present work is Biresin L84 (SIKA, Stuttgart, Germany) with L84T hardener. Irregular-shaped aluminum powder used as a particulate reinforcement was produced by grinding an aluminum ingot. The Al powder grain size varies from 10 to 63 μm after sieving (Manufactured by SIKA, Germany). The mass mixture ratio between the epoxy and the aluminum powder—called AL-EP—was 55:45, and this was used to obtain an optimum strength [25]. In order to minimize the air bubbles formed inside the specimens during the curing process, all mixing and casting processes were carried out in a vacuum chamber (MCP 5/01 Vacuum Casting Machine, MCP HEK GmbH, Germany) with a degassing time of 30–45 min. The AL-EP material was poured in the casting mold. The mold material was then cooled in this chamber for 16–24 h.

A square plate of polypropylene (PP) with dimensions of 50 × 50 × 4 mm was chosen as the master model (substrate) for this work due to the ease of forming [26]. Electric arc spraying (TAFA ARCJET 9000, Hobart Tafa Technologies, United States) was then used to apply a coating onto the PP substrate. The spraying motion was controlled using an industrial robot (MOTOMAN UP50N ROBOT, Yaskawa Electric Corporation, Japan). The coating material was an intermetallic compound of Zn alloy containing 94% zinc and 6% aluminum (Tafaloy 204M, Hobart Brothers, United States). The goal was to produce a coating thickness of approximately 300 μm onto the PP substrate ready for back-casting in order to produce

Table 1 The constant parameters of arc spraying.

Spraying parameters	
DC current	100 A
DC voltage	30 V
Compressed-air pressure	0.4 MPa
Stand-off distance	152 mm
Spray angle	90 degrees
Surface speed (robot)	160 mm s^{-1}

a rapid mold. The optimized parameters of the arc spraying of the Zn-alloy coating were kept constant, as shown in Table 1. These parameters could provide a deposition efficiency (%DE) of the Zn-alloy coating on the PP material of around 52.73% [26]. In the last step, this coating was then backfilled by AL-EP material to create a mold shell, as can be seen by the above-mentioned AL-EP casting procedures.

2.2 Specimen preparation

The vacuum casting technique achieved an even surface of 0.22–0.30 μm *Ra* on the AL-EP casting surface (AL-EP-C). Due to that, the separation of the epoxy and Al reinforcement formed on the top surface during casting (named the EP surface layer). The AL-EP machined surface (AL-EP-M) was prepared from a CNC turning; this machining process is widely used in the fabrication of rapid mold materials. The optimized conditions of machining were then used to achieve a surface roughness of around 1.6–2.2 μm *Ra*. The as-sprayed Zn alloy coating on the AL-EP backing (ZM) was then polished to obtain a surface roughness of 0.05–0.1 μm *Ra* and a remaining coating thickness of 240–280 μm. The average porosity of the coating was lower than 1% that on the planar direction of polished surfaces measured using an image analysis technique (Image Pro Plus V 5.1). The hardness of the AL-EP-C, AL-EP-M, and ZM were 23.8 ± 0.1, 25.5 ± 0.2, and 97 ± 4 HB, respectively. Specimens for the erosion test were cut in dimensions of $30 \times 30 \times 6$ mm. Specimen densities were measured using a water immersion technique (using a 0.0001 g precision balance, GF-200, A&D Co., Japan) and were determined as $1.53 \pm 0.08 \times 10^3$, $1.54 \pm 0.2 \times 10^3$, and $3.82 \pm 0.05 \times 10^3$ kg m^{-3}, respectively. According to ASTM D2734-09 [27], the average porosity measurements of the AL-EP-C and AL-EP-M were 3.75% and 4.36%, respectively. A steel mold material, PM (AISI 420 with supplied condition of annealing (225 ± 4 HB), manufactured by ThyssenKrupp AG, Germany), was also studied for comparison in the erosion test. All four materials, PM, AL-EP-C, AL-EP-M, and ZM were also subjected to the characterization test described in Section 2.3.

2.3 Slurry jet erosion testing

2.3.1 Test rig

In order to create a wear situation similar to a conventional injection molding machine, a horizontal slurry jet erosion test with a custom design was carried out. The test rig consists of a hopper and feeder of erodent particles, a water container, a water-erodent mixing unit, a nozzle unit (inside diameter of 8 mm), a proportional-integral-derivative (PID) feedback control unit of compressed air pressure, a sample holder, and a slurry receiver tray (see Fig. 1). The erodent particles were conveyed via the feeder belt. The erodent mass flow rate (M_E) can be adjusted via the speed of the conveyor belt and the distance between the belt and the hopper.

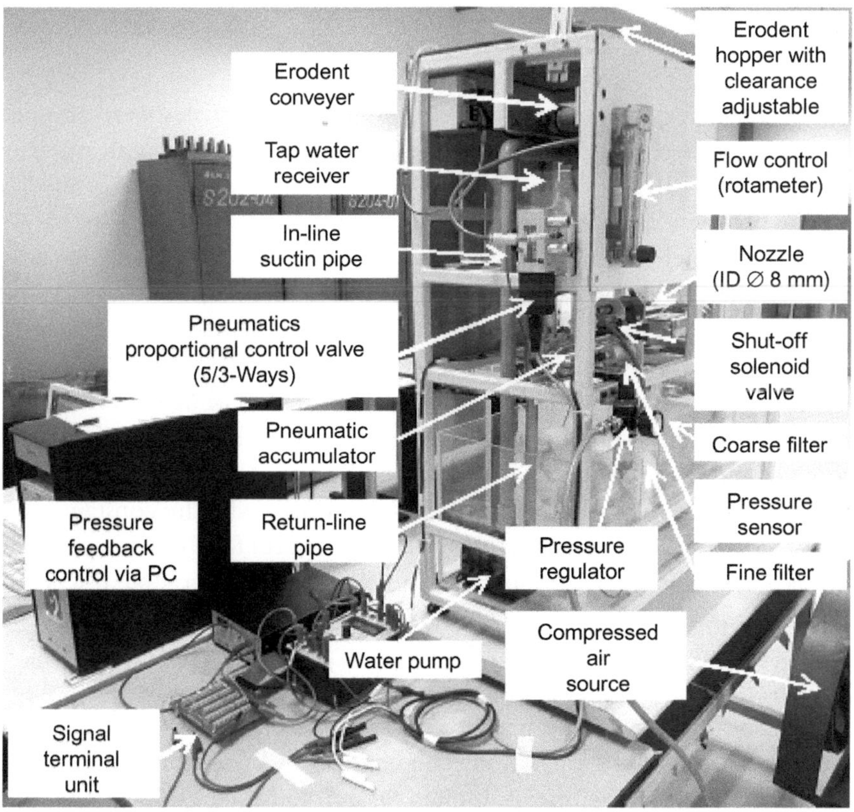

Fig. 1 The slurry jet erosion test rig.

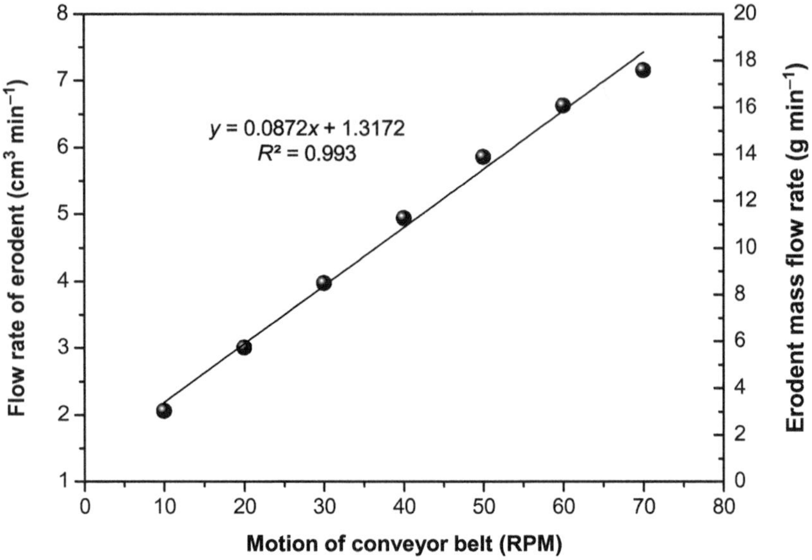

Fig. 2 The flow rate of erodent.

The M_E values are shown in Fig. 2. The erodent particles were dropped by gravity at the end of the conveyor, and were then mixed with water in a mixing unit.

The flow rate of the water (F_W) can be varied from the pump speed and the flow channel of a needle valve, as shown in Fig. 3. This mixing ratio provided the erodent mass fraction (M_F).

The M_F is a ratio of the M_E and F_W values, which represents the slurry concentration (%wt), as shown in Fig. 4. In order to prevent premature degradation, the nozzle was made from hardened steel of 60 ± 2 HRC. The wear damage of the nozzle will lead to inaccuracy in the slurry velocity and wear measurements. The slurry was transported to the nozzle by the force of gravity. It was then accelerated to the target surface by the constant pressure of a compressed air jet, thereby utilizing the PID feedback control. The PID as a closed-loop control unit with an electropneumatic system was implemented as a means to maintain the constant pressure of the compressed air to carry the slurry of the nozzle.

The test set up is described in the following steps: (1) Set the compressed air pressure of 6 bar via the pressure regulator service unit; (2) control the pressure flow rate using the volumetric flow control (Rotameter, RMB-54 SSV series, Dwyer Instruments, United States), and then carry via the proportional control valve (MPYE 5-1/8 HF, Festo GmbH, Germany) to the

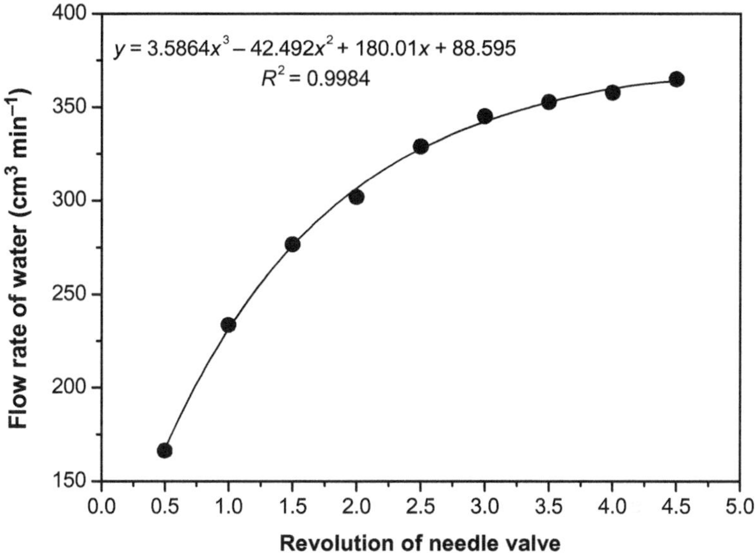

Fig. 3 The flow rate of tap water.

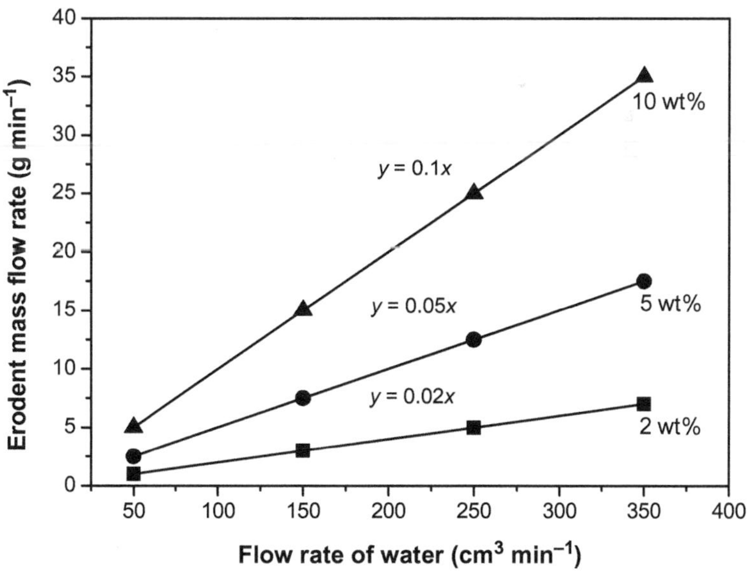

Fig. 4 The diagram of erodent mass fraction.

reservoir (CRVZS-0.4 160234, Festo GmbH, Germany); and (3) turn on or off to set the test duration of a carrier air jet via the shut-off solenoid valve (YQ21M1-5GZ-C6, SMC Pneumatics, Japan) to the nozzle unit with the PC control. During the wear test, the pressure drop signal was indicated by the pressure sensor (SDE-10–10V/20 mA, 19562, Festo GmbH, Germany), and the different pressure signal was then compensated using the PID control unit (see Fig. 1).

The procedure for a feedback control system is shown in Fig. 5. The test rig would set the compressed air pressure during the 0–10 bar (setpoint) using the LabVIEW software on the PC. Their signals convert from the pressure to the voltage via data acquisition (DAQ). The PID control unit utilizes the pressure compensator and controller, which was recognized by the PID autotuning algorithm. There are proportional gain (KP = 1), integral gain (KI = 0.01), and derivative gain (KD = 0). The controlled signals are amplified and optimized using the pulse width modulation method (0–10V) for the proportional way valve. Thereafter, the reservoir as the pneumatic accumulator is sustained by the constant pressure for the nozzle. When the consumed pressure during the wear test is lower than that on the setpoint, the pressure sensor will respond with the signal of pressure losses to the DAQ unit and the PID control unit, respectively. The PID control unit preserves the steady pressure as a means to keep the constant impact velocity of the slurry through the erosion test.

The slurry velocities were indicated using a high-speed videography (i-SPEED 3 series, Olympus UK, United Kingdom (UK)), which was fixed at the suitable shutter speed of 30,000 fps for the setpoint of the pressure between 0.55 to 1.0 bar and 40,000 fps for the pressure setting between 1.15–1.70 bar (see Fig. 6). The average velocities of the slurry are shown in Table 2.

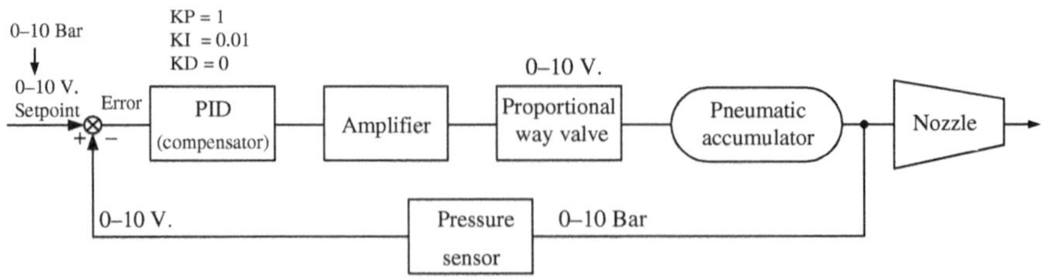

Fig. 5 The schematic view of the pressure feedback control system.

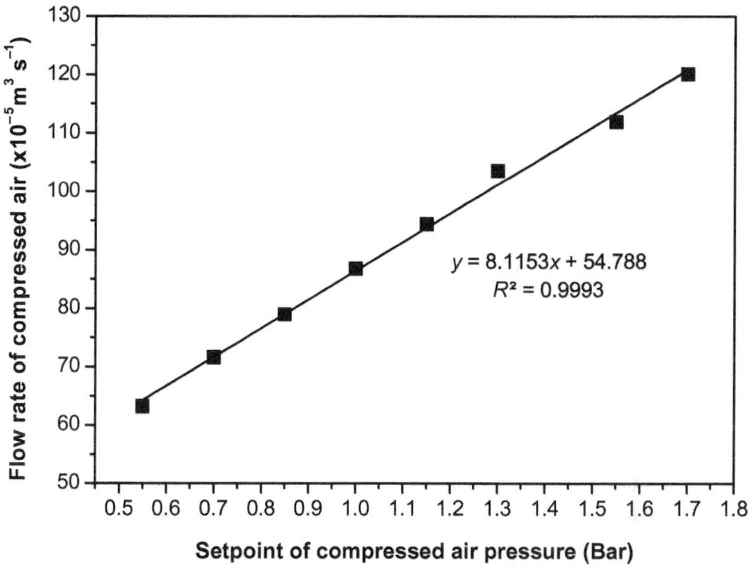

Fig. 6 The diagram for setting up a compressed air flow rate.

Table 2 The average of the slurry velocities.

Setpoint of compressed air pressure (bar)	Velocity of slurry droplet (m s^{-1})
0.7	11.0603
0.85	12.5034
1.55	28.0521

2.3.2 Wear test setup

The parameters for the slurry erosion test were kept constant, except for the impingement angle, which was varied (see Table 3).

2.3.3 Erodent particle

The erodent particles used in this test are a commercial-grade silica abrasive grit (silicon dioxide, Tosa Emery Extra # 180, Ujiden Chemical Industry, Japan). The silica grit was of an angular shape. In the numerical approach described later in Section 3.5, the geometry of the particle was specified as cubic. The erodent particles were sieved to obtain a size range between 90 and 105 μm. A laser diffraction particle size analyzer (MASTERSIZER-S, Malvern Instruments, UK) was used to determine the particle size distribution. Two grit samples were used for

Table 3 The conditions of slurry jet erosion testing.

Testing conditions [28]	
Erodent size (d_{50})	99.515 μm
Erodent mass flow rate (M_E)	$5.47\,g\,min^{-1}$
Flow rate of water (F_W)	$301.9084\,cm^3\,min^{-1}$
Erodent mass fraction (M_F)	1.81%wt
Erodent volume fraction (C)	6.7869×10^{-3}
Flow rate of compressed air (Q) at 0.85 bar	$78.91840 \times 10^{-5}\,m^3\,s^{-1}$
Slurry droplet velocity (v) at 0.85 bar	$12.5034\,m\,s^{-1}$
Test duration (T)	1500 s
Total erodent volume on impact (V_e)	$8.0342 \times 10^{-4}\,m^3$
Stand-off distance refer to impingement of 90 degrees	12 mm
Angles of impingement	30, 45, 60, 75, 90 degrees

the measurement of the median diameter (d_{50}). They were 100.04 and 98.99 μm, with an average value of d_{50} at 99.515 μm [28]. The density of the silica grit was $2.65 \times 10^3\,kg\,m^{-3}$.

2.3.4 Wear measurements

Each sample was cleaned prior to the wear test in an acetone solvent to remove any possible contaminants. In order to avoid any excessive wear at the edges of the workpiece during the wear test, each side of the samples was covered using a pressure-sensitive tape (3M, United States). After the wear test, the workpiece was ultrasonically cleaned in acetone to remove any loose particle on the worn surfaces of the specimens. The workpiece was then placed in an oven at 50°C for 60 min in order to remove any excess humidity on the surface. The mass losses were measured as the weight difference between the specimens before and after the wear tests using a high-precision balance (GF-200, A&D, Japan). The average mass loss was used to determine the erosion rates (see Fig. 7). The worn surfaces were examined, and their roughnesses were measured using an LSCM instrument (LSCM, laser scanning confocal microscopy, 3D Measuring Laser Microscope OLS4000 LEXT, Olympus, Japan). To ensure a minimum error in the roughness measurement, the fixture device was mounted and fixed in position on a stage of the LSCM for all specimens. The wear characteristic of the specimens was examined using a scanning electron microscope (SEM). The maximum wear of the specimens was also to indicate the chemical compositions using energy dispersive X-ray spectroscopy (EDS, JEOL JSM-5410, Oxford Instruments, Japan).

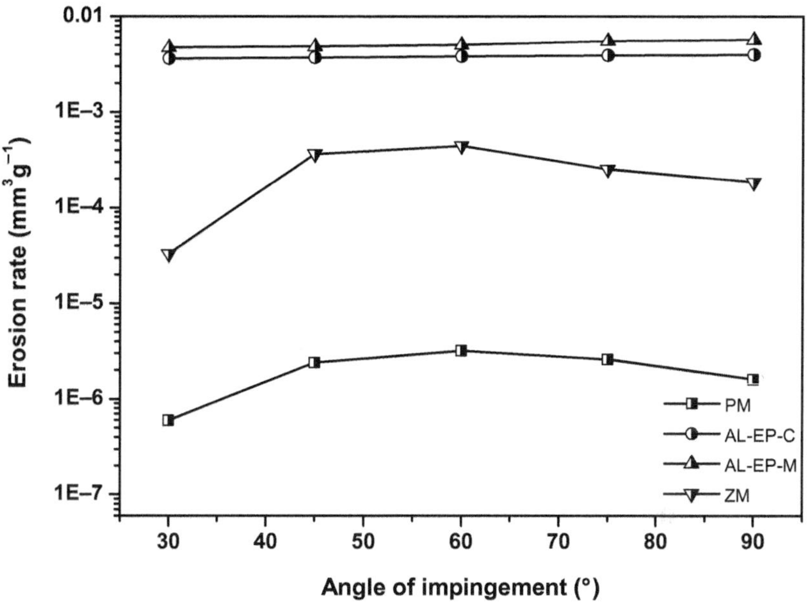

Fig. 7 The volumetric wear of tested materials.

2.3.5 Erosion wear index

A combined Wood [29] and Jha et al. [30] modeling of kinetic energy determination was suggested in Ref. [28]. The total kinetic energy before impact is determined by:

$$E_{KT} = \left(\rho_e QCTv^2 \sin^2\theta\right)/2 \quad (1)$$

where ρ_e, Q, C, T, and v, are the density of the erodent particles, the flow rate of the slurry, the erodent volume fraction, the test duration, and the velocity of the erodent particles at each angle of impingement (θ), respectively.

The calculation of the dimensionless of the specific volumetric wear, W, is [29],

$$W = V / QCT \quad (2)$$

when V is the volumetric wear of the sample.

Eq. (1) shows that the wear of the material depends on the whole accelerated particles at impact, but all erodent particles cannot be forced onto the material surface in a real situation. Thus, the energy transfer during material failure would be estimated as the number of wear losses, namely, the kinetic energy of the worn material, E_{KW}, as shown in Eq. (3), where ρ_s is the density of the sample.

$$E_{KW} = \left(W \rho_s QCTv^2 \sin^2 \theta \right) / 2 \tag{3}$$

A semiempirical approach was used to estimate the slurry jet erosion wear index of the tested materials. The erosion wear index is the ratio of the kinetic energy of the worn material (E_{KW}) and the total kinetic energy (E_{KT}), and would be determined using Eq. (4), which is dimensionless. It indicates that during $0 \leq$ erosion wear index ≤ 1, even if the highest wear occurred in the workpiece, then the index value equals 1.

$$\text{Erision wear index} = W\left(\rho_s / \rho_e \right) \tag{4}$$

3 Results and discussion

3.1 Erosion rate

Fig. 7 shows the erosion rate of the four material groups as a function of the impingement angles at impact. The erosion rates of the AL-EP-C and the AL-EP-M were higher at the normal impact (90 degrees) than that found in inclined impacts (30–75 degrees). The average erosion rate of the AL-EP-M increased by a maximum of $5.69 \times 10^{-3}\,mm^3\,g^{-1}$ while the AL-EP-C had a value of $3.97 \times 10^{-3}\,mm^3\,g^{-1}$. As the angle of impact increases, the erosion rate also increases due to the brittle nature of the AL-EP material experienced during impact [13]. The maximum wear was observed at the impingement angle of 60 degrees for PM and ZM, as described by the nature of the ductile failure [19, 20] and the average erosion rates were 3.19×10^{-6} and $4.44 \times 10^{-4}\,mm^3\,g^{-1}$, respectively.

3.2 Eroded surface observations by LSCM

A contact profilometer is widely used in surface roughness measurements for hard materials, but this method using a stylus probe is not suitable for very soft materials such as polymeric materials. The LSCM technique is a noncontact measurement that can be used to identify any changes in the surface of the soft tool materials in this work. The surface roughness measurement obtained from the LSCM used an evaluation length of 693 μm and a cut-off distance of 80 μm, and the roughness measurements of *Ra*, *Rq*, *Rz*, *Rt*, and *Rsm* were done according to ISO 25178-1 [31].

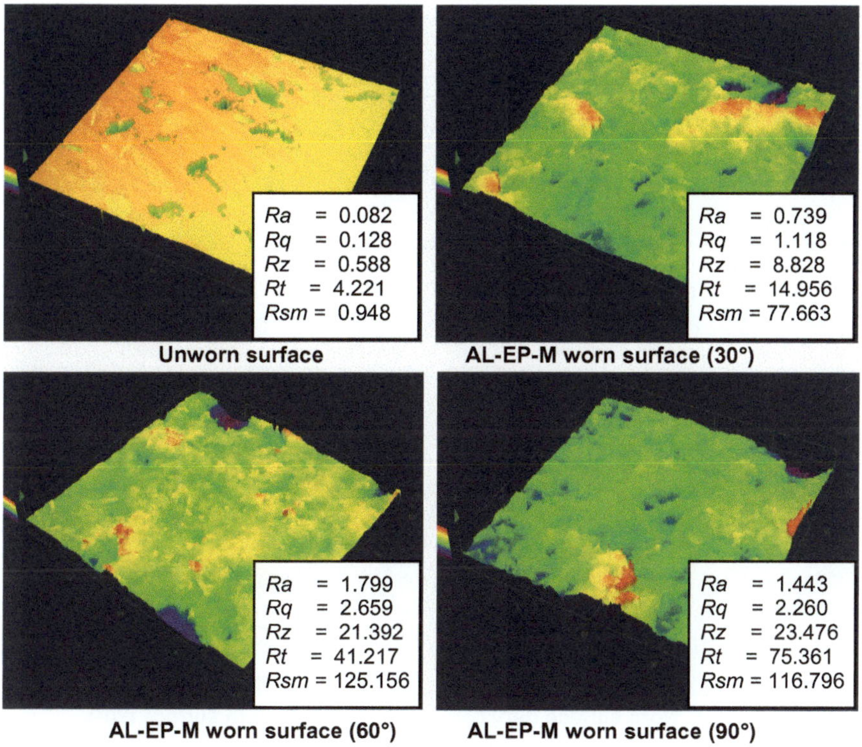

Fig. 8 LSCM observation of the eroded AL-EP-M surfaces.

In Fig. 8, the maximum wear depth of the AL-EP-M was indicated by the *Rt* value for each impingement angle of 30, 60, and 90 degrees, and had values of 14.956, 41.217, and 75.361 μm, respectively. The unworn surface of the AL-EP-M had a value of 4.221 μm *Rt*. The AL-EP-M exhibited a different behavior at inclined impacts (30 degrees, 60 degrees) and normal impact as shown in the *Rq* value. This is believed to be due to the presence of the Al filler on the worn surfaces.

The eroded AL-EP-C at the normal impact saw a decrease in *Ra* values (see Fig. 9) when compared to the AL-EP-M (Fig. 8). This finding can be attributed to the effect of the Al pull-out in the AL-EP-M during the wear tests. The layered structure of the epoxy (without the Al filler) obtained in the AL-EP-C can enhance the wear resistance, thus decreasing the *Rz* value.

Fig. 10 shows that the *Rq* parameter had the highest value of 0.302 μm at 60 degrees of the ZM worn surfaces. This is confirmed by comparing the maximum wear volume as received at 60 degrees of impact (Fig. 7).

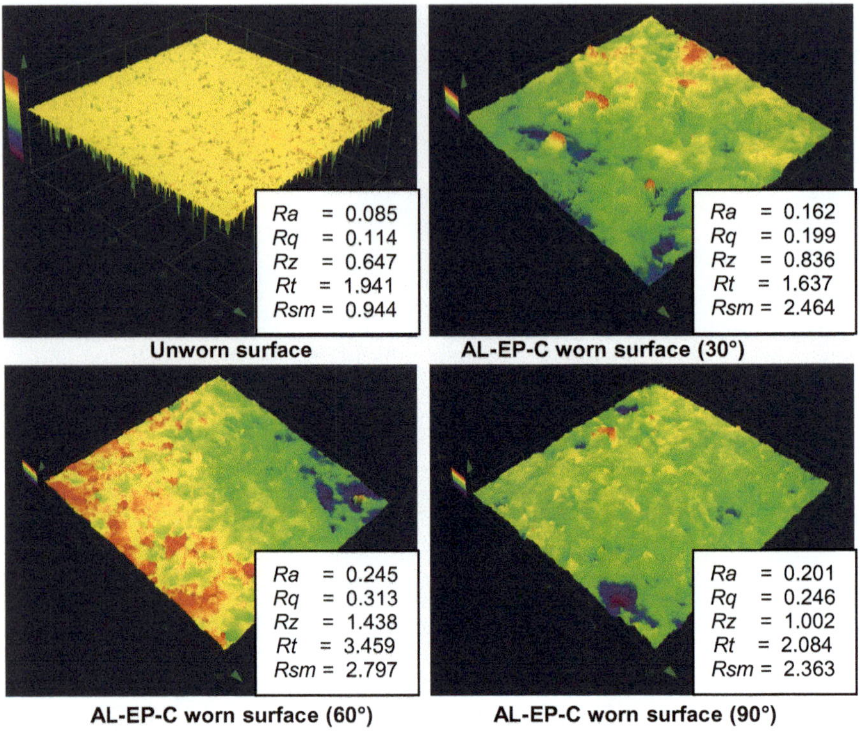

Fig. 9 LSCM observation of the eroded AL-EP-C surfaces.

3.3 Eroded surface observations by SEM

Fig. 11A shows the surface of the PM. The maximum wear at the impingement angle of 60 degrees appeared as the microcutting wear and microploughing wear were experienced.

The maximum wear at the normal impact on the AL-EP-M is shown in Fig. 11B. The surface cracks due to the debonding mechanisms of the epoxy and the reinforced Al particles were observed as microfatigue wear. The reasons for this behavior are related to the fracture toughness (K_{IC}) ratio, as indicated by less than 25% of the matrix and reinforcement. This ratio leads to an increase in crack nucleation and crack propagation mechanisms at the interfacial surface, as described in Ref. [32]. When comparing the K_{IC} values of the epoxy matrix (0.45 MPa m$^{1/2}$) and the Al reinforcement (23 MPa m$^{1/2}$), the fracture toughness ratio of the AL-EP material was computed to be 1.96%. It is attributed mainly to the ease of detachment of Al particles from the epoxy matrix, which may lead to

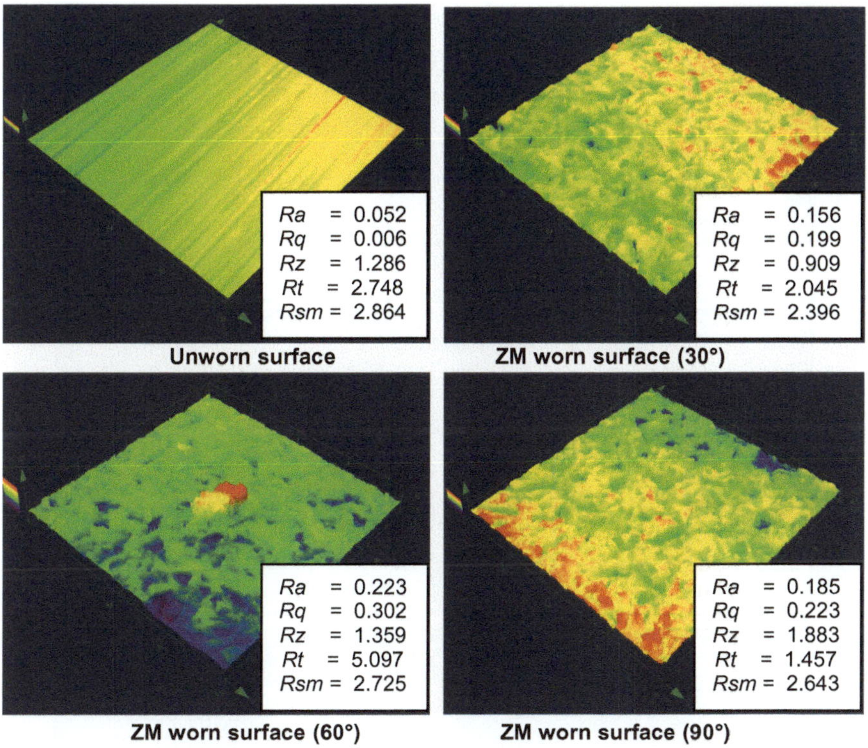

Fig. 10 LSCM observation of the eroded ZM surfaces.

increased fatigue mechanisms on the AL-EP-M. The surface cracks experienced during machining on the AL-EP-M [33] can also induce fatigue wear from repeated impacts of the erodent.

Fig. 11C shows the worn surface of the AL-EP-C. The failure at the normal impact was also found to be the microindentation crater wear. This wear behavior is observed mainly at the EP surface layer of the AL-EP-C (Fig. 9). The wear size of the indentation was found to be at least 5 μm; however, signs of microcracks due to fatigue were not observed. It might be noted that the slurry wedge is not to be penetrated due to the homogeneity of the EP surface layer.

Fig. 11D shows a maximum volumetric loss at 60 degrees of the impingement angle during the wear test of the ZM. The wear areas dominantly obtained were seen by microploughing wear rather than microcutting wear. This is attributed to a ductile failure at the higher inclined angle of impact combined with the effect of angular-shaped

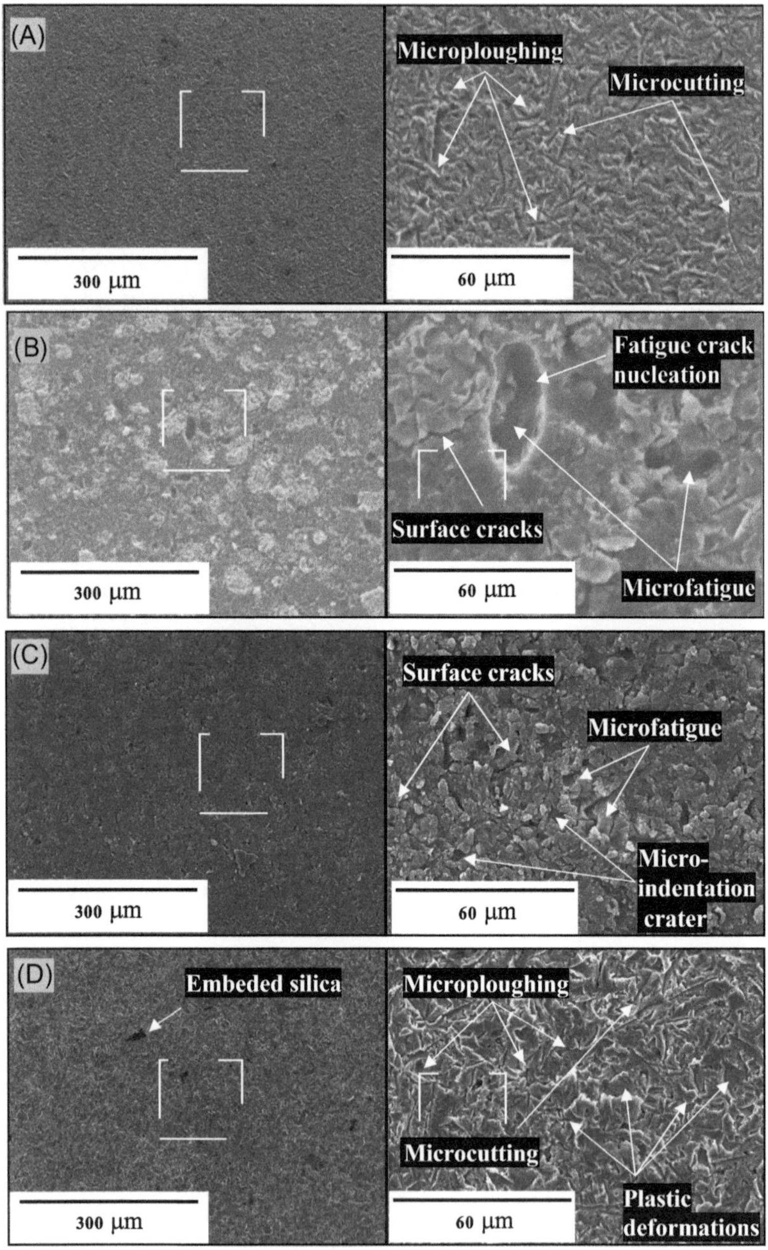

Fig. 11 SEM observation of the eroded surfaces for (A) PM material, (B) AL-EP-M material, (C) AL-EP-C material, and (D) ZM material.

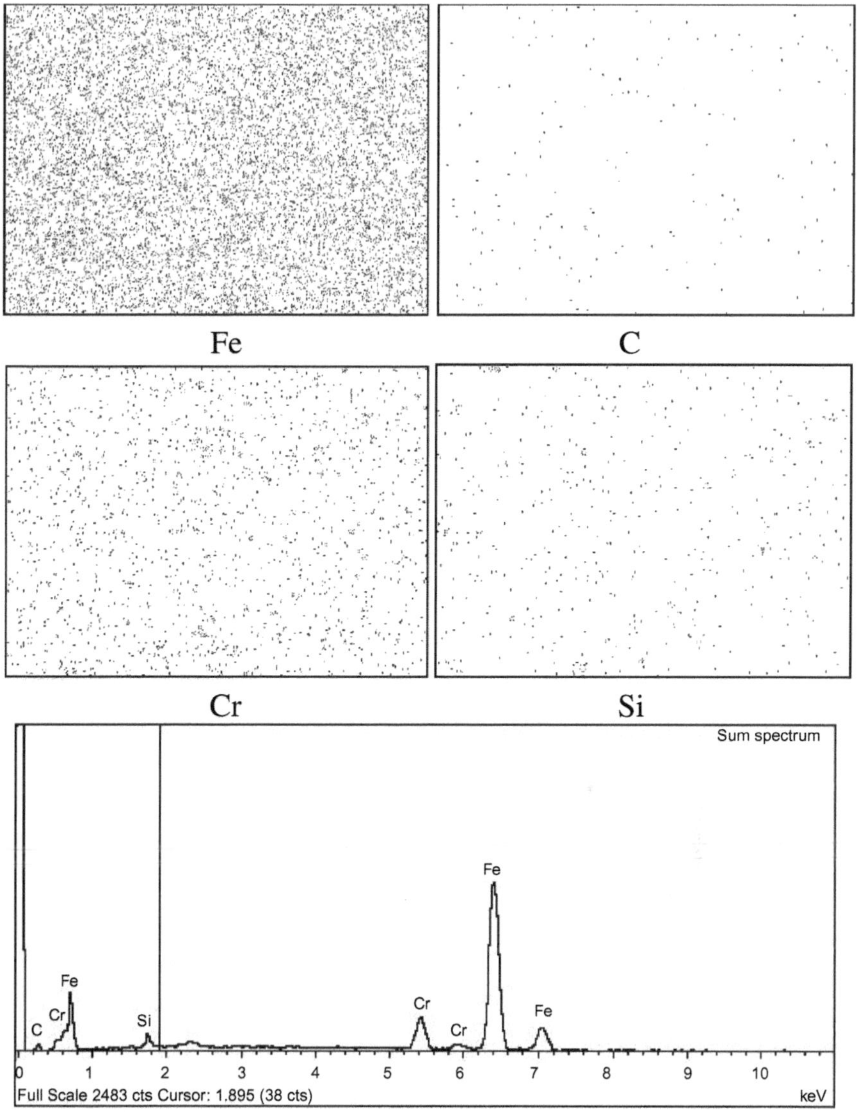

Fig. 12 EDS mapping and spectrum analysis of the PM material.

particles on the ploughing mechanism [23, 24]. The ZM worn surface was also characterized by plastic deformation of the soft metal coating.

In Figs. 12–15, the tested specimens were studied for changes in chemical compositions during wear tests, as seen by EDS. In Fig. 12, the EDS spectrum analysis of the PM was found to be 2.12%wt Si, in

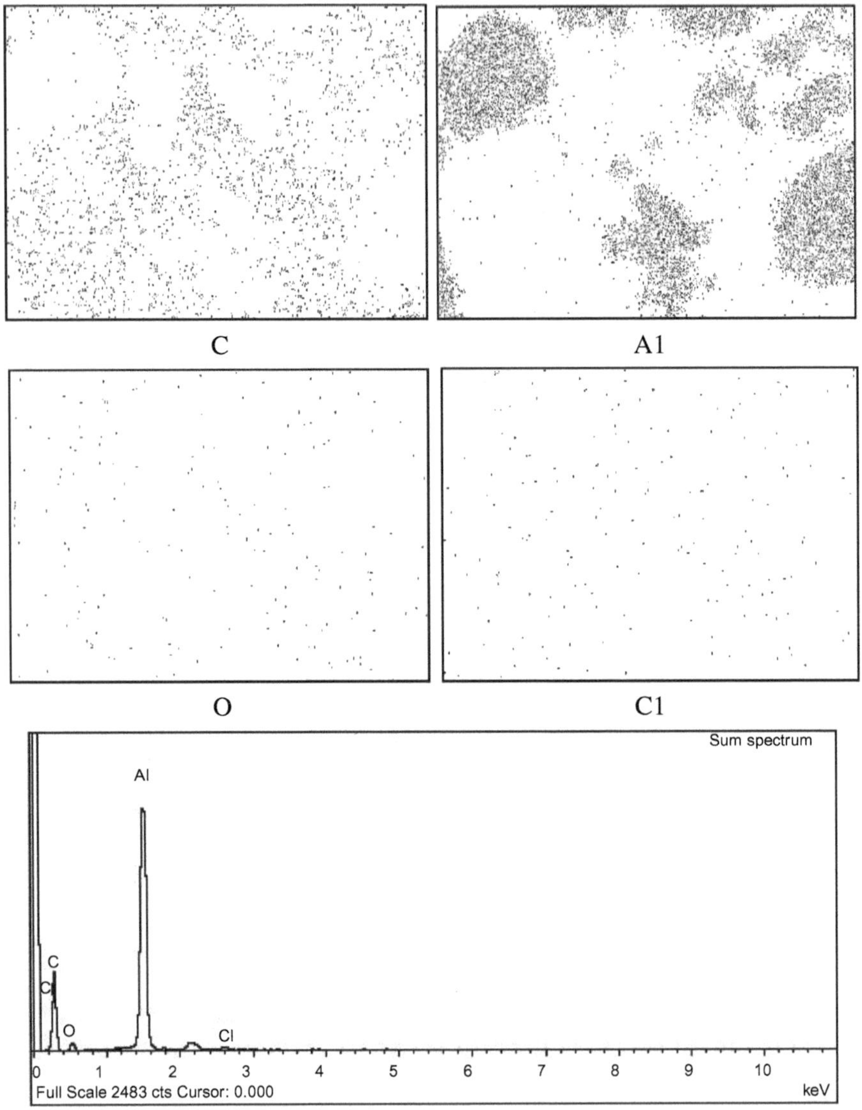

Fig. 13 EDS mapping and spectrum analysis of the AL-EP-M material.

comparison to 1%wt Si in the original PM material. Si is likely to be from the silica erodent, and the embedded silica promoted the reduction in wear rates.

The EDS mapping of the AL-EP-M is shown in Fig. 13, which represents the Al being surrounded by the epoxy matrix. The Cl content in the AL-EP-M was found to be ~0.36%wt, which implies that the AL-EP can

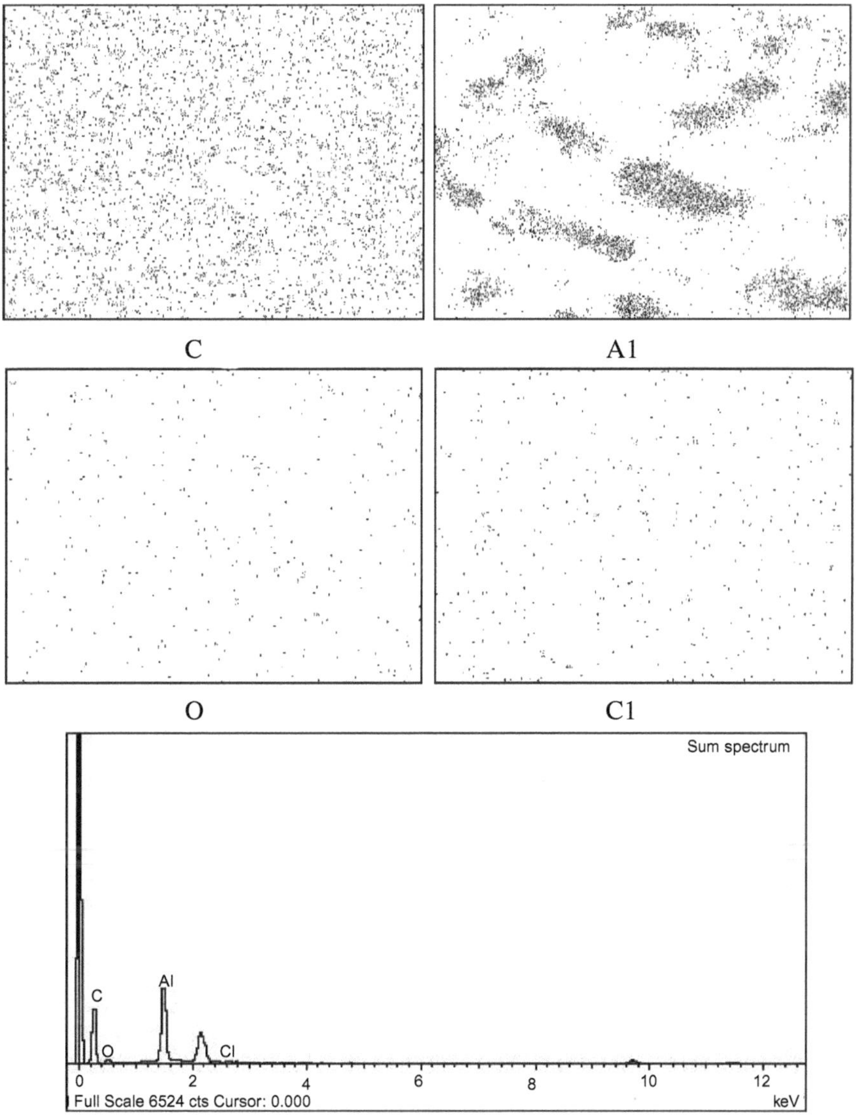

Fig. 14 EDS mapping and spectrum analysis of the AL-EP-C material.

absorb the liquid easier than the ZM (Fig. 15). If the Cl is predominating higher, it can also induce erosion-corrosion wear and polymer degradation. The EDS map information of the AL-EP-C is similar to that of the AL-EP-M, as shown in Fig. 14. In Fig. 15, the oxygen content in the ZM can promote the formation of a zinc oxide layer in the coating, which can reduce the wear rate of the Zn alloy coatings.

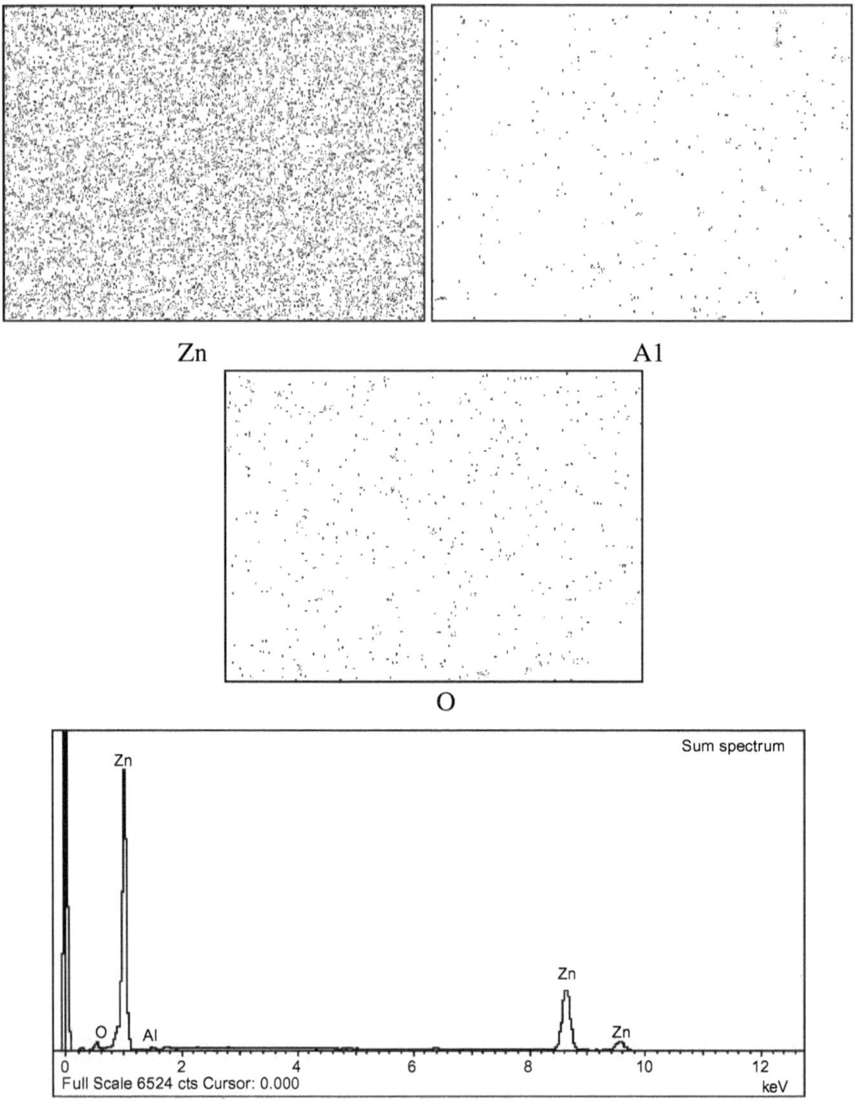

Fig. 15 EDS mapping and spectrum analysis of the ZM material.

3.4 Wear mechanisms

Impact wear is one of the main erosion mechanisms. In order to investigate the effects of the repeated impacts on erosion mechanisms, the schematic views were illustrated to describe the wear mechanisms, as shown in Fig. 16 (A) AL-EP-C, (B) AL-EP-M, and (C) ZM, respectively. All illustrations assumed the stagnation line of the repeated impacts close to the horizontal direction of the test.

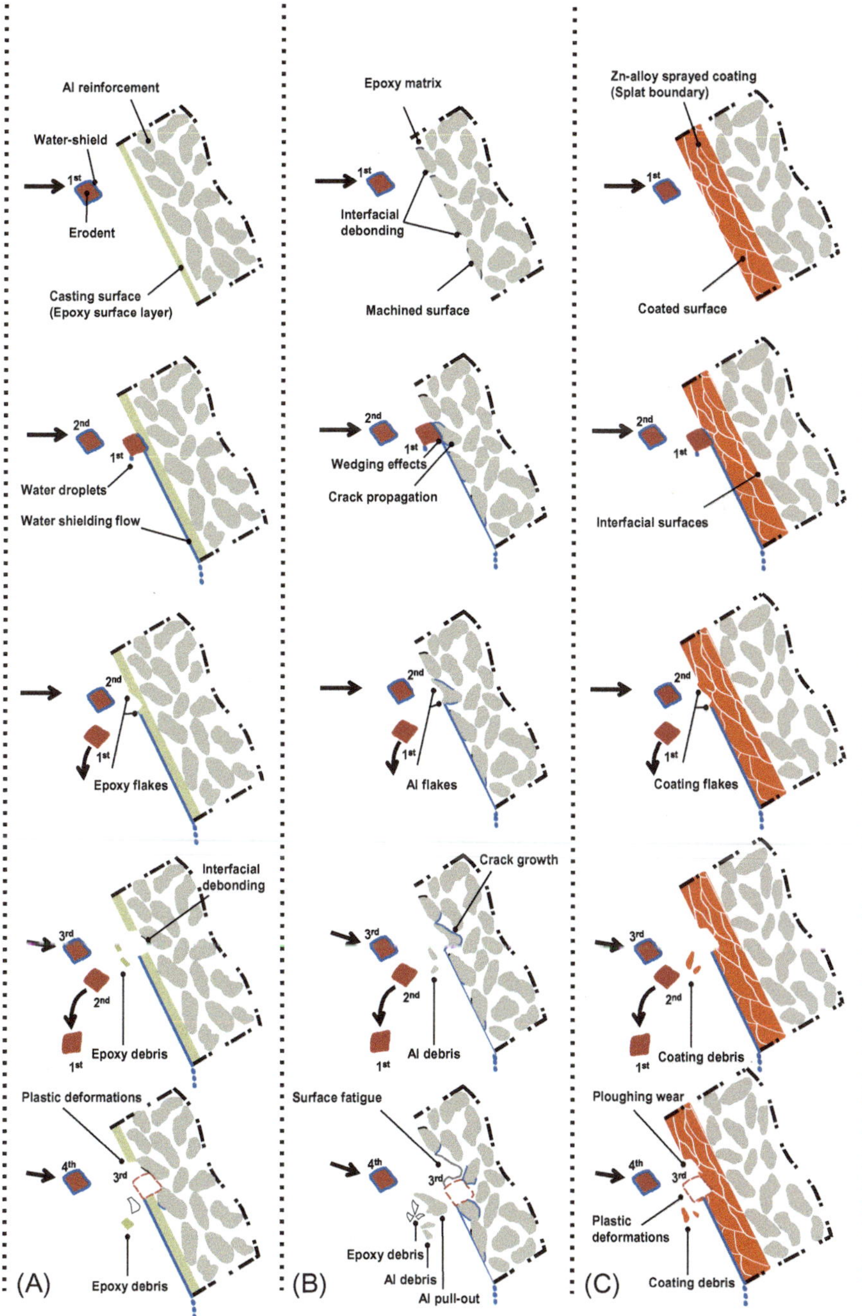

Fig. 16 Schematic views of wear mechanisms for (A) AL-EP-C material, (B) AL-EP-M material, and (C) ZM material.

Fig. 16A shows that the plastic deformation forms at the EP surface layer. Depending on the angle of impact, the homogenous structure of the epoxy resin is mainly affected by the microploughing wear or the microindentation crater wear.

Unlike a byproduct from the casting technique used in the AL-EP-C, the AL-EP-M was fracturing during machining and rising in crack nucleation from the interfacial debonding of the epoxy resin and the Al particles. The repeated impacts of the erodent consequently caused the crack propagation at the interface [13]. The crack growth was increased by the hydrostatic pressure due to the wedging effects of the slurry flow. As a result of fatigue, the Al pull-out occurred as wear debris (see Fig. 16B).

Fig. 16C shows the wear mechanisms of the ZM. As previously mentioned, the ductile Zn coating obtained low porosity; hence, the effect of fatigue is minimized due to the low defects. The ploughing wear predominantly occurred [24], as can be observed by the rim formation in the coating [34]. Moreover, some wear modes were obtained as the plastic deformation of the soft metal. In this failure mode, a bulge formation was readily formed in the soft metal coating where the sharp corner of the erodent penetrated or the embedding of silica inclusion occurred. Principally for the metallic materials (ZM and PM), it is to be noted that the effects of erodent rebound shielding and the water film of the slurry droplets significantly decreased the effect of the cutting wear at a low angle of impingement [24, 34, 35]. The metal coating thickness was also large enough to carry the loading by means of load support from the coating [36], so it can be assumed that the coating can resist the wear from higher impact loading.

In terms of all the wear mechanisms, it can be noted that the erosive wear mode in the AL-EP-M material rapidly occurred by the fatigue mechanisms of the brittle material. The machined surface defects also caused the interfacial debonding of the Al reinforcement and the epoxy resin matrix; hence, the low strength at interfaces led to increased crack nucleation and crack propagation. Particularly, the crack growth and the Al-reinforced particle pull-out are predominately performed by the fluid wedge under the penetration from the hydrostatic pressure into the cracks, and is gradually accelerated in the wear-out of the epoxy matrix. The AL-EP-C material was slightly increased in erosion rates. This is because the superior resistance to crack extension is enhanced from the homogeneity of the EP surface layer (the backfilled AL-EP casting technique). In addition, the damage of the ductile material is formed by the ploughing wear at the higher inclined angle of impact. The failure of the thermal-sprayed coating composite tool

(ZM material) is also caused by the plastic deformation due to the very soft properties of the Zn-based coating.

3.5 Erosion wear index results

In order to describe the effect of impingement angles during impacts on erosion wear properties, the conservation of kinetic energy was performed [24]. In Fig. 17, the erosion wear index was determined using Eq. (4), and the minimum wear index was observed at 30 degrees for all tested materials. The maximum wear index, both in AL-EP-C and AL-EP-M, exhibited similar behavior of brittle fracture at the normal impact (90 degrees), and had values of 6.068×10^{-5} and 8.762×10^{-5}, respectively. Meanwhile, the highest wear indices of both the PM and ZM were found in the higher inclined angle of impact (60 degrees), and had values of 2.505×10^{-7} and 1.697×10^{-5}, respectively. The higher erosion wear indices of impact can be interpreted as higher erosive wear. Also, the highest erosion wear index for each specimen can also be clarified that is readily considered to a brittle failure in AL-EP materials, and is responsible for a ductile failure in the coating material. The findings from the erosion testing can be used as the guideline for plastic injection molding design. For example, the excessive erosion at the normal impact from both the AL-EP-C and the AL-EP-M materials

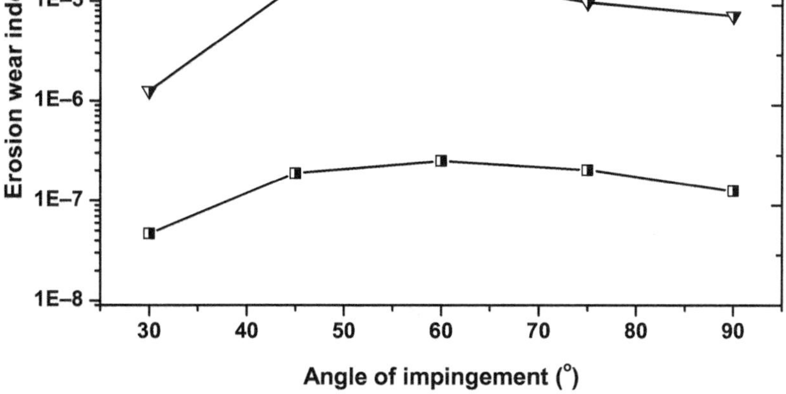

Fig. 17 The erosion wear index of tested materials.

and the increase of wear at the higher inclined angle of impact experienced in the ZM material should be neglected to arrange the runner of a melted polymer in the mold design.

4 Conclusions

Zn-alloy coating increased the erosion resistance of aluminum-filled epoxy composites while producing better surface properties, thereby achieving a low erosion rate ($4.44\times10^{-4}\,mm^3g^{-1}$) at the impingement of 60 degrees, an optimum worn surface roughness (0.223 μm *Ra*), and a low erosion wear index (1.697×10^{-5}). As a result of low wear rates, this coating was slightly removed under common wear modes of ploughing and plastic deformation due to a ductile failure behavior. When compared to the aluminum-filled epoxy composites, the surface cracking and microfatigue modes due to a brittle failure behavior were dominant to increase wear rates.

Acknowledgment

This work was financially supported by the Thailand Graduate Institute of Science and Technology, National Science and Technology Development Agency (the Contract No. TGIST-33-11-50-032D).

References

[1] E. Radstok, Rapid tooling, Rapid Prototyp. J. 5 (4) (1999) 164–168.
[2] C.K. Chua, K.H. Hong, S.L. Ho, Rapid tooling technology: part 1. A comparative study, Int. J. Adv. Manuf. Technol. 15 (1999) 604–608.
[3] S. Jetley, K.D. Low, A rapid tooling technique using a low melting point metal alloy for plastic injection molding, J. Ind. Technol. 22 (3) (2006) 2–8.
[4] C.M. Cheah, C.K. Chua, H.S. Ong, Rapid moulding using epoxy tooling resin, Int. J. Adv. Manuf. Technol. 20 (2002) 368–374.
[5] K. Altaf, A. Majdi Abdul Rani, V. Raghavan, Prototype production and experimental analysis for circular and profiled conformal cooling channels in aluminium filled epoxy injection mould tools, Rapid Prototyp. J. 19 (4) (2013) 220–229.
[6] S.I. Chung, Y.G. Im, H.D. Jeong, T. Nakagawa, The effects of metal filler on the characteristics of casting resin for semi-metallic soft tools, J. Mater. Process. Technol. 134 (2003) 26–34.
[7] S. Ma, I. Gibson, G. Balaji, Q.J. Hu, Development of epoxy matrix composites for rapid tooling applications, J. Mater. Process. Technol. 192-193 (2007) 75–82.
[8] A.J. Pontes, M.P. Queirós, P. Bártolo, A.S. Pouzada, A study on design and performance of hybrid moulds for injection moulding, in: K. Kuzman (Ed.), The 5th International Conference on Industrial Tools, Celje, Slovenia, April 12–15, 2005, pp. 143–146.
[9] A. Phuchamnong, J. Soparat, K. Ananon, P. Viriyaratanasak, Application of CAE technique in epoxy resin tooling fabrication, in: I. Preechawyttipong (Ed.), The 23rd Conference of Mechanical Engineering Network of Thailand, Chiang Mai Province, Thailand, November 4–7, 2009, pp. 1–9. (in Thai).

[10] N.P. Karapatis, J.-P.S. van Griethuysen, R. Glardon, Injection molds behavior and lifetime characterization, in: D. Bourell, J. Beaman, R. Crawford, S. Fish, C. Seepersad (Eds.), The Eighth Solid Freeform Fabrication (SFF) Symposium, August 11–13, Texas, USA, 1997, pp. 317–324.
[11] P. Engelmann, K. Hayden, P. Guichelaar, R. Dealey, M. Monfore, Injection mold wear mechanisms and mold design, Plast. Eng. 57 (2001) 40–50.
[12] P.V. Vasconcelos, F.J. Lino, A.M. Baptista, L.R.J. Neto, Tribological behaviour of epoxy based composites for rapid tooling, Wear 260 (2006) 30–39.
[13] N.-M. Barkoula, J. Karger-Kocsis, Review process and influencing parameters of the solid particle erosion of polymer and their composites, J. Mater. Sci. 37 (2002) 3807–3820.
[14] D.E. Crawmer, Thermal spray processes, in: J.R. Davis (Ed.), Handbook of Thermal Spray Technology, ASM International, Ohio, 2005, pp. 60–62.
[15] L. Pawlowski, The Science and Engineering of Thermal Spray Coatings, second ed., John Wiley & Sons, Ltd., West Sussex, 2008. (Chapter 3).
[16] J.I. Segal, R.C. Cobb, Optimizing arc-sprayed metal tooling for injection molding, in: G. Bennett (Ed.), The First National Conference on Rapid Prototyping and Tooling Research, November 6–7, Buckinghamshire, United Kingdom, 1995, pp. 189–199.
[17] C.K. Chua, K.H. Hong, S.L. Ho, Rapid tooling technology: part 2. A case study using arc spray metal tooling, Int. J. Adv. Manuf. Technol. 15 (1999) 609–614.
[18] I.D. Wimpenny, J.G. Gibbons, Metal spray invar tooling for composites, Aircr. Eng. Aerosp. Technol. 72 (5) (2000) 430–439.
[19] P.S. Grant, S.R. Duncan, A. Roche, C.F. Johnson, Scientific, technological, and economic aspects of rapid tooling by electric arc spray forming, J. Therm. Spray Technol. 15 (4) (2006) 796–801.
[20] H. Zhang, G. Wang, Y. Luo, T. Nakaga, Rapid hard tooling by plasma spraying for injection molding and sheet metal forming, Thin Solid Films 390 (2001) 7–12.
[21] H. Wu, B. Lu, C. Guo, R. Wei, Structural analysis on the arc spraying rapid tooling for sheet forming, Int. J. Adv. Manuf. Technol. 77 (2014) 2277–2287.
[22] B.K. Prasad, O.P. Modi, Slurry wear characteristics of zinc-based alloys: effects of sand content of slurry, silicon addition to alloy system and traversal distance, Trans. Nonferrous Met. Soc. Chin. 19 (2009) 277–286.
[23] O.P. Modi, B.K. Prasad, A.K. Jha, Influence of alumina dispersoid and test parameters on erosive wear behaviour of a casting zinc-aluminium alloy, Wear 260 (2006) 895–902.
[24] M. Abedini, H.M. Ghasemi, Erosion and erosion–corrosion of Al-brass alloy: effects of jet velocity, sand concentration and impingement angle on surface roughness, Trans. Nonferrous Met. Soc. Chin. 27 (11) (2017) 2371–2380.
[25] S. Kerdnoonwong, A. Wongto, C. Chungchoo, Effects of shape and grain size of aluminum powder on the mechanical strength of aluminum filled epoxy resin, in: S. Hankla (Ed.), The 21th Conference of Mechanical Engineering Network of Thailand, October 17–19, Chonburi Province, Thailand, 2007, pp. 132–142. (in Thai).
[26] P. Nunwong, C. Chungchoo, S. Sirivedin, P. Niranatlumpong, Parametric study and material behaviours of rapid mould making by thermally sprayed coating process, in: P. Rattanadecho (Ed.), The 22nd Conference of Mechanical Engineering Network of Thailand, October 15–17, Pathumthani Province, Thailand, 2008, pp. 87–92. (in Thai).
[27] ASTM D2734, Standard Test Methods for Void Content of Reinforced Plastics, ASTM Stand., (2009) 1–3.
[28] P. Nunthavarawong, Slurry erosion model for wear performance prediction of tool steel, Adv. Mater. Res. 717 (2013) 194–199.
[29] R.J.K. Wood, The sand erosion performance of coatings, Mater. Des. 20 (1999) 179–191.
[30] K.A. Jha, J. Mantry, S. Satapathy, A. Patnaik, Erosive wear performance analysis of jute-epoxy-SiC hybrid composites, J. Compos. Mater. 44 (13) (2010) 1623–1641.

[31] ISO 25178, Geometric Product Specifications (GPS)—Surface Texture: Areal—Part 1: Indication of Surface Texture, ISO Stand. (2016)1–25.
[32] Y.G. Lee, C.K.H. Dharan, R.O. Ritchie, A physically-based abrasive wear model for composite materials, Wear 252 (2000) 322–331.
[33] R.W. Lanz, S.N. Melkote, M.A. Kotnis, Machinability of rapid tooling composite board, J. Mater. Process. Technol. 127 (2002) 242–245.
[34] V. Javaheri, D. Porter, V.-T. Kuokkala, Slurry erosion of steel—review of tests, mechanisms and materials, Wear 408–409 (2018) 248–273.
[35] Y. Zhang, Y.-B. Cheng, S. Lathabai, Erosion of alumina ceramics by air- and water-suspended garnet particles, Wear 240 (2000) 40–51.
[36] K. Holmberg, H. Ronkainen, A. Matthews, Wear mechanisms of coated sliding surfaces, in: D. Dowson, C.M. Taylor, T.H.C. Childs, M. Godet, G. Dalmaz (Eds.), Thin Film in Tribology, Tribology Seriesvol. 25, Elsevier, Amsterdam, 1993, pp. 399–407.

CHAPTER 10

Tribology of thermosetting composites

Yunhai Ma[a,b], Yucheng Liu[a,b], and Jin Tong[a,b]
[a]Key Laboratory of Bionic Engineering (Ministry of Education, People's Republic of China), Jilin University, Changchun, People's Republic of China
[b]College of Biological and Agricultural Engineering, Jilin University, Changchun, People's Republic of China

1 Introduction

Thermosetting composites mainly include phenolic resin composites, unsaturated polyester resin composites, epoxy resin composites, polyimide resin composites, bismaleimide resin composites, cyanate ester resin composites, organic silicon resin composites, and melamine formaldehyde resin composites. In fact, phenolic resin composites are widely used in the automobile, electronics, aviation, aerospace, and national defense industries. In this chapter, we mainly present the tribological properties of the phenolic resin-based friction materials. Phenolic resin-based friction materials are mainly used in brakes, clutches, and friction driving devices of all kinds of transportation and mechanical equipment [1, 2]. The friction materials must achieve a particular set of desirable performances, including excellent wear resistance, stable friction coefficients, low fade, low cost, lack of noise or vibration, reliable strength, no environmental pollution, and no damage to the human body [3–5]. Because no single material can satisfy all the above requirements, friction materials are designed as multiphase composites, which include an excess of 10 ingredients [6]. The tribological and morphological properties of the friction materials are generally determined by these ingredients, which are classified into four categories: binder resins, reinforcements, property modifiers, and space fillers [7]. In the design of brake friction materials, both the composition and the manufacturing technique should be comprehensive considered.

Among the various ingredients used in friction materials, binder resins are often considered the heart to determine the various aspects of brake performance [8]. In fact, binder resins could bind the rest of the ingredients in the material matrix and maintain the integrality of the structure of the friction materials in the braking process. Phenolic resin, which is

Tribology of Polymer Composites
https://doi.org/10.1016/B978-0-12-819767-7.00010-4

synthesized through the reaction of phenol and formaldehyde in alkaline or acidic conditions, has commonly been used in friction materials as a binder for decades due to its low cost, good wetting capability, and good combination [2, 9]. However, phenolic resin also has various negative problems such as poor thermal resistance, high hardness, and high modulus. Moreover, the use of phenolic resin makes friction materials with excessive wear at elevated temperatures, high brittleness, high temperature fade, high noise, and other brake performance-related issues [10, 11]. In the last decade, considerable efforts have been extensively investigated by several researchers to look for a substitution for phenolic resin. Shin et al. [12] investigated the influence of different resins on the heat resistance, wear rate, and friction coefficient of the multiphase friction materials. They reported that the friction material containing boron phosphorous-modified phenolic resin had excellent friction stability, wear, and fade resistance at elevated temperatures in comparison to other binder resins. Kuroe et al. [13] prepared a lignin-modified phenolic resin and studied the friction properties of the friction materials. The study concluded that the modified resin could reduce the thermal fade of the friction material at elevated temperatures. Cai et al. [2] examined the influence of nitrile rubber and boron-modified phenolic resin on the thermal and tribological properties of the friction materials. The work revealed that a friction material with a 1:3 ratio of resins had preferable thermal resistance and a high friction coefficient under the conditions of different loads and speeds. Park et al. [14] aimed at replacing phenol resin and a cashew nut shell liquid with different types of lignin in friction material pads, and then improving the friction coefficient and wear resistance. However, nowadays, phenolic resin or modified phenolic resin is still the best choice as a binder to prepare commercial friction materials due to the problems of cost and other friction performances of the new resins.

Wet granulation, which is the process where powder particles gradually enlarge into larger permanent agglomerates, is widely used in many industrial sectors, such as pharmaceutical, detergent, zootechnical, fertilizer, and material manufacturing [15–17]. Wet granulation can make the material easier for handling and downstream processing and the fiber reinforcements are evenly distributed in the material matrix, reducing the abrasion of the mating surface, decreasing the powder content, and increasing the porosity in the final product [15, 18]. In wet granulation, a liquid phase as the "binder" is sprayed into the powder in a tumbling granulator. In fact, the types, viscosity, volume flow rate, add method, and liquid-solid surface tension determine the properties of the final granule and product [16].

After an extensive literature review, limited studies were found on the influence of wet granulation technology on the tribological and morphological properties of the phenolic resin-based friction materials. In this chapter, the brake friction materials were prepared by wet granulation, and then the influences of the final granule size and phenolic resin content on the tribological and morphological properties of the friction materials based on wet granulation were systematically investigated. Moreover, the effects of the granule size and resin content on the wear mechanism were also discussed based on the scanning electron microscope (SEM) and laser scanning confocal microscopy (LSCM) images.

2 Experimental work

2.1 Materials and sample preparation

2.1.1 Materials

Experiment 1The friction materials consisted of a 13 wt% binder (phenolic resin), a 27 wt% reinforcing ingredient (compound mineral fiber), and 60 wt% other fillers (alumina, iron powder, antimony sulfide, petroleum coke, artificial graphite, friction powder, vermiculite foam, precipitated barium sulfate, and hard zinc sulfate). Distilled water was selected as the liquid phase (binder) to be used in wet granulation in Experiment 1.
Experiment 2The friction materials were mainly composed of NBR-modified phenolic resin, $BaSO_4$, CF, abaca fiber, Al_2O_3, graphite, shale, purple sands, porous iron powder, tire powder, water granulated slag, vermiculite, Sb_2S_3, husk powder, and frictional powder. Among the above ingredients, CF, abaca fiber, and husk powder were natural raw materials, and the water granulated slag and tire powder belong to industrial waste. The composition of the friction materials is presented in Table 1. Corn starch paste was selected as the liquid phase (binder) to be used in wet granulation and a 5 wt%

Table 1 The ingredients of the brake friction materials (wt%) [19].

	Ref	FG-13	FG-11	FG-9	FG-7	FG-5
Phenolic resin	11	13	11	9	7	5
$BaSO_4$	11	9	11	13	15	17
Other components[a]	78	78	78	78	78	78

[a] Other components included fibrous ingredients (CF and abaca fiber): 15 wt%, lubricant (graphite): 8 wt% and the rest (Al_2O_3, shale, purple sands, porous iron powder, tire powder, water granulated slag, vermiculite, husk powder, frictional powder): 55 wt%.

(wt/wt) corn starch paste was prepared by dissolving 21 g of corn starch powder in 399 g of 85°C distilled water that needed to be stirred constantly with uniform distribution.

2.1.2 Sample preparation

The granules were produced by using a JF805R typed laboratory tumbling granulator (Wanda Machinery, China). The experimental steps involved in wet granulation that consisted of mixing, spraying of the binder, drying of the wet granular mass, and screening are shown in Fig. 1. The wet granular mass was dried using a drying oven and then was screened using a grading sieve. The granules were poured into a special mold and then were pressed using a JFY50 typed compression molding machine for 30 min at 160°C and a normal pressure of 40 MPa. The obtained sample blocks should undergo heat treatment after hot pressing in order to cure resin and remove the residual stress of the composites. The heat treatment process is as shown in Fig. 2. Friction material samples with standard sizes of 25 mm × 25 mm × 6 mm were prepared for the following tests.

In Experiment 1, the friction material samples with granule sizes of 1–3 mm, 3–5 mm, 5–8 mm, 8–10 mm, and 1–10 mm were labeled as FG13, FG35, FG58, FG810, and FG110, respectively. In Experiment 2, according to the test results of Experiment 1, the granule size of 1–10 mm was selected to prepare the friction materials. The friction material samples containing 5–13 wt% phenolic resin content were labeled as FG-13, FG-11, FG-9, FG-7, and FG-5, respectively. The first letter "F" refers to "friction material," the second letter "G" represents "granulation," and the number is the granule size range in Experiment 1 and the resin content in Experiment 2. For

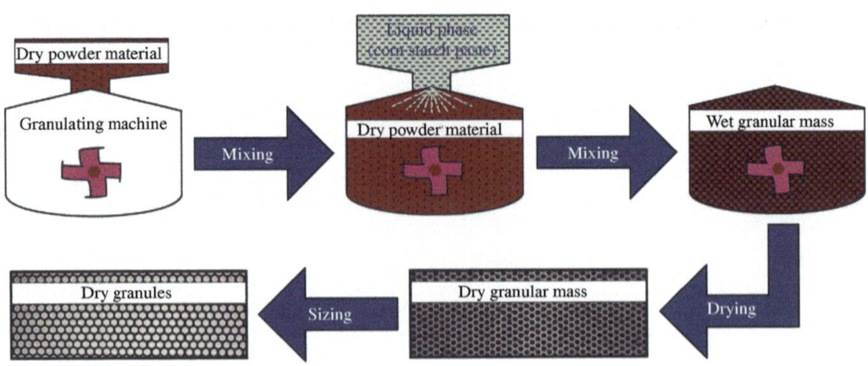

Fig. 1 Schematic diagram of wet granulation.

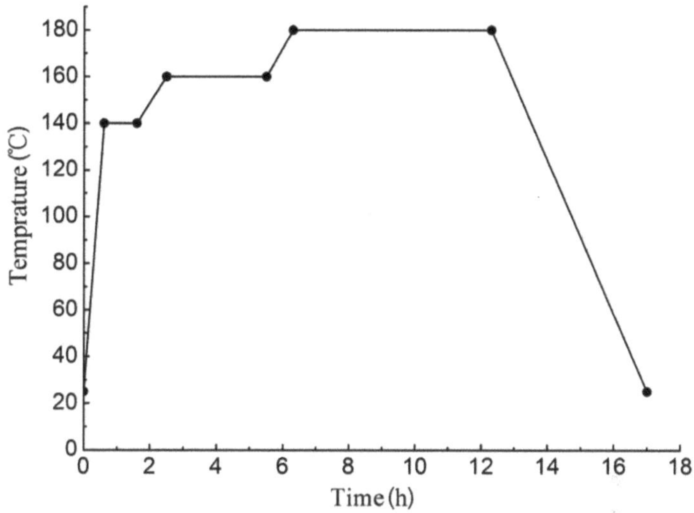

Fig. 2 Heat treatment of the brake friction materials.

comparison, the friction material sample prepared directly from the powder was used as the reference (FG in Experiment 1 and Ref in Experiment 2).

2.2 Tribological property evaluation

The friction-wear tests were performed using a JF150D-II typed constant speed tester (Wanda Machinery, China). The testing machine schematic is illustrated in Fig. 3. The rotating disc with HB 180–HB 220 hardness was chosen as the counter pair. It kept a constant speed of 480 rpm during the tests, and was driven with an electric motor. The rotating disc was heated

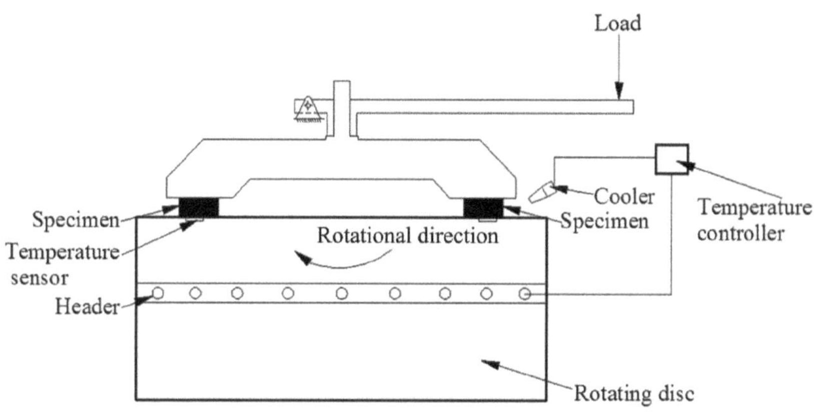

Fig. 3 Brief drawing of the friction-wear test machine.

using a thermoelectric couple and the decrease in disc temperature mainly depended on the cooling water. The temperature was monitored using a temperature sensor and was controlled by using a temperature controller. The normal pressure on the friction material samples always remained at 0.98 MPa. Before each test, all testing samples should be polished with abrasive papers at room temperature. At the fade test, the rotating disc was rotated for 5000 revolutions, subsequently increasing the constant temperature of 100°C, 150°C, 200°C, 250°C, 300°C, and 350°C. At the recovery test, the rotating disc was rotated for 1500 revolutions, subsequently decreasing the constant temperature of 300°C, 250°C, 200°C, 150°C, and 100°C. The friction coefficient and wear rate were defined as follows [20–22]:

$$\mu = \frac{f}{P}, \tag{1}$$

$$\Delta V = \frac{1}{2\pi r}\frac{A}{n}\frac{d_1 - d_2}{f}, \tag{2}$$

where f is the mean force of friction (N); P is the normal pressure (N); r is the center-to-center distance between the rotation and the friction material sample (r = 150 mm); n is the revolutions (n = 5000); A is the superficial area of each sample (A = 625 mm^2); and d_1–d_2 are the thickness differences of the friction material samples before and after the experiments (cm), respectively.

2.3 Morphological property evaluation

The worn surfaces of the tested friction material samples were inspected by a TESCAN VEGA3 model SEM (TESCAN, Czech Republic) with an Oxford INCAx-sight model energy dispersive X-ray spectroscopy machine (EDX, Oxford Instruments, United Kingdom) at 20 kV. Moreover, the worn surfaces of the tested samples were measured by using an OSL3000 model 3D LSCM (Olympus Company, Japan). The worn surface roughness of Ref, FG-13, FG-11, FG-9, FG-7, and FG-5 were measured using a hommel roughness tester (MarSurf LD120, Mahr, German).

3 Experimental results and discussion

3.1 Effect of final granule size on the tribological and morphological properties

3.1.1 Wear properties of friction materials

The friction coefficients of the samples FG, FG13, FG35, FG58, FG810, and FG110 during the fade and recovery tests are presented in Fig. 4 [15].

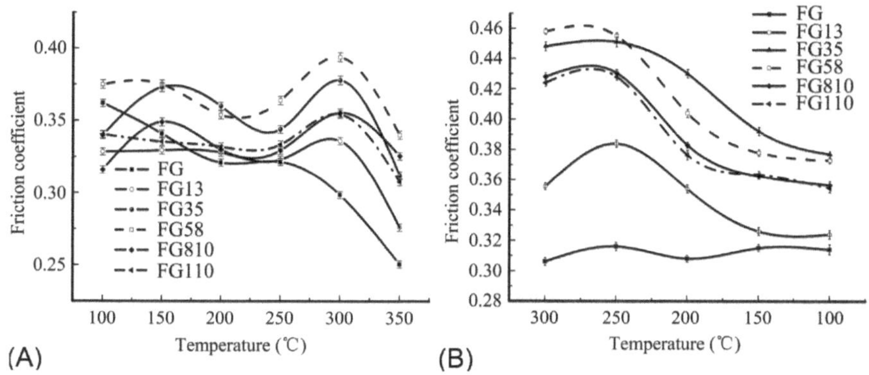

Fig. 4 Friction coefficient of samples FG, FG13, FG35, FG58, FG810, and FG110. (A) In fade tests, and (B) in recovery tests.

Fig. 4A shows that the friction coefficient of sample FG prepared directly from the powder was decreased with the increasing of the test temperature during the fade test. The friction coefficient significantly decreased, especially for the test temperatures higher than 250°C, which presents serious heat fading. The appearance of heat fading was attributed to the thermal decomposition of the organic components (e.g., phenolic resins and fibers) at elevated temperature, which was responsible for weakening the fiber-matrix bond.

Interestingly, the behavior of the other samples was quite different. Specifically, the samples FG13, FG35, FG58, and FG810 with a homogeneous size of the granule presented an increasing of the friction coefficient between the test temperatures of 100°C and 150°C. The friction coefficient of these samples decreased from 150°C to 200°C and then increased from 200°C to 300°C. A drastic decreasing of friction performance was recorded when the test temperature was further increased (Fig. 4A).

The improved friction performance of the samples obtained by granulation with respect to those obtained directly from the powder (sample FG) was attributed to an improved thermal stability of the resins. Sample FG110 obtained by the granule size of 1–10 mm had hybrid behavior. In fact, the friction coefficient of sample FG110 slightly dropped from 100°C to 250°C, and then increased between 250°C and 300°C.

Fig. 4B shows that the friction coefficient of the samples obtained by granulation decreased with the decreasing of the test temperature during the recovery test. In fact, the friction coefficient of these samples (FG13, FG35, FG58, FG810, FG110) was always higher than that of the sample prepared from the powder (sample FG).

The wear rates of samples FG, FG13, FG35, FG58, FG810, and FG110 are presented in Fig. 5 [15]. As can be seen from Fig. 5, the wear rate of the friction materials was greatly affected by the experimental temperature, and the wear rate of all friction materials was increased with the increasing of the experimental temperature. This phenomenon is in agreement with the research results of Öztürk et al. [23]. The wear rate of sample FG was noticeably higher than that of other samples obtained by granulation, especially at a high elevated temperature. In fact, the application of granulation could yield a decreasing of the wear rate from 8.62% to 61.27%. Sample FG810 had the best wear-resistance property in terms of the lowest wear rate, followed by sample FG35. The results of the wear rates indicated that the friction materials prepared with granulation have an excellent wear-resistance property.

3.1.2 Worn surface analysis

The worn morphologies such as primary and secondary plateaus, wear debris, adhesive pit, and grooves can provide useful information to investigate the worn mechanism of the friction materials [24–29]. The worn surface of the tested friction materials was observed by SEM. The typical SEM micrographs of the worn surfaces of the tested friction materials are shown in Fig. 6 [15].

Fig. 6A shows the worn surface of sample FG. It can be seen that the worn surface is very rough, and the material matrix is also very loose and not well blended together. Large flake debris (see Fig. 6A) was detached from the material matrix and probably constituted the main cause of adhesive wear. Moreover, the worn surface in Fig. 6A also provide the evidence of voids and bare fibers, which indicated that sample FG has a high wear

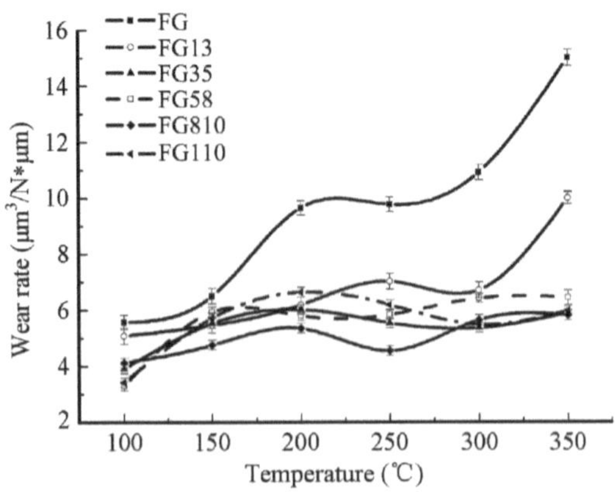

Fig. 5 Wear rate of samples FG, FG13, FG35, FG58, FG810, and FG110.

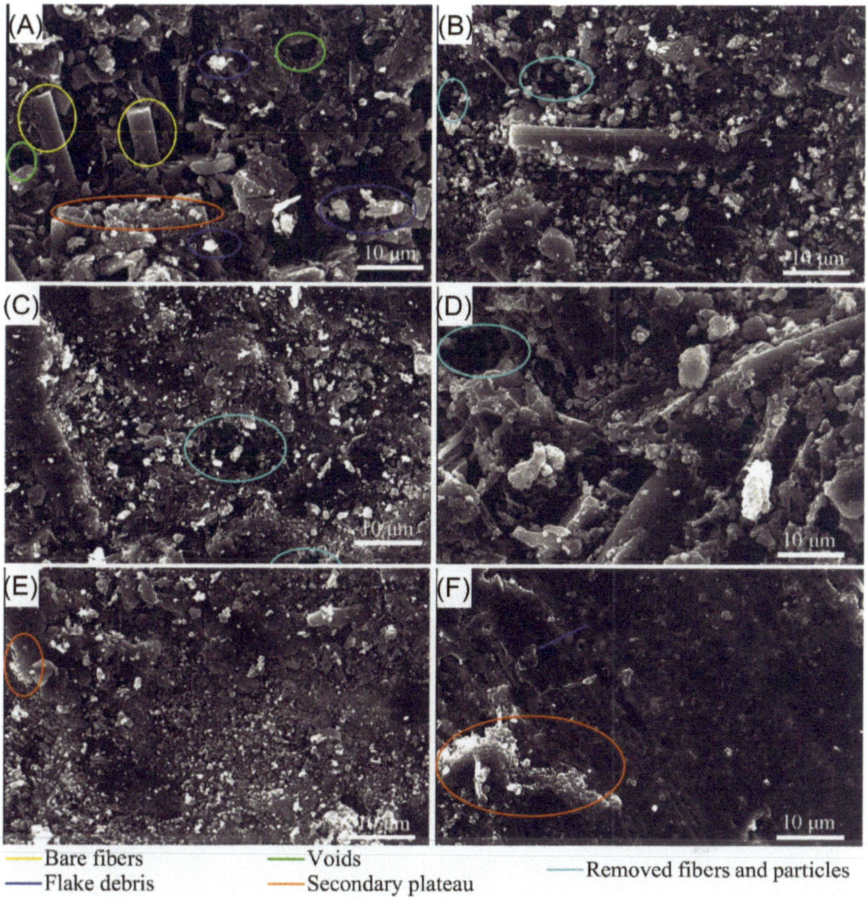

Fig. 6 SEM micrographs of friction material after friction and wear tests. (A) FG; (B) FG13; (C) FG35; (D) FG58; (E) FG810; and (F) FG110.

rate and the worst wear-resisting property. The worn surface of samples FG13, FG35, and FG58 are presented in Fig. 6B–C. These worn surfaces were damaged. In fact, the elements of the material matrix (such as fibers and particles) were removed and then some wear debris was presented in the worn surface, which shows a very heterogeneous and rough worn surface. Fig. 6E shows the surface morphology of sample FG810. Its worn surface was very smooth, and a few secondary plateaus were also presented on the worn surface. In fact, the matrix material in the friction materials was well blended together and very little wear debris was presented on the worn surface of the friction materials, which reflects that sample FG810 yielded the highest wear resistance behavior. Fig. 6F shows the worn surface of sample FG110. It can be seen from Fig. 6F that some medium-sized wear debris and thick secondary plateaus were presented on the worn matrix surface. However, its worn surface is very smooth in comparison to sample FG110, and few bare fibers and wear debris were presented on the worn surface. Kumar et al. [30] also concluded that the formation of secondary contact patches could improve the stability of the friction coefficient and decrease the wear rate of the friction materials.

A three-dimensional (3D) profile of the worn surfaces of samples FG, FG13, FG35, FG58, FG810, and FG110 is shown in Fig. 7; the worn surface roughness is presented in Table 2. In fact, the roughness of the sample worn surface is an important parameter that could quantitatively characterize the tribological properties of the friction materials [31]. It can be seen that with the increasing of the particle size in the samples prepared with granulation, the surface 3D morphology became more and more rough (Fig. 7) and the

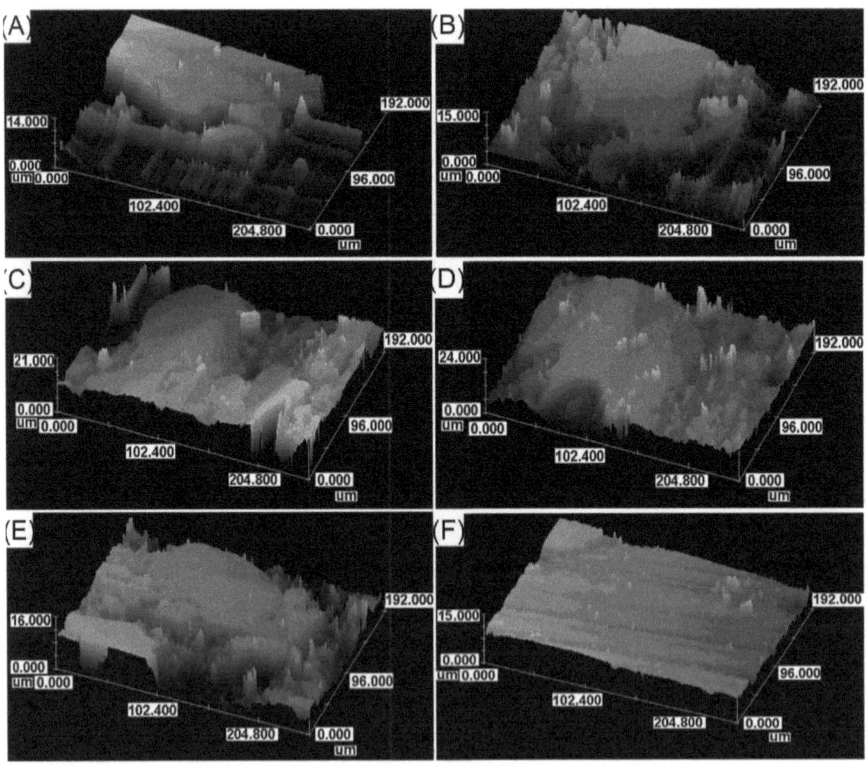

Fig. 7 Three-dimensional profile of friction material after friction and wear tests. (A) FG; (B) FG13; (C) FG35; (D) FG58; (E) FG810; and (F) FG110.

Table 2 Worn surface roughness of samples FG, FG13, FG35, FG58, FG810, and FG110.

Samples	**FG**	**FG13**	**FG35**	**FG58**	**FG810**	**FG110**
Ra (μm)	1.252 ± 0.015	1.679 ± 0.24	1.337 ± 0.018	1.346 ± 0.015	0.701 ± 0.011	0.478 ± 0.009

worn surface roughness (Ra) was also increased. Sample FG prepared without granulation showed a moderate rough worn surface. Sample FG110 had the very lowest worn surface roughness.

3.2 Effect of phenolic resin content on the tribological and morphological properties

3.2.1 Friction performance analysis

The friction coefficient of the brake friction materials relates directly to the stability and reliability in the braking process of the motor vehicle [32]. A friction test was performed at a constant speed of 480 rpm to evaluate the influence of the decreasing resin content on the change in the friction coefficient. Fig. 8 shows the friction coefficient of samples Ref, FG-13, FG-11, FG-9, FG-7, and FG-5 in each temperature at the fade and recovery tests. In general, the friction coefficient of the friction materials was generally impacted by the test temperature, the composition of the composite material, the normal pressure, and the rotating speed [7]. In the present study, the normal pressure and speed were kept constant during the above tests, and the effects of the test temperature and composition on the friction performance were systematically investigated. In fact, the test temperature remarkably affects the adhesion and the deformation resistance of the composite materials [33]. The component of the composite material determines the combination reaction of the raw material and its microstructure, which eventually leads to the change in the friction coefficient during the friction test.

In Fig. 8A, during the fade test, the friction coefficient of the friction material samples was increased and then decreased rapidly by increasing the test temperature from 100°C to 350°C. The increasing of the friction

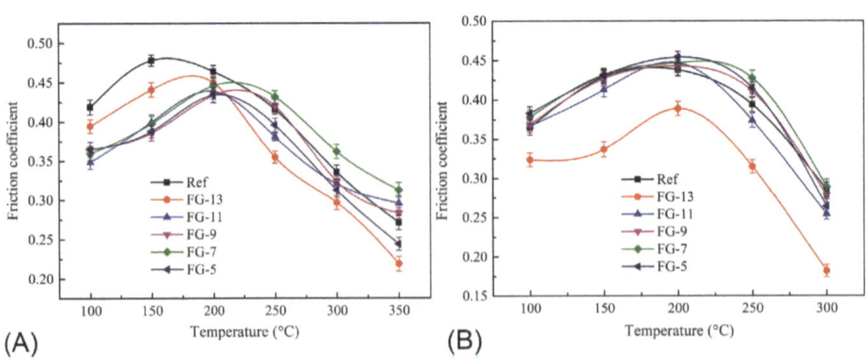

Fig. 8 Friction coefficient at fade test (A) and at recovery test (B).

coefficient was ascribed to the incomplete matching of the sample and the disc, and the glass transition of the phenolic resin in the matrix of the friction material [34]. The decreasing of the friction coefficient was due to the organic component thermal decomposition at elevated temperatures [35]. The organic components mainly included phenolic resin, CF, abaca fiber, and husk powder in the present work. Interestingly, the friction coefficient of sample FG-13 was relatively low and that of sample FG-7 was relatively high in comparison to other samples when the temperature exceeded 250°C. A possible explanation is that a relatively high resin content in the matrix of sample FG-13 led to the higher fade resistance. Moreover, compared to the sample Ref, the friction coefficient of samples FG-11, FG-9, and FG-7 tended to be stable at elevated temperatures. It indicated that the application of wet granulation could increase the friction coefficient and fade resistance of the friction materials due to increasing the glass-rubber transition temperature and the better positive hybrid effect of the resin and corn starch paste binder at high temperatures. As concluded by Natarajan et al. [36], the higher friction coefficient is important to the commercial brake friction materials. Overall, the friction coefficient of the tested friction materials met the requirements in the GB 5763-2008.

In Fig. 8B, during the recovery test, the friction coefficient of the friction material samples was increased with decreasing the test temperature from 300°C to 200°C and then was decreased rapidly with decreasing the test temperature from 200°C to 100°C. The cause of the increasing of the friction coefficient at high test temperatures was that some wear debris acted as the third body, damaged the disk and composite matrix [37]. Moreover, the decreasing of the friction coefficient at a lower temperature was attributed to the change in the rheology of the wear debris between the friction interface and the topography of the contact area [38]. It should be pointed out that the friction coefficient for sample FG-13 was always lower than that for samples with lower resin content, and sample FG-13 presented the worst friction properties with the lowest friction coefficients of 0.182–0.389. It indicated that sample FG-13 does not satisfy the requirement of GB 5763-2008 and its formulation cannot be applied to produce commercial brake friction materials.

3.2.2 Wear performance analysis

Wear resistance determines the working life of the friction materials during service; it is determined by the ingredients and manufacturing technique [39]. In the present study, the wear resistance of the samples with different

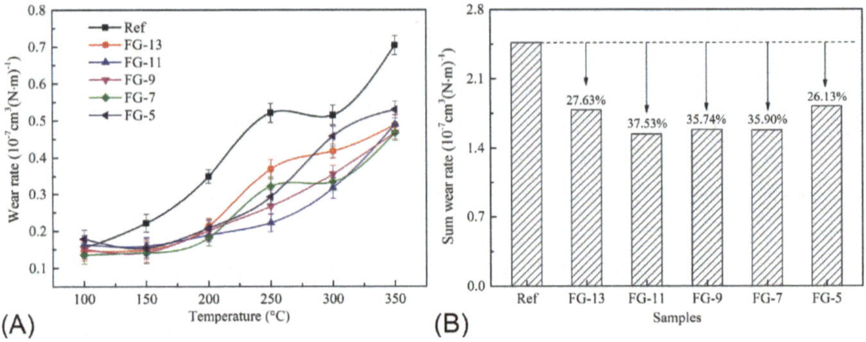

Fig. 9 (A) Wear rate and (B) sum wear rate of the friction materials.

resin contents was compared by measuring the wear rate and the sum wear rate. The wear tests demonstrated that the wear rate of the samples was remarkably affected by the manufacturing technique and resin content; moreover, it was increased with increasing the test temperature (Fig. 9A). It is interesting to note that the wear rate of the friction materials prepared by wet granulation was relatively lower than that of sample Ref prepared directly from the powder. The results of the sum wear rate presented in Fig. 9B also indicated that the friction materials prepared by wet granulation had a noticeably lower wear rate, and its sum wear rates were reduced by 26.13%–37.53% in comparison to that of the sample Ref. Furthermore, among the friction materials prepared by wet granulation, the samples FG-11, FG-9, and FG-7 had the best performance in terms of lowest wear rates, and samples FG-13 and FG-5 gave the highest sum wear rates. In other words, wet granulation applied to prepare the friction materials could improve the wear rate and wear resistance.

In conclusion, improvements in the wear resistance of the friction materials prepared by wet granulation were attributed to the formation stage, size, liquid phase, and particle distribution. Wet granulation is a complicated process that involves a series of simultaneous physical changes: nucleation, agglomeration, and breakage [40]. This increased the internal compaction degree of the granules and promoted the formation of a hardened hull with a high-density liquid phase binder/powder layer on the surface of the granules [41]. Additionally, along with the gradual formation of large granules during wet granulation, the interparticle cohesion forces such as surface tension, capillary pressure, and viscous forces between the powder particles were increased [42]. This decreased the particle-particle segregation [43], which changed the bonding form between the powder particles and also changed the microstructure of the granules in the friction materials. On the other hand, a certain amount of corn starch paste as a binder is required to promote the formation of a liquid bridge between the fine particles and the high cohesion of the granules [44]. Furthermore, the appropriate phenolic resin content in the friction materials could also effectively improve the interface bonding strength between the granules during the braking process. The better positive hybrid effect of the corn starch paste binder and resin not only provided highly adhesive strength for the thermo-forming of the friction material, but also decreased the resin content.

3.2.3 Worn surface analysis

The worn morphologies such as primary and secondary plateaus, wear debris, adhesive pit, and grooves can be applied to analyze the worn mechanism that governed the wear during the performed tests [29, 45, 46]. Thus, the worn surfaces of the tested friction materials were observed by SEM observation and 3D profiles were also reconstructed. The typical SEM micrographs and 3D profiles of the worn surfaces of the tested samples are shown in Figs. 10 and 11, respectively.

As observed in Fig. 10A, a large amount of wear debris as well as a larger adhesive pit could apparently be found, and grooves and contact plateaus also appeared on the worn surface of the sample Ref. Moreover, the 3D profile presented a relatively worse and rougher worn surface than other samples. In fact, the manufacturing technique determines the wear resistance of the friction materials and then affects its worn mechanism [39]. The traditional manufacturing technique applied in the fabrication of sample Ref could not make powder materials blend well together and also could not change the internal structure of the friction materials. Thus, plenty of wear debris was easily produced on the worn surface, and the hard asperities in the wear debris embedded the friction surface. Then, certain parallel shallow grooves presented, indicating the abrasive wear as the wear mechanism of the sample Ref. Meanwhile, a larger adhesive pit formed on the friction surface under the shearing force and probably constituted the main cause of adhesive wear. Furthermore, primary plateaus and secondary plateaus appeared on the worn surface. These plateaus were not bonded well to the friction material matrix, which then could not improve the wear resistance of sample Ref. The main wear mechanisms of sample Ref are abrasive wear and adhesive wear. The severe abrasive wear mechanism and the adhesive wear mechanism along with weak secondary plateaus in the sample Ref are introduced as the main reasons for the higher wear rate and worse wear resistance (Fig. 9).

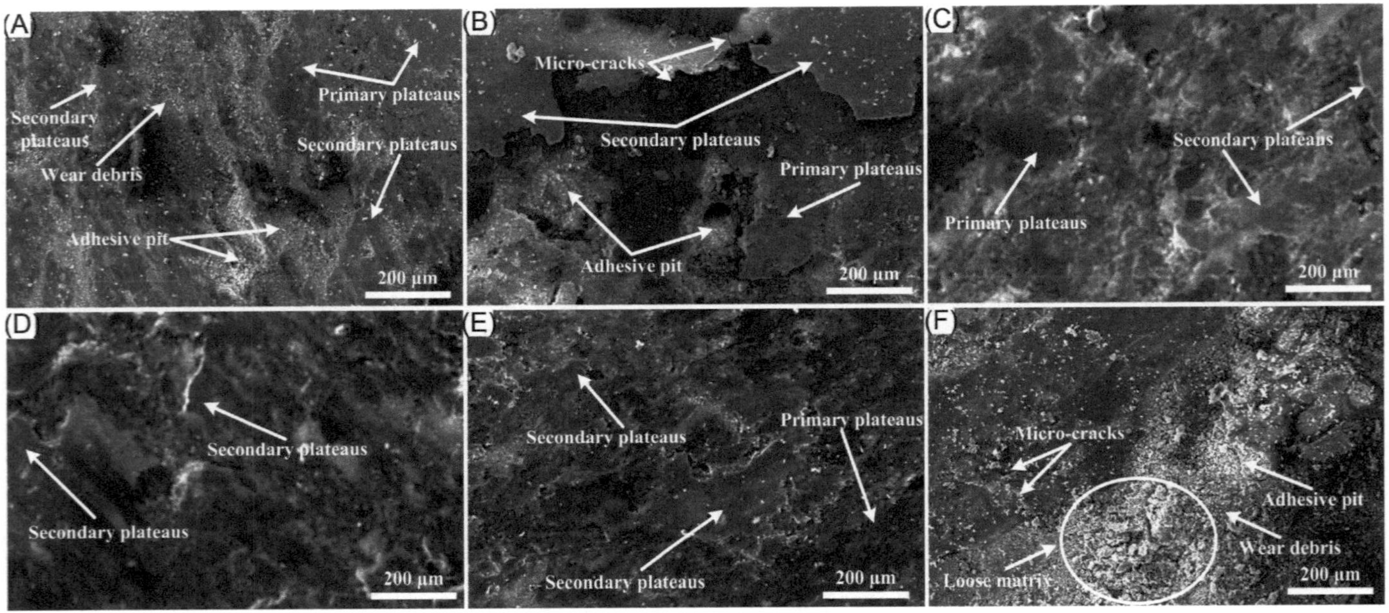

Fig. 10 SEM micrographs of friction material worn surfaces. (A) Ref; (B) FG-13; (C) FG-11; (D) FG-9; (E) FG-7; and (F) FG-5.

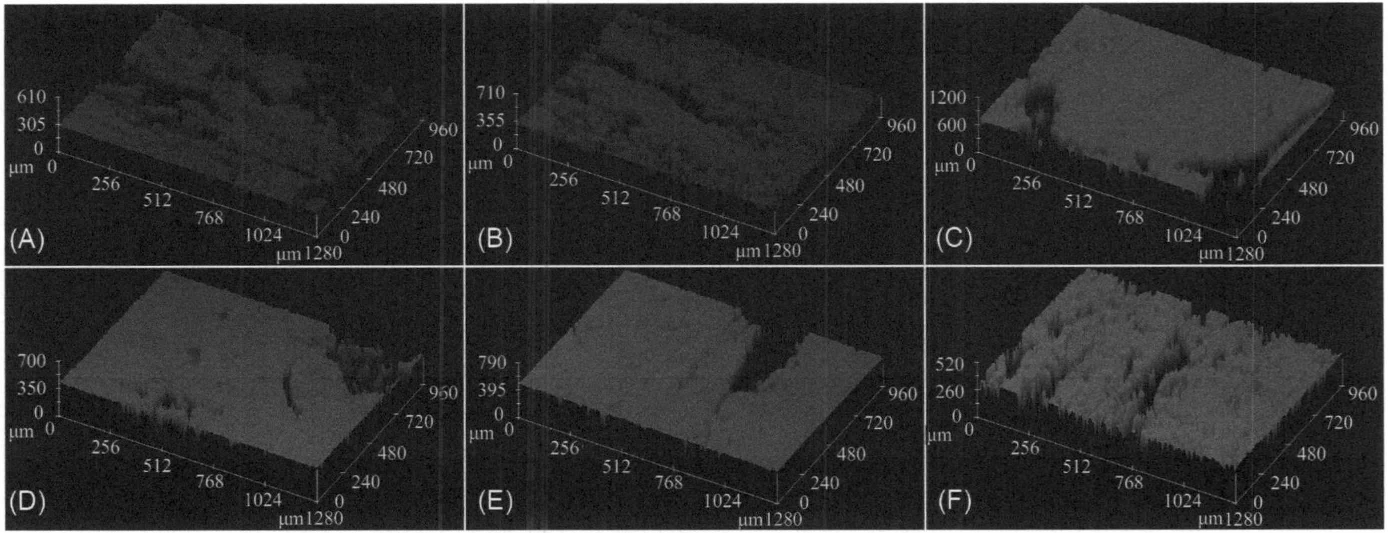

Fig. 11 Three-dimensional profile of friction material worn surfaces. (A) Ref; (B) FG-13; (C) FG-11; (D) FG-9; (E) FG-7; and (F) FG-5.

Figs. 10B and 11B are the worn surface micrograph and 3D profile of sample FG-13, which presented a relatively rougher worn surface. Large flake debris detached from the friction surface and then a larger adhesive pit formed on the friction surface. This was mainly due to the fact that too much resin in the friction material could easily soften the matrix at elevated test temperatures. Large pieces of uncompacted contact plateaus were also presented on the surface, which could be easily peeled off from the worn surface under the action of friction and shear forces. Therefore, the plateaus could not play a protective role for the friction surface. Moreover, microcracks also could be clearly observed. The main reason for microcrack formation is the action of unstable forces and temperature fields, showing typical fatigue wear characteristics. The adhesive wear and fatigue wear are the main wear mechanisms of specimen FG-13.

Fig. 10C–E correspond to samples FG-11, FG-9, and FG-7, respectively. It could be clearly observed that their worn surfaces were relatively smooth, and their 3D profiles in Fig. 11C–E also presented the smooth worn surface. Meanwhile, large primary plateaus and secondary plateaus were also observed on the worn surface. In general, the primary plateaus originate from a lower removal rate of the mechanically stable and wear-resistant components of the matrix materials (hard particles) [47, 48]. The primary plateaus acted as barriers tended to stop and stick together the finer wear debris on the worn surface with such barriers, and then the wear debris was compacted, promoting the formation of secondary plateaus [49–51]. Many studies showed that the formation of the secondary contact plateaus on the worn surface plays a crucial role in improving the friction and wear behaviors of the brake friction materials [52–56]. This is consistent with the results of the friction-wear properties of the friction materials prepared by wet granulation with adequate resin content in Figs. 8 and 9. The improvement of the friction-wear properties following wet granulation was associated with the fact that the corn starch paste could form a liquid bridge between the fine particles and the formed granule, which could be considered a relatively independent unit with high cohesion and a hard shell. Moreover, the adequate resin content could effectively improve the interface bonding strength between the granules of the friction materials. Furthermore, in the current work, the granules with a hardened hull belong to the mechanically stable and wear-resistant components in the matrix materials. Also, they can promote the formation of primary plateaus, which provide the nucleating point for secondary plateaus.

In Figs. 10F and 12A, a large amount of wear debris and a larger adhesive pit were presented on the worn surface of sample FG-5. Microcracks and a loose material matrix could also apparently be observed, showing a relatively rough worn surface that could also be found on the 3D profile of sample FG-5 in Fig. 11F. Too little resin in the composite matrix could not provide enough adhesion between the granules, and then the granules easily peeled off from the friction surface, forming a loose matrix. Moreover, microcracks, wear debris, and a larger adhesive pit appeared successively under the action of the normal pressure, shear forces, and frictional heat, which indicated a significant decreasing of the wear resistance of sample FG-5. In the present study, wear debris that presented on the worn surface of sample FG-5 was inspected using EDX spectroscopy and the obtained results are presented in Fig. 12 and Table 3. Besides a lot of C, O, and Fe, a certain amount of Si, Ca, Al, Mg, K, and S could be detected from the wear debris. The above elements mainly come from the rotating disc and friction material matrix. Therefore, abrasive wear, adhesive wear, and fatigue wear are the main wear mechanisms of sample FG-5.

The worn surface roughnesses (*Ra*) of Ref, FG-13, FG-11, FG-9, FG-7, and FG-5 are presented in Table 4. It showed that the increase in worn surface roughness follows the order FG-11 < FG-9 < FG-7 < FG-13 < FG-5 < Ref. Sample FG-11 showed the lowest roughness ($Ra = 1.395\,\mu m$), following by sample FG-9 ($Ra = 1.582\,\mu m$) and sample FG-7 ($Ra = 1.696\,\mu m$) while sample Ref presented the highest roughness ($Ra = 2.509\,\mu m$). The results are basically consistent with the tribological and morphological analysis results. In general, the higher roughness is mainly attributed to the severe damage of the worn surface of the friction material samples in the dry tests, which directly shows the increasing of *Ra* values. It indicated that the worn surface of the friction materials prepared by wet granulation was relatively smooth and the friction materials had better wear resistance in comparison to sample Ref prepared directly from the powder.

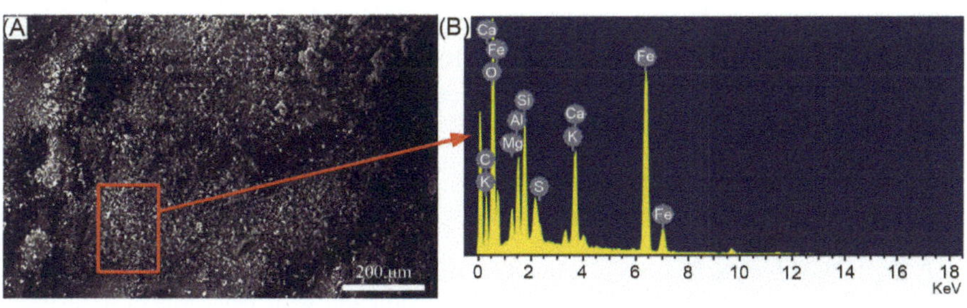

Fig. 12 Typical worn micrograph (A) and EDX spectrum (B) of sample FG-5.

Table 3 Atomic and weight percentage of wear debris on the worn surface of FG-5.

Element	C	O	Fe	Si	Ca	Al	Mg	K	S
Weight (%)	15.68	41.16	26.46	5.37	4.50	3.78	1.95	0.93	0.17
Atomic (%)	26.61	52.46	9.66	3.90	2.29	2.86	1.63	0.48	0.11

Table 4 Worn surface roughness of Ref, FG-13, FG-11, FG-9, FG-7, and FG-5.

Samples	Ref	FG-13	FG-11	FG-9	FG-7	FG-5
Ra (μm)	2.509	2.191	1.395	1.582	1.696	2.333

4 Conclusion

In this chapter, the influence of the final granule size and the phenolic resin content on the tribological properties of the phenolic resin-based friction materials prepared by wet granulation technology were systematically investigated at each subsequently increasing constant temperature of 100°C, 150°C, 200°C, 250°C, 300°C, and 350°C under constant speed and pressure conditions (480 rpm, 0.98 MPa). The worn surface of the friction materials was observed using SEM and LSCM analysis. The obtained results of the present chapter provide evidence that wet granulation improved the friction wear properties of the friction materials. Wet granulation techniques provide the possibility of improving the friction and wear performances and morphological properties of the friction materials. Moreover, in the material design, both the composition and manufacturing technique of the friction materials should be carefully selected.

Acknowledgments

This project was supported by the National Natural Science Foundation of China (Grant No. 51875242), by the China-EU H2020 FabSurfWAR project (Grant Nos. 2016YFE0112100 and 644971), by the Natural Science Foundation of the Jilin Province of China (Grant No. 20190302129GX), by the Jilin Province Science and Technology Development Plan Item (Grant Nos. 20170101173JC and 20170204015NY), by the 111 project (Grant No. B16020), and by the Jilin Province Development and Reform Commission Plan Item (Grant No. 2018C044-3).

References

[1] Z.J. Ji, W.Y. Luo, K.K. Zhou, S. Hou, Q.F. Zhang, J.Y. Li, H.Y. Jin, Effects of the shapes and dimensions of mullite whisker on the friction and wear behaviors of resin-based friction materials, Wear 406 (2018) 118–125.

[2] B.S. Rajan, M.A.S. Balaji, A.B.M.A. Noorani, M.U.H. Khateeb, P. Hariharasakthisudan, P.A. Doss, Tribological performance evaluation of newly synthesized silane treated shell powders in friction composites, Mater. Res. Express 6 (6) (2019), 065317.
[3] Y.C. Liu, Y.H. Ma, J.T. Yu, J. Zhuang, S.Y. Wu, J. Tong, Development and characterization of alkali treated abaca fiber reinforced friction composites, Compos. Interfaces 26 (1) (2019) 67–82.
[4] T. Singh, A. Tiwari, A. Patnaik, R. Chauhan, S. Ali, Influence of wollastonite shape and amount on tribo-performance of non-asbestos organic brake friction composites, Wear 386 (2017) 157–164.
[5] Y.H. Ma, Y.C. Liu, L.D. Wang, J. Tong, J. Zhuang, H.L. Jia, Performance assessment of hybrid fibers reinforced friction composites under dry sliding conditions, Tribol. Int. 119 (2018) 262–269.
[6] Q. Che, G. Zhang, L. Zhang, H. Qi, G. Li, C. Zhang, F. Guo, Switching brake materials to extremely wear-resistant self-lubrication materials via tuning interface nanostructures, ACS Appl. Mater. Interfaces 10 (22) (2018) 19173–19181.
[7] Z.H. Wang, G.H. Hou, Z.R. Yang, Q. Jiang, F. Zhang, M.H. Xie, et al., Influence of slag weight fraction on mechanical, thermal and tribological properties of polymer based friction materials, Mater. Des. 90 (2016) 76–83.
[8] H. Jang, Brake friction materials, in: S.K. Sinha, B.J. Briscoe (Eds.), Polymer Tribology, Imperial College Press, London, 2009, pp. 521–549.
[9] M. Asim, N. Saba, M. Jawaid, M. Nasir, M. Pervaiz, O.Y. Alothman, A review on phenolic resin and its composites, Curr. Anal. Chem. 14 (3) (2018) 185–197.
[10] B.S. Rajan, M.A.S. Balaji, K. Sathickbasha, P. Hariharasakthisudan, Influence of binder on thermomechanical and tribological performance in brake pad, Tribol. Ind. 40 (4) (2018) 654–669.
[11] A. Gardziella, L.A. Pilato, A. Knop, Phenolic Resins: Chemistry, Applications, Standardization, Safety and Ecology, Springer, New York, 2000.
[12] M.W. Shin, K.H. Cho, W.K. Lee, H. Jang, Tribological characteristics of binder resins for brake friction materials at elevated temperatures, Tribol. Lett. 38 (2010) 161–168.
[13] M. Kuroe, T. Tsunoda, Y. Kawano, A. Takahashi, Application of lignin-modified phenolic resins to brake friction material, J. Appl. Polym. Sci. 129 (1) (2013) 310–315.
[14] J. Park, H. Hwang, J.Y. Kim, J.W. Choi, Applicability of lignin polymers for automobile brake pads as binder and filler materials and their performance characteristics, Environ. Technol. (2018), https://doi.org/10.1080/09593330.2018.1503338.
[15] Y.H. Ma, Y.C. Liu, C. Menon, J. Tong, Evaluation of wear resistance of friction materials prepared by granulation, ACS Appl. Mater. Interfaces 7 (41) (2015) 22814–22820.
[16] V. De Simone, D. Caccavo, G. Lamberti, M. d'Amore, A.A. Barba, Wet-granulation process: phenomenological analysis and process parameters optimization, Powder Technol. 340 (2018) 411–419.
[17] M.F.H. Wolff, V. Salikov, S. Antonyuk, S. Heinrich, G.A. Schneider, Novel, highly-filled ceramic–polymer composites synthesized by a spouted bed spray granulation process, Compos. Sci. Technol. 90 (2014) 154–159.
[18] T.C. Seem, N.A. Rowson, A. Ingram, Z. Huang, S. Yu, M. de Matas, I. Gabbott, G.K. Reynolds, Twin screw granulation—a literature review, Powder Technol. 276 (2015) 89–102.
[19] Y.C. Liu, L.N. Wang, D.X. Liu, Y.H. Ma, Y. Tian, J. Tong, P. Senthamaraikannan, S. Saravanakumar, Evaluation of wear resistance of corn stalk fiber reinforced brake friction materials prepared by wet granulation, Wear 432–433 (2019) 102918.
[20] Y.H. Ma, Y.C. Liu, W.B. Shang, Z.H. Gao, H.B. Wang, L. Guo, J. Tong, Tribological and mechanical properties of pine needle fiber reinforced friction composites under dry sliding conditions, RSC Adv. 4 (69) (2014) 36777–36783.

[21] Y.H. Ma, Y.C. Liu, J. Peng, S.S. Ma, L. Guo, J. Tong, Friction and wear properties of semimetallic friction composites with bionic non-smooth structure, Mater. Res. Innov. 18 (2014) 387–392.
[22] Y.C. Liu, Y.H. Ma, X.M. Lv, J.T. Yu, J. Zhuang, J. Tong, Mineral fibre reinforced friction composites: effect of rockwool fibre on mechanical and tribological behaviour, Mater. Res. Express 5 (9) (2018), 095308.
[23] B. Öztürk, F. Arslan, S. Öztürk, Effects of different kinds of fibers on mechanical and tribological properties of brake friction materials, Tribol. Trans. 56 (2013) 536–545.
[24] Y.H. Ma, Y.C. Liu, S.S. Ma, H.B. Wang, Z.H. Gao, J.J. Sun, J. Tong, L. Guo, Friction and wear properties of dumbbell-shaped jute fiber-reinforced friction materials, J. Appl. Polym. Sci. 131 (17) (2014), https://doi.org/10.1002/app.40748.
[25] W. Osterle, M. Griepentrog, T. Gross, I. Urban, Chemical and micro structural changes induced by friction and wear of brakes, Wear 250–251 (2001) 1469–1476.
[26] A. Patnaik, M. Kumar, B.K. Satapathy, B.S. Tomar, Performance sensitivity of hybrid phenolic composites in friction braking: effect of ceramic and aramid fibre combination, Wear 269 (2010) 891–899.
[27] L.X. Zhong, S.Y. Fu, X.S. Zhou, H.Y. Zhan, Effects of surface microfibrillation of sisal fibre on the mechanical properties of sisal/aramid fibre hybrid composites, Compos. Part A 42 (2011) 244–252.
[28] M.H. Choa, K.H. Cho, S.J. Kim, D.H. Kim, H. Jang, The role of transfer layers on friction characteristics in the sliding interface between friction materials against gray iron brake disks, Tribol. Lett. 20 (2005) 101–108.
[29] T.N. Li, C.X. Zhou, Polyamide/clay nanocomposites, J. Funct. Polym. 1 (2000) 112–118.
[30] M. Kumar, B.K. Satapathy, A. Patnaik, D.K. Kolluri, B.S. Tomar, Hybrid composite friction materials reinforced with combination of potassium titanate whiskers and aramid fibre: assessment of fade and recovery performance, Tribol. Int. 44 (2011) 359–367.
[31] E. Mikael, Friction and Contact Phenomena of Disc Brakes Related to Squeal, Acta Universitatis Upsaliensis, Uppsala, 2000.
[32] Y. Cai, X. Yin, S. Fan, L. Zhang, L. Cheng, Y. Wang, H. Yin, Effects of particle sizes and contents of ceramic fillers on tribological behavior of 3D C/C composites, Ceram. Int. 40 (9) (2014) 14029–14037.
[33] H. Jang, K. Ko, S.J. Kim, R.H. Basch, J.W. Fash, The effect of metal fibers on the friction performance of automotive brake friction materials, Wear 256 (3–4) (2004) 406–414.
[34] Y.C. Kim, M.H. Cho, S.J. Kim, H. Jang, The effect of phenolic resin, potassium titanate, and CNSL on the tribological properties of brake friction materials, Wear 264 (3–4) (2008) 204–210.
[35] F. Eddoumy, H. Kasem, H. Dhieb, J.G. Buijnsters, P. Dufrenoy, J.P. Celis, Y. Desplanques, Role of constituents of friction materials on their sliding behavior between room temperature and 400°C, Mater. Des. 65 (2015) 179–186.
[36] N. Natarajan, S. Vijayarangan, I. Rajendran, Wear behavior of A356/25SiCp aluminium matrix composites sliding against automobile friction material, Wear 261 (2006) 812–822.
[37] Y.C. Liu, J. Xie, N. Wu, L.D. Wang, Y.H. Ma, J. Tong, Influence of silane treatment on the mechanical, tribological and morphological properties of corn stalk fiber reinforced polymer composites, Tribol. Int. 131 (2019) 398–405.
[38] M. Kchaou, A. Sellami, R. Elleuch, H. Singh, Friction characteristics of a brake friction material under different braking conditions, Mater. Des. 52 (2013) 533–540.
[39] H.J. Hwang, S.L. Jung, K.H. Cho, Y.J. Kim, H. Jang, Tribological performance of brake friction materials containing carbon nanotubes, Wear 268 (2010) 519–525.
[40] Y. Chen, J. Yang, R.N. Dave, R. Pfeffer, Granulation of cohesive Geldart group C powders in a Mini-Glatt fluidized bed by pre-coating with nanoparticles, Powder Technol. 191 (2009) 206–217.

[41] E. Rondet, M. Delalonde, T. Ruiz, J.P. Desfours, Identification of granular compactness during the kneading of a humidified cohesive powder, Powder Technol. 191 (2009) 7–12.
[42] S.M. Iveson, J.D. Litster, Growth regime map for liquid-bound granules, AICHE J. 44 (7) (1998) 1510–1518.
[43] A. Castellanos, The relationship between attractive interparticle forces and bulk behavior in dry and uncharged fine powders, Adv. Phys. 54 (2005) 263–376.
[44] M.R.P. Fernandes, A.E. Martinelli, A.N. Klein, G. Hammes, C. Binder, R.M. Nascimento, Production of nickel matrix composites reinforced with carbide particles by granulation of fine powders and mechanical pressing, Powder Technol. 305 (2017) 673–678.
[45] Y.C. Liu, Y.H. Ma, J.J. Che, L.J. Duanmu, J. Zhuang, J. Tong, Natural fibre reinforced non-asbestos organic non-metallic friction composites: effect of abaca fibre on mechanical and tribological behaviour, Mater. Res. Express 5 (5) (2018), 055101.
[46] J. Fei, D. Luo, C. Zhang, H. Li, Y. Cui, J. Huang, Friction and wear behavior of SiC particles deposited onto paper-based friction material *via* electrophoretic deposition, Tribol. Int. 119 (2018) 230–238.
[47] P.C. Verma, R. Ciudin, A. Bonfanti, P. Aswath, G. Straffelini, S. Gialanella, Role of the friction layer in the high-temperature pin-on-disc study of a brake material, Wear 346 (2016) 56–65.
[48] P. Cai, Z. Li, T. Wang, Q. Wang, Effect of aspect ratios of aramid fiber on mechanical and tribological behaviors of friction materials, Tribol. Int. 92 (2015) 109–116.
[49] W. Österle, C. Prietzel, H. Kloß, A.I. Dmitriev, On the role of copper in brake friction materials, Tribol. Int. 43 (12) (2010) 2317–2326.
[50] A. Saffar, A. Shojaei, Effect of rubber component on the performance of brake friction materials, Wear 274 (2012) 286–297.
[51] C. Menapace, M. Leonardi, G. Perricone, M. Bortolotti, G. Straffelini, S. Gialanella, Pin-on-disc study of brake friction materials with ball-milled nanostructured components, Mater. Des. 115 (2017) 287–298.
[52] J.H. Gweon, B.S. Joo, H. Jang, The effect of short glass fiber dispersion on the friction and vibration of brake friction materials, Wear 362–363 (2016) 61–67.
[53] T. Singh, A. Patnaik, R. Chauhan, A. Rishiraj, Assessment of braking performance of lapinus–wollastonite fibre reinforced friction composite materials, J. King Saud Univ. Eng. Sci. 29 (2) (2017) 183–190.
[54] I. Sugozu, I. Mutlu, K.B. Sugozu, The effect of ulexite to the tribological properties of brake lining materials, Polym. Compos. 39 (1) (2018) 55–62.
[55] Y.H. Ma, Y.C. Liu, C.Y. Mao, J.W. Li, J.T. Yu, J. Tong, Effects of structured fibre on mechanical and tribological properties of phenolic composites for application to friction brakes, Polym. Polym. Compos. 26 (4) (2018) 315–324.
[56] M. Amirjan, Microstructure, wear and friction behavior of nanocomposite materials with natural ingredients, Tribol. Int. 131 (2019) 184–190.

CHAPTER 11

Tribological studies of surface-modified sansevieria cylindrica fiber-reinforced polyester composites

C. Bennet, K. Mayandi, J.T. Winowlin Jappes, Nagarajan Rajini, and I. Siva
Department of Mechanical Engineering, Kalasalingam Academy of Research and Education, Krishnankoil, Tamil Nadu, India

1 Introduction

Nowadays, polymer material plays a significant role in engineering applications because of its high strength-to-weight ratio, environmentally friendly nature, and good mechanical properties such as tensile strength, flexural strength, and better wear resistance [1]. Natural fibers provide better or equivalent alternation of synthetic fibers because of their availability in the form of fibers that are taken from various plants in different parts at low cost [2]. The wear behavior of fiber-reinforced polymer composites is easily enriched by the inclusion of short fibers as well as the processing and fabrication of the composites by using short fibers at a very low cost as compared to long fiber-reinforced polymer composites. The fiber length and weight ratio are the two things that affect the wear properties of the composites. Uma Devi et al. [3] investigated the influence of fiber length on the wear properties of pineapple leaf fiber-reinforced polymer composites. Chin and Yousif [4] studied the influence of fiber orientation for bearing application on a kenaf fiber-reinforced polymer composite. It was concluded that the composite made with a normal alignment of fiber related to the sliding direction showed better wear resistance. Chand and Dwivedi [5] studied that the influence of the fiber weight fraction on the wear performance of sisal fiber-reinforced epoxy composites shows an increased friction coefficient with further increases in the fiber weight fraction.

Tribological studies on linen and jute fiber-reinforced unsaturated polyester resin were carried out by El-Sayed et al. [6]. The fibers were oriented in normal, longitudinal, and transverse directions, resulting in increases

Tribology of Polymer Composites
https://doi.org/10.1016/B978-0-12-819767-7.00011-6

in the coefficient of friction value and decreases in the wear rate in all cases. Unal et al. [7] reported pure polytetrafluoroethylene (PTFE) and its composites where with load increment, the friction coefficient decreased. Erosion behavior studies on E-glass fiber-reinforced polyester composites were carried out by Mahapatra et al. [8] and the optimal parameter for the minimum wear rate was reported through the Taguchi approach. The abrasive behaviors of untreated sugarcane fiber-reinforced unsaturated polyester composites have been studied by Mahapatra et al. [9]. It was observed that the fiber length affects the wear effect of the composites. A fiber length of 7–8 mm is identified for the minimum wear of the composites.

The objective of the work is to find the wear performance of the treated sansevieria cylindrica fiber polyester composites fabricated with curing temperature conditions at 60°C. The properties such as specific wear rate and coefficient of friction are determined as a function of the varying load.

2 Materials and experimental details

2.1 Materials used

The composites were fabricated using polyester resin as the matrix and sansevieria cylindrical fiber as the reinforcement material. This polyester resin is commercially and easily available in local markets and is cheap compared to other resins. The polyester resin is procured from the local supplier, Vasavibala Resins. The fibers were harvested from Cumbum, Tamil Nadu, India. By using a retting process and the hand extraction method, the fiber was separated from the stem of the plant. The extracted untreated fiber was thoroughly washed until the primary and secondary walls of the plant were removed from the fiber surfaces.

2.2 Chemical treatment

The untreated raw sansevieria cylindrical fiber (SCFs) surfaces were modified with various chemical treatments such as sodium hydroxide (NaOH), calcium hydroxide ($Ca(OH)_2$, silane, and potassium permanganate ($KMnO_4$). A trichloro vinyl silane solution treatment process was performed after the NaOH treatment process, maintaining the water pH at 5. Finally, the treated fibers were dried at room temperature conditions.

2.3 Fabrication of composites

The sansevieria fiber-reinforced polymer composites (SCFPC) were fabricated using compression molding techniques. Initially, the lower die of the compression mold was cleaned and wax was applied to prevent the sticking of the fiber with the mold. The short sansevieria (SC) fiber was randomly oriented to the bottom die of the mold. The length of the short fiber used was 40 mm and 40 wt.% fiber quantity was used to fabricate the composite slab. After random orientation of the fiber, the polymer resin was poured over the precompressed fiber mat. The polymer resin was prepared using polyester, cobalt napthanete, and methyl ethyl ketone peroxide (MEKP). Initially, the polyester resin was taken in a glass vessel, then curing agents were added to the resin such as MEKP and cobalt napthanete. The curing agent was thoroughly mixed with the resin until the resin became a homogeneous mixture. After pouring the resin to the compression mat, the upper die of the mold was closed and placed in a compression molding machine. The compression pressure was continuously applied until the mold was completely closed; the mold curing temperature was set at 60°C.

2.4 Wear test

The wear tests were performed with a steel circular disc on rectangular composite samples as per the ASTM standard. The experimental procedure can be explained with the following testing parameters as given in detail below. All the composites, which are exhibiting better performance on mechanical properties, are studied to understand and optimize the tribological parameters for suitable engineering applications. The pin-on-disc wear test machine is shown in Fig. 1. Typical specimens are presented in Fig. 2.

Fig. 1 Pin-on-disc wear test machine.

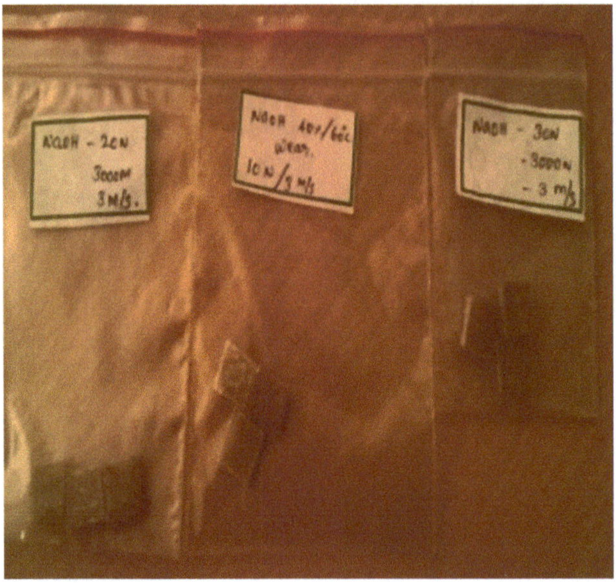

Fig. 2 Wear test specimens.

2.5 Scanning electron microscopy (SEM)

The surface characteristics of worn SCFPC were studied by using the Carl Zeiss model V18 scanning electron microscope and the surface behavior of the SCFPC was examined at different magnifications.

3 Results and discussion

3.1 Specific wear rate

Nowadays, the polymer-based fiber-reinforced composites have attracted the attention of tribologists due to their superior properties such as self-lubrication, ease of fabrication, and corrosion resistance against wear and friction. Specific material loss is the important factor to understand the dry sliding wear characteristics and it is referred to as a loss of volume of material per unit of time. The inclusion of SC in the presence of short fibers in polyester resin can provide improved resistance to the removal of material. The changes in the wear loss of unmodified and chemically modified SCFPC composites are discussed below.

Accordingly, in this work the wear and friction behaviors of SCFPC composites are studied under different chemical treatment conditions at a curing temperature of 60°C. All the trials are performed keeping the randomly oriented fibers in the composites as a parallel face on the circular disk. Before starting every test, the samples are cleaned thoroughly using acetone and the weight measurement is taken at up to four decimal points using an electronic weighing balance. After the test duration, the composite pin is weighed again in the same balance. The change in mass between the initial and final weights is the wear of the material. For every test, a new composite pin is used. The wear loss is determined in terms of specific wear rate (W_s) using the standard formula given in Eq. (1). Finally, the worn-out composite pin is again cleaned with acetone solution before the weight measurement.

$$W_s = \left[\Delta V / (P \times S)\right] \left(\text{mm}^3 / \text{Nm}\right) \tag{1}$$

where ΔV is the change in volume loss, P is the normal load, and S is the distance traveled.

Fig. 3 shows the wear performance of all treated and untreated composites with a 40 mm fiber length and 40 wt.% at a curing temperature of 60°C. The decreasing value of the specific wear rate is noticed for the increasing load. During the initial period of wear, the polyester matrix is subjected to direct contact with the disc and then wears out with the faster rate. At the same time, the minimum magnitude of the load is not enough to see the exposure of the fiber at the contact surface. But, when the load is increased, there is a chance for the fiber to have contact at the counter material. Hence, the abrasive nature of natural fiber can produce more resistance to wear.

Among the treated composites, almost all the composites show an equal amount of wear loss for the varying loading conditions. The better interfacial bonding can be attributed to more resistance to the wear in the sliding direction. However, a large material loss is noticed in the case of $Ca(OH)_2$-treated composites. This may be due to the presence of dominant matrix content at the contact surface.

3.2 SEM analysis of worn out surface

The SEM micrographs of alkali, $Ca(OH)_2$, silane, and the $KMnO_4$-treated worn out composites are shown in Fig. 4, Fig. 5, Fig. 6, and Fig. 7, respectively. Fig. 4A and B shows the micrographs of the worn-out sliding surfaces of the SCFPC composites using NaOH-modified SC fiber under 10 and 30 N loads for a wear distance of 3000 m and at a sliding speed of 3 m/s.

With respect to the NaOH chemical modifications, only the matrix failure is noticed at the contact surface at a lower load, as shown in Fig. 4A, whereas the presence of the fiber end is clearly observed on top of the

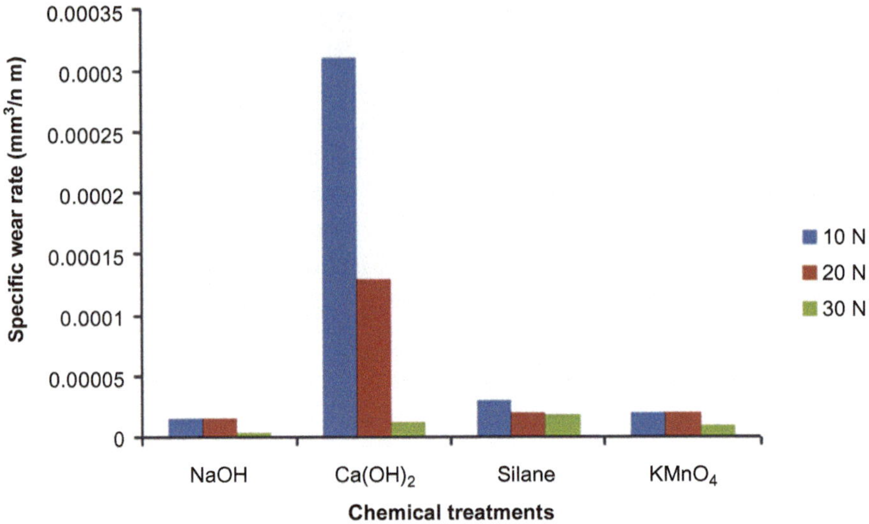

Fig. 3 Effect of chemical treatment on the specific wear rate of treated sansevieria cylindrica fiber (40 mm, 40 wt.%, 60°C) reinforced polyester composites.

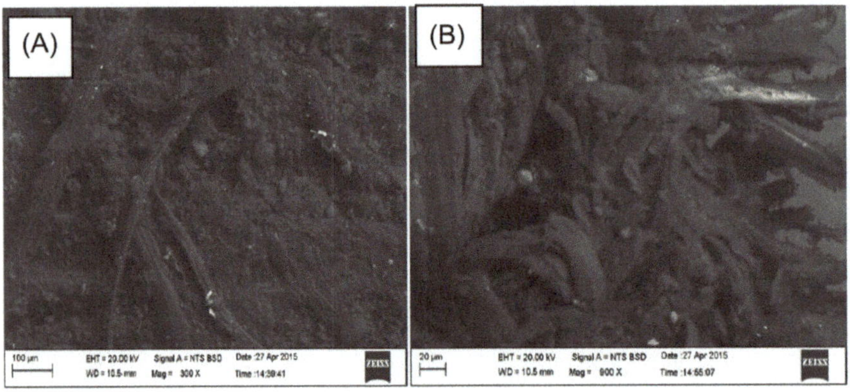

Fig. 4 (A), (B) SEM studies on NaOH-treated sansevieria cylindrica fiber (40 mm, 40 wt.%, 60°C) reinforced polyester composites subjected to a load of 10 and 30 N.

contact surface at the higher load of 30 N. The presence of fiber bundles and the separation of fibers are clearly seen from Fig. 4B, which indicates the removal of the matrix at a higher load. In fact, at the same loading conditions of 10 and 30 N, failures such as the matrix failure and macrocracks are indicated by means of a thick line that appeared to be dominant. It results in major material loss due to the hindering of the fiber surface of the $Ca(OH)_2$-treated SCFPC composites, as seen in Fig. 5A and B.

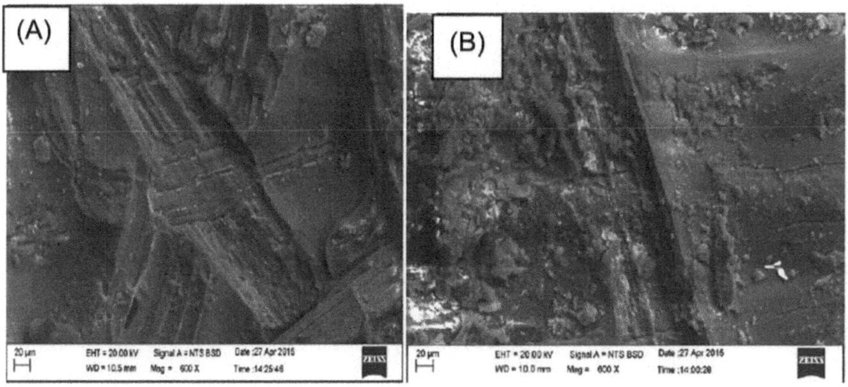

Fig. 5 (A), (B) SEM studies on $Ca(OH)_2$ treated sansevieria cylindrica fiber (40 mm, 40 wt.%, 60°C) reinforced polyester composites subjected to a load of 10 and 30 N.

Fig. 6A and B shows the surface topography of the worn-out surfaces of the SCFPC composites made of silane-treated SC fiber-reinforced polyester composites.

As shown in Fig. 6A and B, only a small amount of difference can be seen on the material removal samples under identical conditions, although the worn surfaces of the silane-treated SCFPC (Fig. 6A) look a bit less wavier than those of the other ones (Fig. 6B). However, the worn-out topography of the silane-modified SCFPC composite is relatively less irregular (Fig. 6B), and the phenomenon of the dominant crack advance at the interface between the mats is stopped. It exhibits more resistance to a loss of material in the composite.

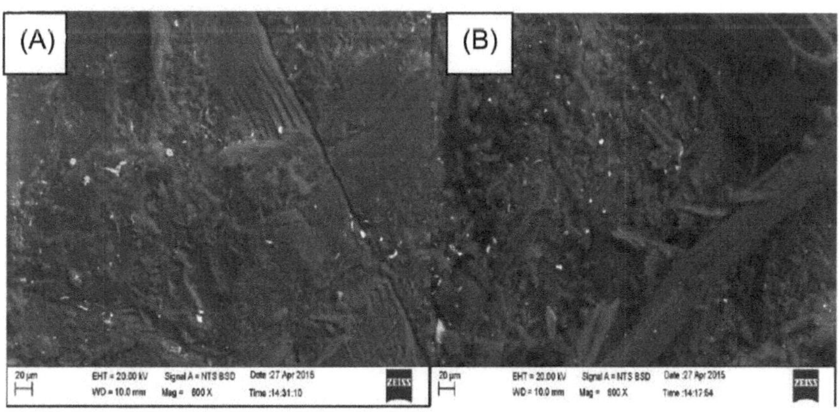

Fig. 6 (A), (B) SEM studies on silane-treated sansevieria cylindrica fiber (40 mm, 40 wt.%, 60°C) reinforced polyester composites subjected to a load of 10 and 30 N.

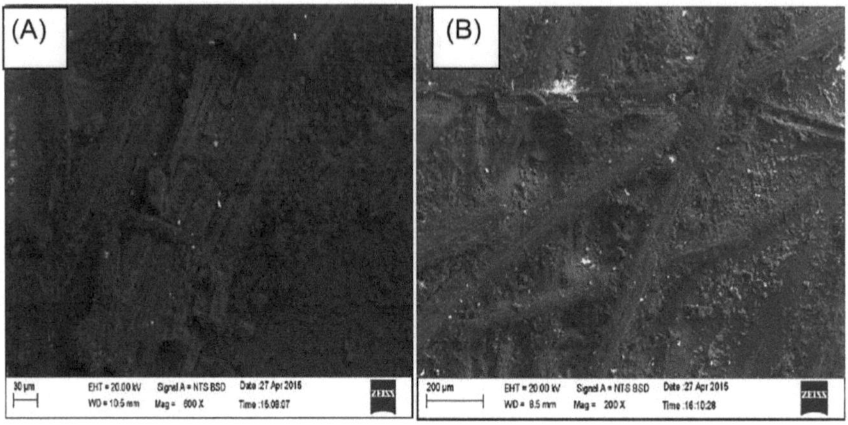

Fig. 7 (A), (B) SEM studies on $KMnO_4$ treated sansevieria cylindrica fiber (40 mm, 40 wt.%, 60°C) reinforced polyester composites subjected to a load of 10 and 30 N.

Fig. 7A and B presents the SEM images of the worn surfaces of the SCFPC composites made of $KMnO_4$-treated SC fiber-reinforced polyester composites. A polished surface is noticed in Fig. 7A and B. This could be the result of good adhesion of the fiber and matrix and it could avoid the damage of the fiber. Due to this better bonding between the fiber and matrix, the explicit exposure of fiber is not present. However, more uneven voids are found due to the removal matrix debris. Probably, due to this reason, the magnitude of adhesion wear gets reduced, and consequently, the wear loss drops to some extent, as shown in Fig. 7A and B.

3.3 Coefficient of friction

This is another important parameter in the tribological studies that needs to be studied to understand the behavior of materials in lieu of addressing the problem in the field of power transmission elements (or) for any other between any two mating components. Apart from all other wear testing parameters such as load, velocity, and distance, the material behavior at the contact surface is found to be significant for addressing the friction characteristics. Hence, it is inevitable to study the friction behavior for a newly developed material.

Fig. 8 shows the coefficient of friction for all the treated SCFPC composites. It is clearly seen from the figure that the friction coefficient is decreasing with the increasing load. This is expected to happen because the increasing load can lead to openings for adhesion wear. The depletion of

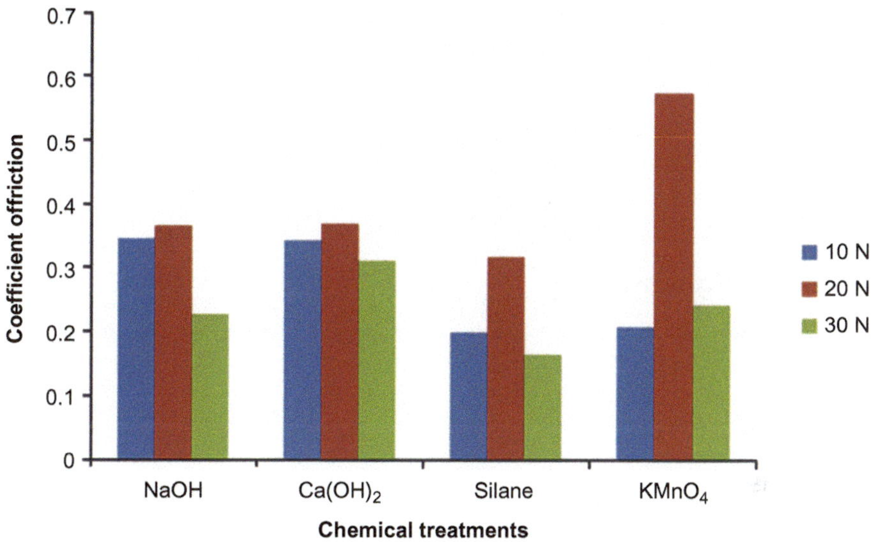

Fig. 8 Effect of chemical treatment on the coefficient of friction of treated sansevieria cylindrica fiber (40 mm, 40 wt.%, 60°C) reinforced polyester composites.

the fiber/matrix adhesion is taking place and the removal of the matrix and fiber debris is adhered to the counterpart. They act as a third body between the composite sample and the counterpart material. However, there is no specific trend found in the friction curve for all the cases, indicating the effectiveness of the randomly oriented composites. Moreover, it also depends on the modifications in the surface texture on the fiber topography due to various chemical treatments because the degree of interaction between the fiber and the matrix is based on the surface energy of the fiber. Hence, it is difficult to achieve a systematic pattern on the friction curve.

4 Conclusions

The following conclusions were obtained from the wear studies of SCFPC.

- In the study of the wear performance of SCFPC, the varying load is found to influence the material loss with respect to sliding distance. The change in magnitude of the material loss is observed in all the chemical treatments that could happen due to the improved adhesion and change in morphology of the fiber surface.
- SEM observations of different wear mechanisms such as matrix failure, crack formation, voids, and wear debris indicate the dominant matrix failure; this could be the reason for the decreasing specific wear rate.

- No specific trend is found in the case of friction characteristic curves for all the treated composites. However, the change in the friction curve is found to be observed with respect to chemical treatments and varying loads. It can be due to the varying degree of adhesion obtained from various chemical treatments and it could cause the formation of a transfer layer in the counter material.

Acknowledgment

The authors express gratitude to the Centre for Composite Materials, Kalasalingam Academy of Research and Education, Krishnankoil, for providing the necessary help to carry out this work.

References

[1] Y. Wang, K.S. Chen, J. Mishler, S.C. Cho, X.C. Adroher, A review of polymer electrolyte membrane fuel cells: technology, applications, and needs on fundamental research, Appl. Energy 88 (2011) 981–1007.
[2] L. Ranakoti, M. Pokhriyal, A. Kumar, Natural fibers and bio polymers characterization: a future potential composite material, J. Mech. Eng. 68 (1) (2018) 33–50.
[3] L. Uma Devi, S.S. Bhagwan, S. Thomas, Mechanical properties of pineapple leaf fiber reinforced polyester composites, J. Appl. Polym. Sci. 64 (9) (1997) 1739–1748.
[4] C.W. Chin, B.F. Yousif, Potential of kenaf fibres as reinforcement for tribological applications, Wear 267 (2009) 1550–1557.
[5] N. Chand, U.K. Dwivedi, Influence of fiber orientation on high stress wear behavior of sisal fiber-reinforced epoxy composites, Polym. Compos. 28 (2007) 437–441.
[6] A.A. El-Sayed, M.G. El-Sherbiny, A.S. Abou-El-Ezz, G.A. Aggag, Friction and wear properties of polymeric composite materials for bearing applications, Wear 184 (1995) 45–53.
[7] H. Unal, A. Mimaroglu, U. Kadıoglu, H. Ekiz, Sliding friction and wear behavior of polytetrafluoroethylene and its composites under dry conditions, Mater. Des. 25 (2004) 239–245.
[8] S.S. Mahapatra, V. Chaturvedi, Modelling and analysis of abrasive wear performance of composites using taguchi approach, Int. J. Eng. Sci. Technol. 1 (1) (2009) 123–135.
[9] S.S. Mahapatra, A. Patnaik, A. Satapathy, Taguchi method applied to parametric appraisal of erosion behavior of GF-reinforced polyester composites, Wear 265 (2008) 214–222.

CHAPTER 12

Friction and wear properties of carbon nanotube-reinforced polymer composites

M. Ramesh[a], R. Arun Ramnath[b], and C. Deepa[c]
[a]Department of Mechanical Engineering, KIT-Kalaignarkarunanidhi Institute of Technology, Coimbatore, Tamil Nadu, India
[b]Department of Mechanical Engineering, PSG College of Technology, Coimbatore, Tamil Nadu, India
[c]Department of Computer Science and Engineering, KIT-Kalaignarkarunanidhi Institute of Technology, Coimbatore, Tamil Nadu, India

1 Introduction

CNTs were initially developed by Iijima in the 1990s [1]. CNTs, discovered less than 30 years ago, created great attention among researchers because of their favorable properties and promising applications in many industrial sectors. Compared to carbon fibers, CNTs have excellent strength and modulus as well as a perfect layered structure, ensuring their increasing use as wearable materials in tribology-related fields [2]. In recent years, CNTs have been employed in many research investigations due to their extraordinary mechanical, electrical, and thermal properties. In addition to their shape, higher aspect ratio, flexibility, and lubrication properties, CNTs are unavoidable in tribological applications. Due to their outstanding intrinsic physical properties, particularly mechanical and thermal, CNTs have received much attention in the research community in recent years. In addition, their form, high aspect ratio, and high mobility make them promising candidates for solid lubrication or as a tribological lubricant additive [3–6].

Recently, the focus of research has been on studying the friction and wear properties of natural fiber-reinforced composites. The increase in the wear and friction properties of these composites in mechanical components is due to the tribological loading conditions [7]. Under dry contact

Tribology of Polymer Composites
https://doi.org/10.1016/B978-0-12-819767-7.00012-8

conditions, the wear and friction characteristics were investigated at different operating parameters, such as a 10–30 N load applied, a 1–3 m/s sliding speed, and a 1000–3000 m sliding distance. The results showed that the integration of fibers into the matrix significantly improves the wear behavior. The friction coefficient was reduced by 10%–44% and the real wear rate of the composites was reduced by almost 70% when compared to the matrix [8].

Alkali-treated white/brown coir fiber-reinforced composites were prepared by Valasek et al. [9], and the effect of the surface treatment on the mechanical properties and abrasive composite wear was investigated. This study concluded that coir fiber chemical treatment could improve the interaction with the matrix and increase the strength of the composite enhanced by brown fiber. In addition, the composite containing brown coconut alkali-treated fibers had the best wear resistance as well as high porosity and hardness. The hot-pressing process was used to produce polymer composites reinforced with raw corn stalk fiber. The effect of the concentration of silane solution on the tribological and morphological properties of the manufactured composites was investigated [10]. The results obtained will direct the use of agricultural waste such as maize stalk and the development of a new or improved polymer composite manufacturing technology.

Chin and Yousif [11] tried to use kenaf fibers as bearing reinforcing materials. They concluded from the attempt that the composite based on kenaf fiber exhibited better wear performance in normal orientation and increased wear performance by around 85%. El-Tayeb [12] investigated the friction and wear characteristics of sugarcane- and glass fiber-reinforced polyester composites at various operating parameters such as load, sliding speed, and test duration. This study concluded that the polyester composite enhanced by sugarcane fiber is a promising material that can compete with pure synthetic fiber composites. Hashmi et al. [13] developed cotton composites modified by graphite and observed the friction and sliding wear behavior at different loads and concentrations of graphite applied. They concluded that the actual wear rate decreased when cotton was reinforced and graphite was added, but the friction coefficient increased when cotton was reinforced and significantly decreased when graphite was added in the cotton-based polyester composite. Xin et al. [14] examined the characteristics of friction and wear of sisal fiber-reinforced composites with a specific resin content. They concluded from the analysis that sisal fiber is an ideal replacement for brake pads in cars.

2 CNT composites

Some current studies in the literature indicated that when used as a reinforcement process in polymer-based composites, CNTs have the ability to reduce friction and wear as a protective film and solid lubricant. CNTs were found to have a high tensile modulus and a load-carrying capacity of up to 1TPa and 200GPa [15]. Despite their distinctive properties, CNTs find fewer applications due to certain limitations. They may not be dispersed evenly in composites and a strong interfacial bonding is not found. Due to the strong van der Waals interaction that causes agglomerates, the major limitation of CNTs is their processing difficulty. This reduces the properties and also the composite interfacial bonding [16, 17]. CNTs are used as abrasive materials in composites because of their superior lubricating characteristics like graphite and excellent mechanical properties. In tribological applications, CNTs are commonly favored because of their ability to reduce wear. In order to conduct investigations into the tribological actions of CNTs, it is preferable to begin experiments under dry conditions in order to prevent the substantial influence of lubricants such as oil and grease on the material at work. The two possible methods are the study of a deposited CNT film as the substrate and the study of a composite material [18, 19]. As CNT films operate above the surface, the friction and wear reduction effects are often associated with the ability of CNTs to roll and thus isolate the surface of the rubbing [20].

The effects of friction and wear are considered to be major influencing parameters in the case of fiber-reinforced polymer composites. There are three different categories of composite materials: polymer matrix composites, metal matrix composites, and ceramic matrix composites. Composites of the polymer matrix are favored in applications where light weight without any heat resistance applicable. Where high temperature inertia and higher mechanical properties are desired, ceramic matrices are preferred. The intermediate level of metal matrix composites provides high ductility, high electrical conductivity, and good heat resistivity [21–23]. This chapter is an attempt to explore the friction and wear properties of CNT-reinforced composites and the effect of orientation and the percentage of CNTs in composites. Comparisons between the randomly distributed CNT composites and the aligned CNT enhanced composites, shows a more uniform and higher distribution of CNTs can only be found in aligned CNT enhanced polymer composites.

3 Friction and wear experiments

3.1 Common characteristics of the wear mechanism

CNTs reinforce composites to maintain stable and relatively smaller friction coefficients for the polymer matrix and other carbon matrices. In general, they conduct at higher temperatures, especially when reinforced with carbon matrix composites. When added with a certain ratio, CNTs cause the wear loss to decrease to a minimum. The wear mechanism of the CNT-reinforced composites has some common characteristics, as follows:

(i) CNTs do not induce abrasive wear mechanisms with a very high magnitude of elastic modulus and hardness. CNTs thus weaken the abrasive wear mechanism with a contradictory action and cause a decrease in wear loss. This decline in wear loss is primarily due to the self-lubricating properties and wearable features of CNTs.

(ii) Mostly, CNTs form a uniform layer of self-lubricating film on the worn substrate. This film forming results in a weaker adhesive wearing process for the composites at the same time. This weaker abrasive wear mechanism results in the design of wear mechanisms for fatigue and delamination.

3.2 Wear test set-up

To test the friction and wear characteristics of the produced composites, a pin-on-disk rotating style friction and wear tribo-meter was used. The pin-on-disk schematic diagram is shown in Fig. 1. The counter face is a ground-hardened steel disk with a peak wear diameter of 140 mm and a thickness of 8 mm (specifications: EN-31, 64 HRC, surface roughness

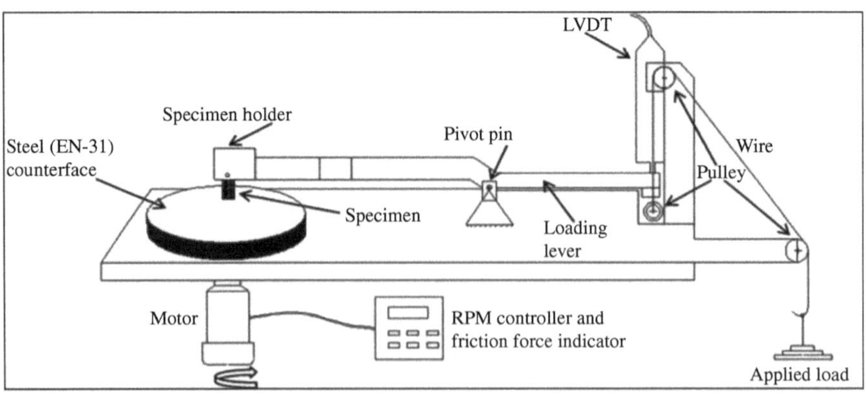

Fig. 1 Schematic arrangement of pin on disk set-up [8].

0.7 mm Ra). Sliding takes place between a stationary pin clamped in the holder of the sample and the spinning disk. The normal load, sliding speed, and wear path diameter may vary depending on the test requirements [8].

3.3 Friction and wear tests

Under ambient conditions, the friction and wear tests for the CNT composites were conducted using ring-on-block-style wear testers. Specimens of 20 × 12 × 6 mm size were manufactured and tested. A stainless steel ring coated with a thin chromium film was used as the counterpart, with a diameter of 40 mm and a length of 12 mm. Experimental studies were carried out under standard atmospheric conditions at a surface speed of 0.42 m/s under charging conditions of 50 and 200 N. The experimental tests were conducted for 65 min and the frictional torque was registered every 5 min. Based on the values of the frictional torque, the coefficient of friction values were determined. For wear testing, the specimen surface and the steel rings were well polished and cleansed with acetone in an ultrasonic bath. The wear values were measured with the aid of an analytical balance of 0.1 mg accuracy [24]. The friction and wear tests of the CNT-based composite materials were conducted with the help of the testing equipment (type: MM-200). The size of the specimens was fabricated as 10 × 6.5 × 8 mm and the rings were made of C45 steel. The size of the steel rings was of diameter 40 mm and length of 8 mm. Fig. 2 illustrates the contact between the ring and the block. The experiments were conducted at a speed of 200 rpm and 50 N with a time duration of 30 min.

At higher temperatures, the friction and wear tests of the well-matched CNT-reinforced carbon composites were performed and the test conditions differed entirely. The tests were carried out with a high temperature

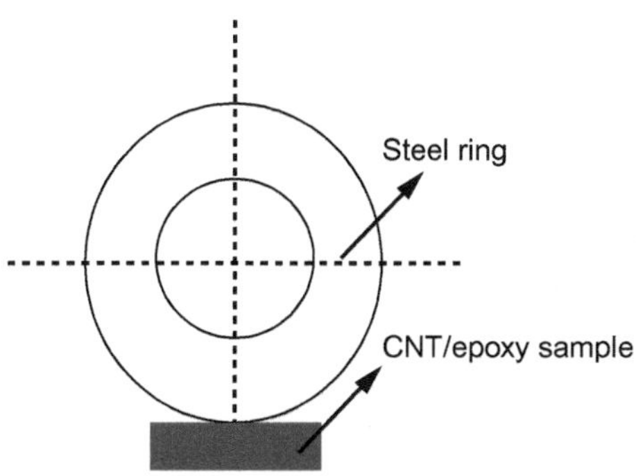

Fig. 2 Experimental set-up for ring-block arrangement of CNT epoxy composites.

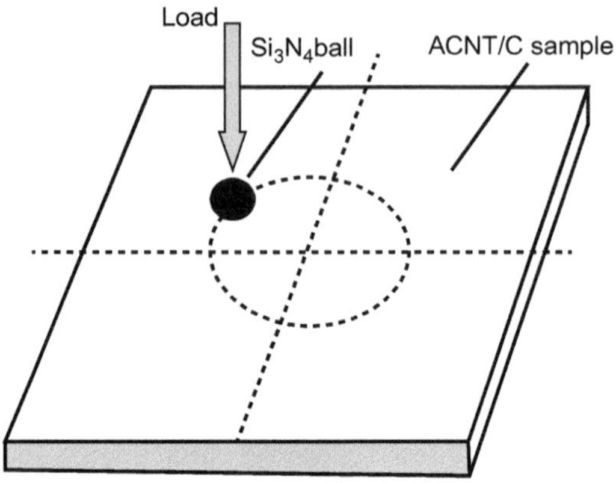

Fig. 3 Experimental setup of ball-disk mode.

friction and wear tester model HT-500 and the ball-disk mode was used [25–27]. The configuration of the ball-disk mode is illustrated in Fig. 3. The tests were conducted at ambient atmospheric conditions and the testing temperature varies from room temperature to a maximum of 500°C. The tests were conducted for a period of 60 min with an applied load of 2 N.

4 Friction and wear properties

It has been shown that the frictional behavior is primarily influenced by the CNTs present in the contact zone, as only minor differences can be observed in the friction coefficient between the samples. Consequently, the known effect of an improved microstructure, resulting in increased hardness of the CNT-enhanced specimens, tends to play a minor role in the reduction of friction compared to the solid lubrication effect induced by the CNTs. However, a continuous supply of CNTs to the tribo-touch can be considered isolated for the reinforced composites, resulting in a long-term reduction in friction relative to the CNT-coated test [23].

4.1 Randomly dispersed CNT epoxy composites

In a three-rolled press, CNTs and epoxy resin are combined in some proportions. This black mixture spent 90 min in vacuum at 130°C. The mixture was then cast into a mold and finally healed for 3 h at 180°C. The density of the epoxy resin was found as 1.21 g/cm^3, which is approximately equal to that of the molded composites, which was 1.2–1.3 g/cm^3. The curing process in molding the CNT composite produces a smooth surface. The specimens with the fractured surface indicated that CNTs has been pulled

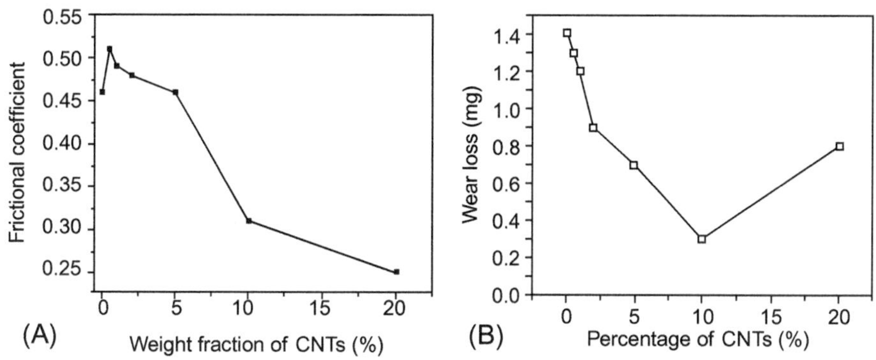

Fig. 4 (A) Coefficient of friction; (B) loss due to wear of CNT composites.

out in the closer range of 1 μm. It can be inferred that the interfacial bonding between the CNTs and the epoxy resin is less.

Fig. 4 illustrates the relationship between the wear losses and the friction coefficient with regard to the composition of the CNTs present in the composites. The results revealed that the friction coefficients vary slightly if the ratio of CNTs is less than 5% by weight and lies within the range of 0.45–0.5. When the CNT ratio increases by weight in the range of 10%–20%, the friction coefficient decreases from 0.31–0.25. The frictional coefficient and wear losses depict a different trend, as shown in Fig. 4. A minimum wear loss value is found if the weight of the CNT ratio is 10%. The CNT-reinforced composite wear loss is one-fifth of the pure epoxy resin. The figure shows the variations in the experimental results on the coefficient of friction and the wear loss influence of CNTs. The topography of the surface shows that the contact surfaces are covered with certain asperity heights. The result of individual contact spots during the initial stages of contact gives the actual contact area as relatively smaller than the geometric contact area.

The strength and hardness of steel is higher than that of epoxy resin, and the asperities on the surface of the epoxy matrix change depending on the load applied when these surfaces come into contact with each other. Relatively higher asperities are broken or damaged by the shear force between the sliding surfaces, and the damaged debris falls off and deforms on the worn surface as a flat film. This results in increased contact area and a smooth worn surface is formed. With an increase in contact area, the attractive interaction forces between the contact surfaces increases significantly. CNTs have an outstanding self-lubricating property, other than their higher modulus values. Also, the coefficient of friction values of CNTs are highly

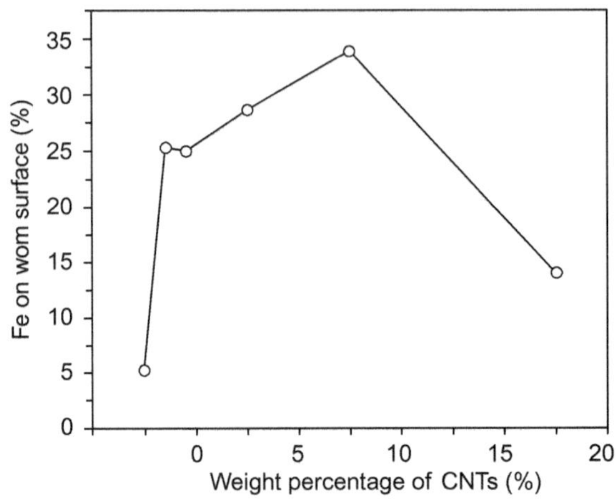

Fig. 5 Fe content variation with regard to changes in CNT content.

determined by the matrix material. CNTs are wearable and protrude from the surfaces. Fig. 5 shows the behavior of a worn surface with the variations in weight percentage of CNTs.

The results showed that there is a higher ratio of Fe content to the worn surface with the smaller amount of CNTs. At the same time as the weight increase of 10% of CNTs, a smooth frictional film layer is formed between the sliding surfaces; this film helps to self-lubricate. With an increase in CNT content of 20%, no significant differences are observed in the worn and unworn surfaces. A higher self-lubricating property means a lower coefficient of friction is achieved. It can be inferred on the basis of these findings that the effect of CNTs on wear loss is almost identical to that of the friction coefficient. Frictional film has a major impact on wear loss reduction. When the ratio of CNTs exceeds a certain proportion, the cross-linked structure of epoxy resin is to some extent destroyed, resulting in a reduction in composite shear strength.

4.2 Aligned CNT epoxy composites

Aligned CNTs are produced in a tubular furnace via the chemical vapor deposition process. Inert gases such as xylene, argon, and hydrogen are pyrolyzed and deposited on a quartz substratum by a catalyst at 700–800°C using ferrocence [28–30]. Fig. 6 shows the morphology of the aligned CNTs. The morphology of the composite surface shows that the CNTs grew in a vertical direction to that of the quartz substrate with a layer formation

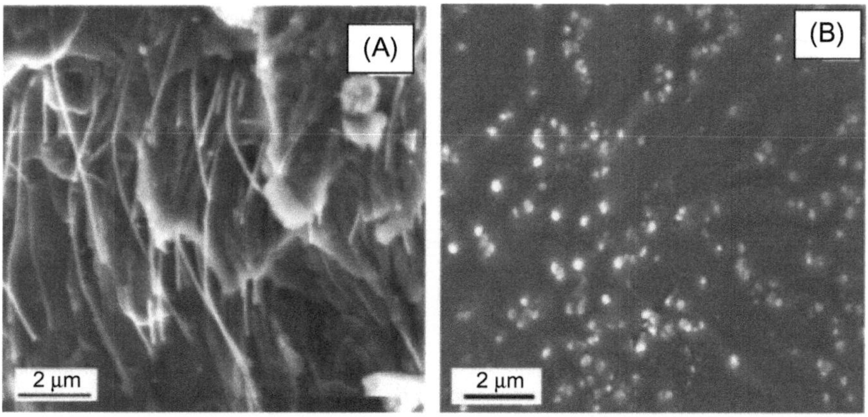

Fig. 6 Morphology of aligned CNT epoxy composite: (A) intersection in the parallel direction; (B) upper surface.

of 4–6 mm thick that is clearly visible in the images. The diameter of the CNTs ranges from 40 to 60 nm with an initial density of 0.10–0.15 g/cm^3.

The nanotubes are formed closely and uniformly dispersed to form a regularly shaped porous block. The pores on the surface are very small, with a diameter of 113 nm. Due to the high wettability, strong interfacial bonding is observed in the epoxy composites integrated in the CNT. Balanced CNT composites are stronger because of their spatial properties compared to randomly distributed CNT epoxy composites. Epoxy was therefore used to fill the apertures of the reinforced CNT composites in accordance [31]. Surface morphology showed that the CNTs were well aligned with the matrix and that there was no deformation. The density of the well-balanced, CNT-enhanced epoxy composites was found within a range of 1.2–1.3 g/cm^3.

In order to investigate further the tribology of the epoxy matrix, well-aligned CNTs were used as reinforcing material. Experimental tests were conducted to determine whether the wear and friction characteristics were similar to those of randomly dispersed CNTs. Fig. 7 illustrates the variations in the friction coefficient of various specimens at varying test durations. From the figure, it may be observed that the friction coefficient and the wear loss of the side surface are very much less when compared with that of the epoxy matrix and the upper surface of the well-aligned CNTs. Theoretical evaluations indicated that the CNTs contribute nearly 5% in weight of the composites, and this is almost equivalent to that of randomly oriented CNT-reinforced composites. In case the surface on the upper side

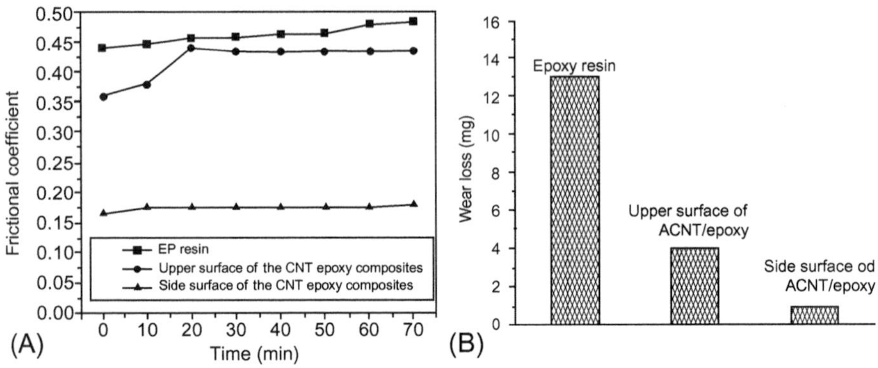

Fig. 7 Tribological properties of well-aligned CNT composites: (A) variation of coefficient of friction with time; (B) wear loss.

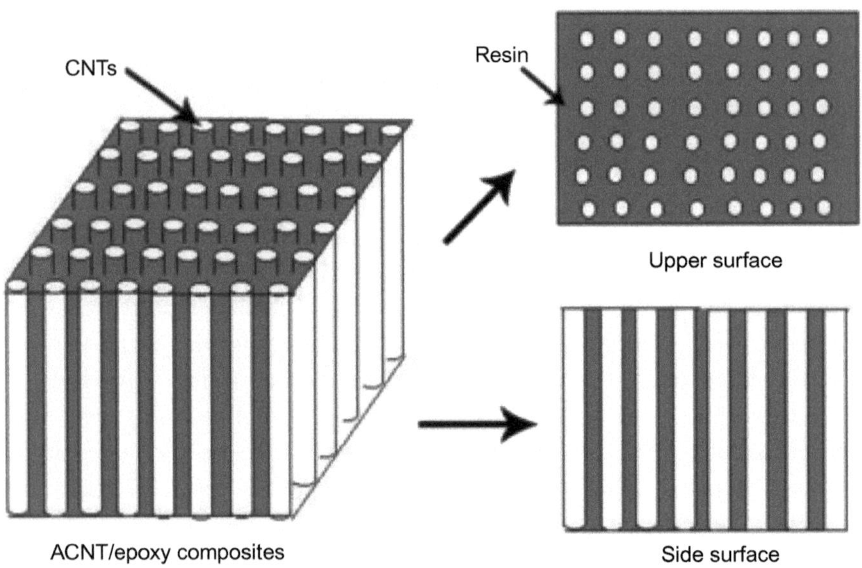

Fig. 8 Projection of CNTs on the upper and the side surface.

exhibits wearing behavior, the actual area of contact between the CNT and the steel counterpart is limited. Due to higher strength values, CNTs are pulled out quickly than other fractured segments. Fig. 8 demonstrates the influence of CNTs on the upper and the side surfaces. The wear and friction values between the side and top surfaces vary with one other because the actual sliding area between the CNT and the mold is larger. Therefore, lower coefficient of friction and wear losses are expected.

4.3 Aligned CNT carbon composites

Apart from epoxy-based resin, hydrocarbon gas infiltrates into any small hole in a certain vacuum. Hence, each CNT is coated with a layer of pyrocarbon after chemical vapor infusion without any damage to the orientation to the aligned CNTs. With the addition of carbon as a matrix material, self-lubrication properties are induced into the composites. Self-lubrication is a common characteristic of carbon as a matrix material [32, 33]. After the chemical vapor infusion process, the resultant density was found as 1.3–1.5 g/cm^3. Due to their enhanced self-lubricating properties, carbon materials are found in a wide range of applications as a self-lubricating material as well as a sealing material. Carbon and its composites are found as a standard material for the development of aircraft disk brakes. CNTs exhibit better mechanical and thermal properties when compared with carbon fibers. An experimental study conducted by aligning CNT carbon composites was fabricated to analyze their friction and wear properties. The fabrication process densifies the preform efficiently and induces good interfacial strength between the CNTs and the carbon matrix. Fig. 9 depicts the friction and wear test results of the composites.

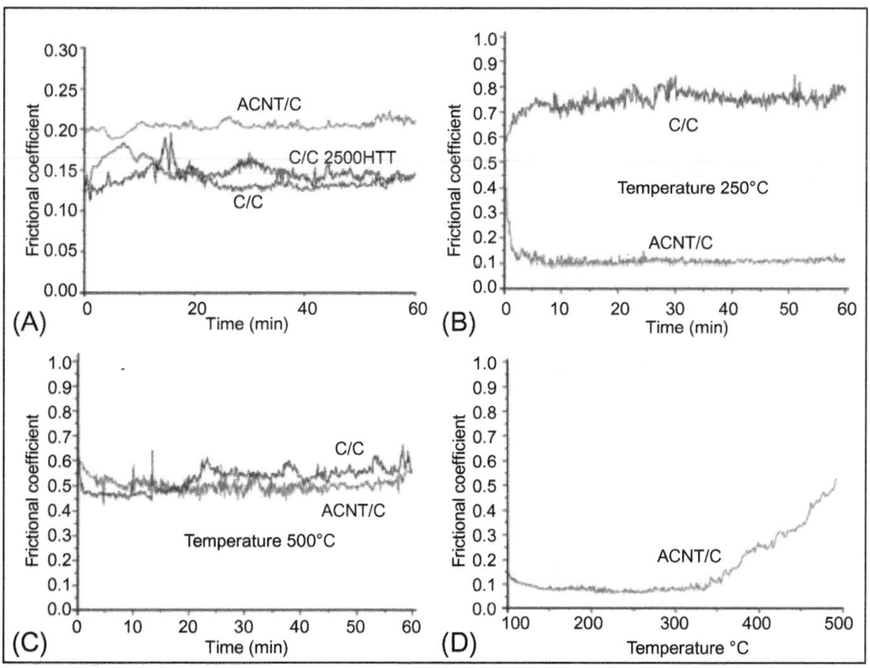

Fig. 9 Friction test results of carbon and aligned CNT composites at: (A) room temperature, (B) 250°C, (C) 500°C, and (D) from room temperature to 500°C.

The figure shows that the coefficient of friction is located at room temperature as 0.2, which is slightly higher than that of the carbon composites. Higher coefficient variations in friction values are observed when the measurements are conducted at higher temperature conditions. The friction coefficient of combined CNT and carbon composites at a temperature of around 250°C is within the range of 0.10–0.15, whereas the carbon composite values are around 0.70–0.80. The friction coefficients of two composites are in a similar range of 0.45–0.60 when the measurements are carried out at 500°C. Generally, smaller changes in the coefficient of friction from room temperature to 350°C of the aligned CNT/carbon composites imply more stable tribological properties of aligned CNT/carbon composites than carbon composites. The width of the friction track was considered as the evaluation criterion for the qualitative analysis of the wear loss. The wear loss is too low to measure and is therefore assumed to be the size of the crack. Relative to carbon composites at room temperature and 250°C, the wear loss of matched CNT/carbon composites is lower. Around 500°C, there are no sudden changes and the wear loss is almost the same as shown in Fig. 10.

Once the tests are carried out at room temperature, they are broken and compressed into small particles due to loading the asperities on the sliding surfaces. The plowing friction effect occurred in this case to some degree, resulting in relatively higher friction coefficients. For aligned CNT composites, the flaky debris and smooth surface without large pores result in a stable frictional coefficient and low wear loss. Water or vapor that serves as a lubricant during friction was desorbed when the experiments were conducted at 250°C. This triggered desorption leads to a higher friction coefficient as well as wear loss. In lubricating the frictional cycle, the frictional film, if formed, plays a dominant role. Because of this ideal frictional film from the aligned CNT composites, the friction coefficient values are lower and stable. Similarly, no standard film layer is created for carbon composites to cover the worn surface. In the case of carbon composites, friction due to the plowing phenomenon and adhesion can coexist [34, 35]. Based on the morphology review, it is found that the friction film is packed with layers of delaminated film and broader cracks on the worn surface of carbon composites. In the case of carbon composites, the presence of larger cracks results in higher and unstable friction coefficients.

On both specimens, when measured at 500°C, larger friction cracks were found, suggesting that the oxidation occurred mainly on the damaged surface on the film's nanosized debris and the interface between the

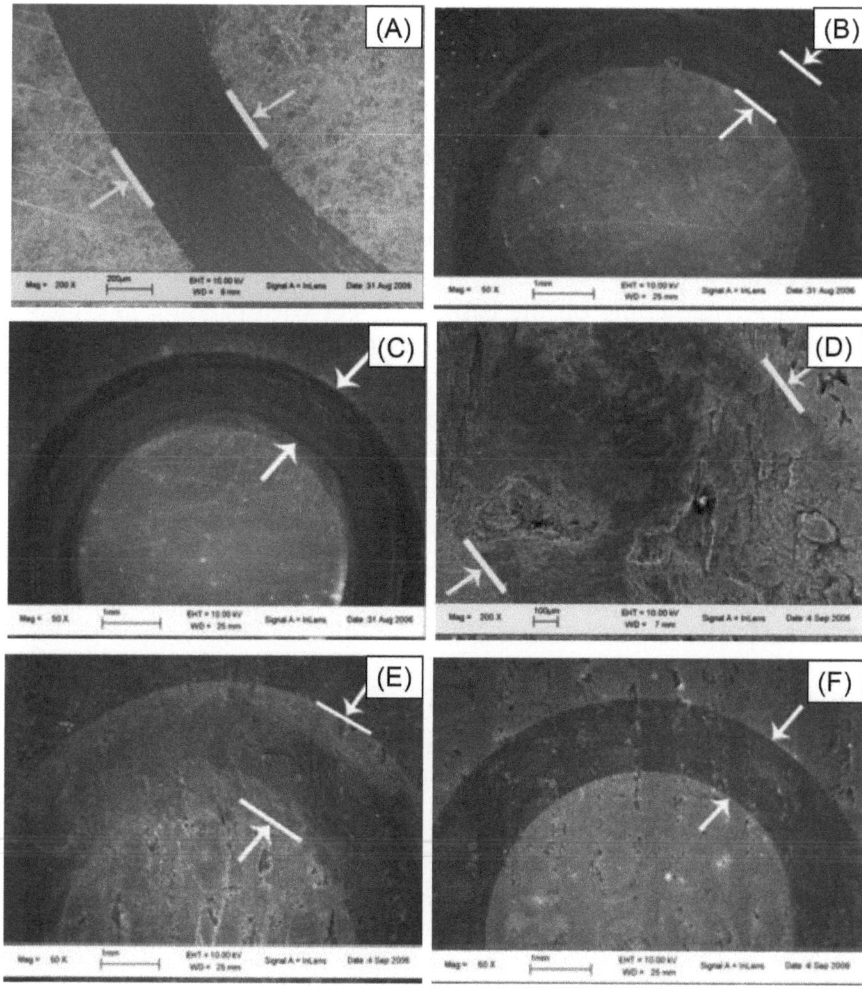

Fig. 10 Morphology analysis of worn surface tested at room temperatures, 250°C, and 500°C (A–C) aligned CNT/C composites; (D–F) carbon composites.

reinforcement and the matrix. The frictional film would therefore be impaired by oxidation, and higher friction coefficient values in CNT/carbon composites are observed. The oxidation of carbon composites improves the process of film forming on the entire worn-out surface rapidly. More integrated friction film results in a lower frictional coefficient for carbon composites. The tribo-chemical wear mechanism is a secondary factor in the wear loss of both the samples at 500°C with the presence of the oxidation process. Apart from the tribo-chemical wear mechanism, the friction film fatigue and the delamination wear mechanism are the two main factors that have a major impact on the wear loss and the tribological properties of CNT composites.

4.4 CNT-reinforced nickel matrix composites

No chemical reaction between nickel (Ni) and CNTs can be observed when using Ni as the component of the matrix. In addition, Ni only forms metastable carbides under very specific conditions and is therefore ideal for strengthening on the basis of CNTs with parasitic interphases [36–38]. Compared to unreinforced CNT-coated bulk Ni specimens in terms of friction and wear, CNT-reinforced Ni matrix composites gain significant knowledge of the tribological effect of CNTs and the underlying tribo-mechanism. Finally, it can be mentioned that CNTs can withstand the accumulated stress and maintain their structural state for the given strain to some extent. A comprehensive study using complementary analytical methods, including Raman spectroscopy and electron microscopy, is used to understand the mechanisms involved in friction and wear [23].

4.5 Microstructure properties

Microstructural analyses were conducted by means of microscopic analysis and the microhardness was measured by using a microhardness tester. The degree of graphitization present in CNT composites was determined by X-ray diffractometer analysis. The worn surfaces of the composites were analyzed using the Raman effect and the X-ray energy dispersive spectrum. The tests were conducted at room temperature using a microscopic Raman spectrometer. The experiments were conducted for 30 s and three different trials were taken. The microscopy images of wear debris are presented in Fig. 11. Large volumes of wear debris appeared across the counter face of the smooth polylactic acid. Most of the debris, as shown in the figure, was in a plate-like shape. Yet relatively smaller wear debris with low volume was obtained on the counter face in the case of composites, which supports the composites' low wear level. Such debris includes wearing fibers and polymer. Throughout sliding, the sliding layer is rubbed against both longitudinal and transverse fibers in the mat and small and large fiber particles are removed. In general, this fiber debris and the thin polymer film covering the fiber particles are moved to the counter face. Originally, the material was removed from the composite's sliding face in the form of a thin film and then dispersed across the counter face. However, some of the debris collected

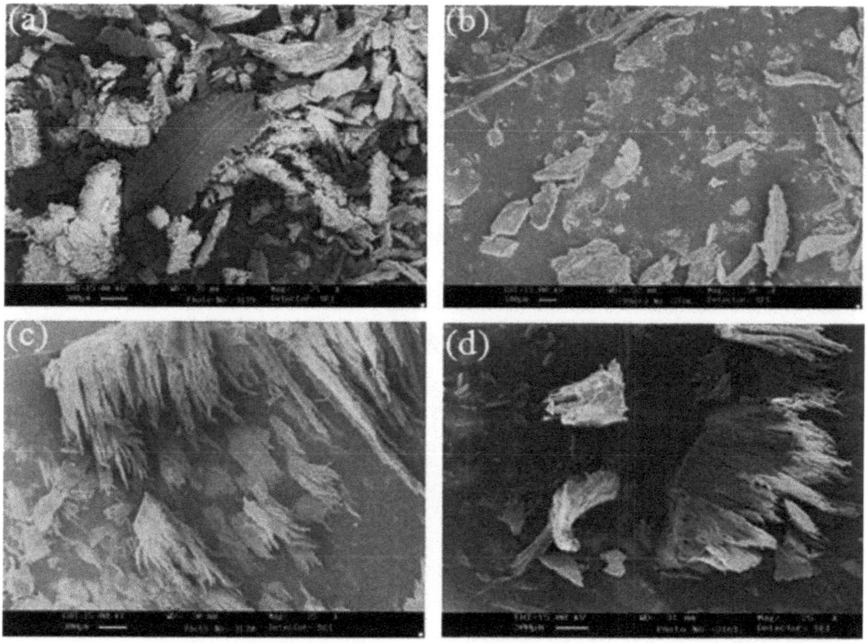

Fig. 11 SEM images of wear debris at an applied load of 30 N and at a sliding speed 3 m/s: (A) polylactic acid, (B) polylactic acid/grewia optiva, (C) polylactic acid/nettle, and (D) polylactic acid/sisal.

one after another during testing at the composite specimen's trailing edge and formed a bunch of debris, as shown in Fig. 11C–D [8].

The wear track SEM images observed after 200 and 20,000 sliding cycles on the Ni reference and the corresponding wear tracks for the CNT-coated sample are shown in Fig. 12 [23]. It can be identified from the illustrations that the main wear mechanism is plowing for both tracks. This is apparent because the opposing body's hardness exceeds the Ni surface's hardness. In comparison, an increased wear track size after a wear track of 20,000 sliding cycles is seen very clearly when compared to 200 cycles. This is a consequence of the continuous indentation of the Al_2O_3 indent in the specimen during the test, resulting in a greater contact area and thus increasing the wear track size. No extreme wear of the Ni substratum can be found in the case of the 200 sliding cycles. The Ni surface is still polished, apart from the minor scratches that could have been caused by the ball's larger asperities. It should be remembered that individual grains can still be identified after the tribological test, which is only possible for highly polished surfaces. Within the scratches described, CNTs that are confirmed by Raman spectroscopy can be found.

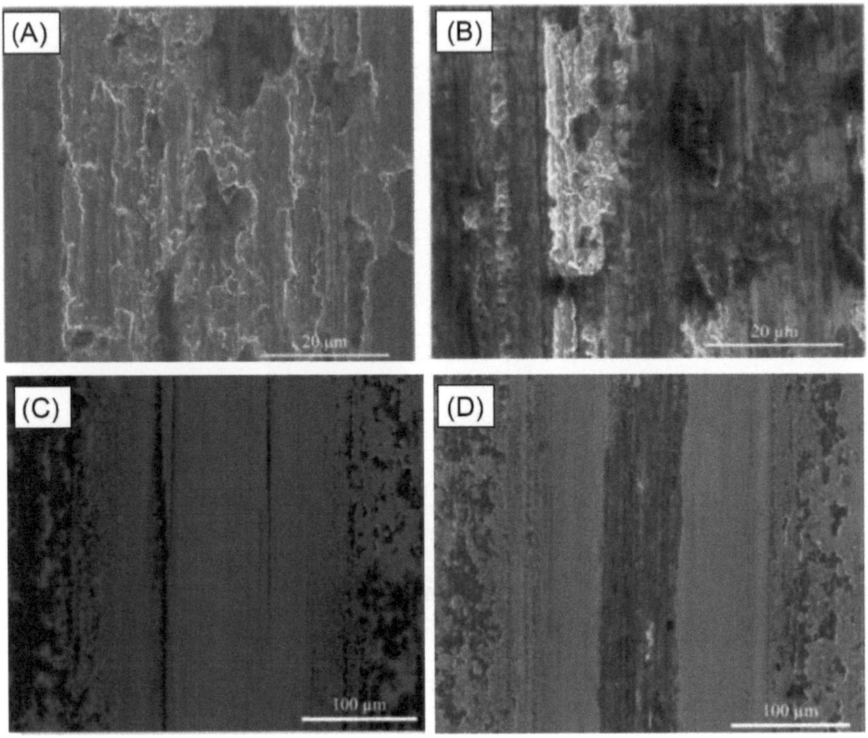

Fig. 12 Wear tracks of Ni-coated samples (A, B) and the CNT-coated samples (C, D) after 200 and 20,000 sliding cycles [23].

For 20,000 sliding cycles, the images can distinguish two places in the wear tracks very quickly. On the one side, a small area with extreme wear and marked plowing as well as signs of oxidation can be found in the middle of the wear line. On the other side, a larger wear track with some small scratches and a similar appearance after 200 sliding cycles compared to the wear track is seen clearly. It is clear that there is no improvement in wear track width when comparing the wear track width between 200 and 20,000. Unlike the Ni comparison, however, the width of the region with extreme wear is decreased, which may be due to the CNTs that prevent direct contact of the ball's flanks with the Ni substratum.

5 Conclusion

This chapter on the tribological characteristics of CNT-reinforced matrix composites focuses on the effect of CNTs on the friction and wear characteristics when reinforced with polymeric composites as well as carbon matrix composites. Randomly dispersed and well-aligned CNTs reinforced with different matrix materials are discussed in detail. The wear and frictional characteristics of CNT-reinforced composites are studied on the basis of experimental tests, morphology analysis, and microstructural analysis. From the literature, it is revealed that CNTs maintain stable and consistent low coefficients of friction and wear loss when investigated at higher temperatures and reinforced with carbon as a matrix material. This is mainly due to CNT's self-lubricating properties and their influence on weakening the mechanism of abrasive wear. It is concluded from the chapter that CNTs are preferred in engineering applications because of their higher values of elastic modulus, hardness, and self-lubricating properties to reduce wear loss.

References

[1] S. Iijima, Helical microtubules of graphitic carbon, Nature 354 (1991) 56–58.

[2] G. Qianming, L. Dan, L. Zhi, Y. Xiao-Su, L. Ji, Tribology properties of carbon nanotube-reinforced composites, Tribol. Interf. Eng. Ser. 55 (2008) 245–267.

[3] K.-T. Lau, D. Hui, The revolutionary creation of new advanced materials-carbon nanotube composites, Compos. Part B 33 (2002) 263–277.

[4] R.S. Ruoff, D.C. Lorents, Mechanical and thermal properties of carbon nanotubes, Carbon 33 (1995) 925–930.

[5] Y. Lu, P.K. Liaw, The mechanical properties of nanostructured materials, JOM 53 (2001) 31–35.

[6] P. Kim, L. Shi, A. Majumdar, P. McEuen, Thermal transport measurements of individual multiwalled nanotubes, Phys. Rev. Lett. 87 (2001) 215502.

[7] N.S.M. El-Tayeb, B.F. Yousif, T.C. Yap, Tribological studies of polyester reinforced with CSM 450-R-glass fiber sliding against smooth stainless steel counterface, Wear 261 (2006) 443–452.

[8] P.K. Bajpai, I. Singh, J. Madaan, Tribological behavior of natural fiber reinforced PLA composites, Wear 297 (2013) 829–840.

[9] P. Valasek, R. D'Amato, M. Muller, A. Ruggiero, Mechanical properties and abrasive wear of white/brown coir epoxy composites, Compos. B Eng. 146 (2018) 88–97.

[10] Y. Liu, J. Xie, N. Wu, L. Wang, Y. Ma, J. Tong, Influence of silane treatment on the mechanical, tribological and morphological properties of corn stalk fiber reinforced polymer composites, Tribol. Int. 131 (2019) 398–405.

[11] C.W. Chin, B.F. Yousif, Potential of kenaf fibres as reinforcement for tribological applications, Wear 267 (2009) 1550–1557.

[12] N.S.M. El-Tayeb, Study on the potential of sugarcane fibers/polyester composite for tribological applications, Wear 265 (2008) 223–235.

[13] S.A.R. Hashmi, U.K. Dwivedi, N. Chand, Graphite modified cotton fibre reinforced polyester composites under sliding wear conditions, Wear 262 (2007) 1426–1432.

[14] X. Xin, C.G. Xu, L.F. Qing, Friction properties of sisal fibre reinforced resin brake composites, Wear 262 (2007) 736–741.

[15] M.F. Yu, O. Lourie, M. Dyer, K. Moloni, T.F. Kelly, R.S. Ruoff, Strength and breaking mechanism of multi-walled carbon nanotubes under tensile load, Science 287 (2000) 637–639.

[16] R. Haggenmueller, H.H. Gommans, A.G. Rinzler, J.E. Fischer, K.I. Winey, Aligned single-wall carbon nanotubes in composites by melt process methods, Chem. Phys. Lett. 330 (2000) 219–225.

[17] K.-T. Lau, S.-Q. Shi, Failure mechanism of carbon nanotube/epoxy composites pretreated in different temperature environments, Carbon 40 (2002) 2965–2966.
[18] S.R. Dong, J.P. Tu, X.B. Zhang, An investigation of the sliding wear behavior of Cu-matrix composite reinforced by carbon nanotubes, Mater. Sci. Eng. A 313 (2001) 83–87.
[19] A. Peigney, C. Laurent, E. Flahaut, A. Rousset, Carbon nanotubes in novel ceramics matrix nanocomposites, Ceram. Int. 26 (2000) 677–683.
[20] W.X. Chen, J.P. Tu, Z.D. Xu, Tribological behavior of carbon-nanotube-filled PTFE composites, Tribol. Lett. 15 (3) (2003) 275–278.
[21] D.S. Lim, J.W. An, H.J. Lee, Effects of carbon nanotube addition on the tribological behavior of carbon-carbon composites, Wear 252 (5–6) (2002) 512–517.
[22] W.X. Chen, J.P. Tu, L.Y. Wang, H.Y. Gan, Z.D. Xu, X.B. Zhang, Tribological application of carbon nanotubes in a metal-based composites coating and composites, Carbon 41 (2) (2003) 215–222.
[23] L. Reinert, S. Suarez, A. Rosenkranz, Tribo-mechanisms of carbon nanotubes: friction and wear behavior of CNT-reinforced nickel matrix composites and CNT-coated bulk nickel. Lubricants 4 (2016) 11, https://doi.org/10.3390/lubricants4020011.
[24] Z. Jin, K.P. Pramoda, G. Xu, S.H. Goh, Dynamic mechanical behavior of melt-processed multi-walled carbon nanotube/poly (methyl methacrylate) composites, Chem. Phys. Lett. 337 (2001) 43–47.
[25] J.C. Kearns, R.L. Shambaugh, Polypropylene fibers reinforced with carbon nanotubes, J. Appl. Polym. Sci. 86 (2002) 2079–2084.
[26] K.T. Kim, S.I. Cha, S.H. Hong, Hardness and wear resistance of carbon nanotube reinforced Cu matrix nanocomposites, Mater. Sci. Eng. A 449–451 (2007) 46–50.
[27] T.W. Scharf, A. Neira, J.Y. Hwang, J. Tiley, R. Banerjee, Self-lubricating carbon nanotube reinforced nickel matrix composites, J. Appl. Phys. 106 (2009) 013508.
[28] B.K. Yen, T. Ishihara, On temperature-dependent tribological regimes and oxidation of carbon-carbon composites up to 1800°C, Wear 196 (1996) 254–262.
[29] K.T. Lau, H. David, Effectiveness of using carbon nanotubes as nano-reinforcements for advanced composite structure, Carbon 40 (9) (2002) 1605–1606.
[30] Q.M. Gong, Study on Manufacture Process and Properties of C/C Composites Used for Aircraft Brakes, (PhD thesis) Central South University, PR China, 2002.
[31] B.K. Yen, Influence of water vapor and oxygen on tribology of carbon materials with sp2 valence configuration, Wear 192 (1–2) (1996) 208–215.
[32] M. Endo, Y.A. Kim, Y. Fukai, T. Hayashi, M. Terrones, H. Terrones, M.S. Dresselhaus, Comparison study of semi-crystalline and highly crystalline multiwalled carbon nanotubes, Appl. Phys. Lett. 79 (10) (2001) 1531–1533.
[33] C.A. Cooper, D. Ravich, D. Lips, J. Mayer, H.D. Wagner, Distribution and alignment of carbon nanotubes and nano fibrils in a polymer matrix, Compos. Sci. Technol. 62 (2002) 1105–1112.
[34] J. Tan, T. Yu, B. Xu, Q. Yao, Microstructure and wear resistance of nickel-carbon nanotube composite coating from brush plating technique, Tribol. Lett. 21 (2006) 107–111.
[35] A.D. Moghadam, E. Omrani, P.L. Menezes, P.K. Rohatgi, Mechanical and tribological properties of self-lubricating metal matrix nanocomposites reinforced by carbon nanotubes (CNTs) and graphene—a review, Compos. B Eng. 77 (2015) 402–420.
[36] S. Suarez, F. Soldera, O.C. Gonzalez, D. Acevedo, F. Mucklich, Thermomechanical behavior of bulk Ni/MWNT composites produced via powder metallurgy, Adv. Eng. Mater. 14 (2012) 499–502.
[37] S. Suarez, F. Lasserre, F. Mucklich, Mechanical properties of MWNT/Ni bulk composites: influence of the microstructural refinement on the hardness, Mater. Sci. Eng. A 587 (2013) 381–386.
[38] S. Suarez, F. Lasserre, O. Prat, F. Mucklich, Processing and interfacial reaction evaluation in MWNT/Ni bulk composites, Phys. Status Solidi 211 (2014) 1555–1561.

CHAPTER 13

Sustainable tribology: Processing and characterization of multiscale thermoplastic composites within hydropower applications

Prabakaran Saravanan[a] and Nazanin Emami[b]
[a]Assistant Professor, Department of Mechanical Engineering, BITS-Pilani, Hyderabad, Telengana, India
[b]Polymer Composites and Biotribology Group, Division of Machine Elements, Luleå University of Technology, Luleå, Sweden

1 Introduction

Hydroelectric power is the most preferred way of energy production among all renewable energy sources due to constant and reliable energy production throughout all seasons with precise control over the demand and load. Hydrodynamic sliding bearings, that is, journal bearings, thrust bearings, and shaft bushings, are commonly used to support the rotating parts such as turbines and others. However, there are challenges to the effective operation of hydropower turbine bearings. Hydropower turbine bearings have to be operated at low speed and extremely high load conditions, which pushes the operation of the bearing that supports the turbine to extreme contact pressure conditions (over ~30 MPa) [1, 2]. Also, longevity of operation (~40 years) is a main requirement because the maintenance of these massive machines affects production line. It is a significantly time-consuming and expensive process and it depends on the wear durability of the material pairs as well as effective lubrication. Therefore, failure of the bearing due to high friction and wear is inevitable if suitable lubrication is not employed in place along with the right material. That leads to the next challenge of this application, that is, the kind of lubrication to be used. Conventional fossil fuel-based oil lubrication has a significant impact on the river environment and ecosystems such as flora and fauna due to oil leakage and spillage into rivers. Fig. 1A shows a digital photo of actual oil spillage into a river from a hydropower station. Fig. 1B presents breakdowns of total oil spillages into marine environments annually [3], and the lubricant leakage is ~10%.

Tribology of Polymer Composites
https://doi.org/10.1016/B978-0-12-819767-7.00013-X

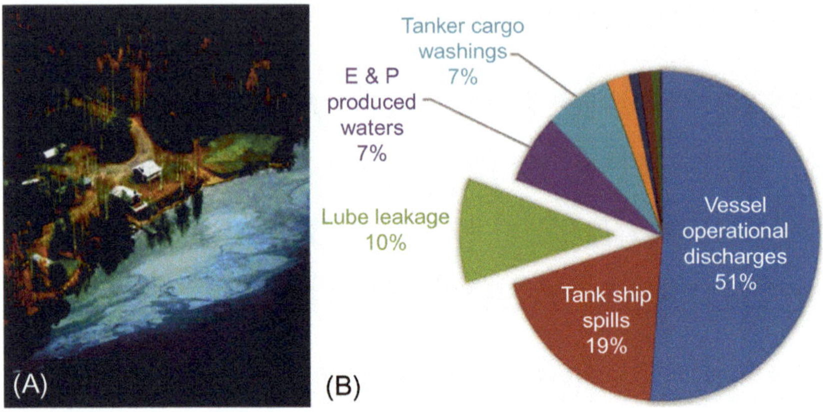

Fig. 1 (A) Digital image of oil spillage from a hydropower station to a river. (B) Split of annual oil inputs to the marine environment [2]. *(Reproduced with permission from A. Golchin, Polymeric Materials for Bearing Applications: Tribological Studies in Lubricated Conditions (Doctoral Thesis), Luleå University of Technology (Online open access), 2015.)*

Therefore, more environmentally friendly, sustainable, biodegradable, and nontoxic lubricants have to be developed for environmental preservation. Water or water-based lubricants have many advantages such as abundance, highly economical nature, sustainability, nontoxicity, and high heat capacity. Despite these advantages, water is considered a poor lubricant for harsh engineering applications because of it's low viscosity, solvent nature (corrosiveness), high volatility, and many other physical properties. Table 1 provides a comparison between water and other common liquid lubricants [4].

Table 1 Physical Properties of common liquid lubricants at atmospheric pressure.

	Dynamic viscosity (η) ($\times 10^{-3}$Pa s)			**Pressure-viscosity coefficient (α) ($\times 10^{-9}$ Pa^{-1})**		
Lubricants	**30°C**	**60°C**	**100°C**	**30°C**	**60°C**	**100°C**
High VI—light machine oil	38	12.1	5.3	–	18.4	13.4
High VI—heavy machine oil	250	50.5	12.6	25.0	21.3	17.6
Medium VI—spindle oil	18.6	6.3	2.4	20	16	13
Low VI—heavy machine oil	310	44.2	9.4	34.6	26.3	19.5
Water	0.80	0.47	0.28	0	0	0

Data taken from G.W. Stachowiak, A.W. Batchelor, Engineering Tribology, 2013.

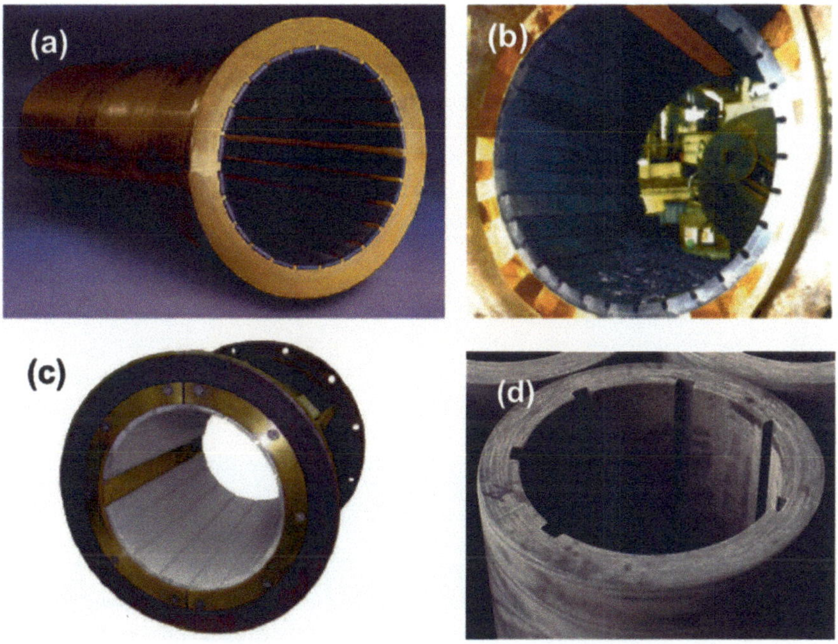

Fig. 2 Commercially available polymer bearings for water-lubricated conditions. (A) Romor; (B) Vesconite; (C) Thordon SXL; and (D) Trelleborg Orkot TXM marine [2]. *(Source: https://www.tss.trelleborg.com/.)*

However, keeping the environment as the primary focus, the material selection can be made more wisely to make the water lubrication possible. Polymers, particularly thermoplastics and their composites, are the most promising materials for water-based lubrication compared to metal, as corrosion becomes an important issue. Fig. 2 shows some of the commercial polymer bearings for water-lubricated applications. Based on the scientific knowledge accumulated over the years on the tribology of polymers in water-lubricated conditions, the important points about thermoplastic composite development, manufacturing, and characterization are discussed.

2 Thermoplastics and thermoplastic composites in hydropower applications

2.1 Thermoplastic polymers of interest

Fig. 3 shows a pyramid of thermoplastic polymers that is classified by structural order (i.e., amorphous and crystalline), application, operating temperature range, and performance. The polymers used to develop multiscale

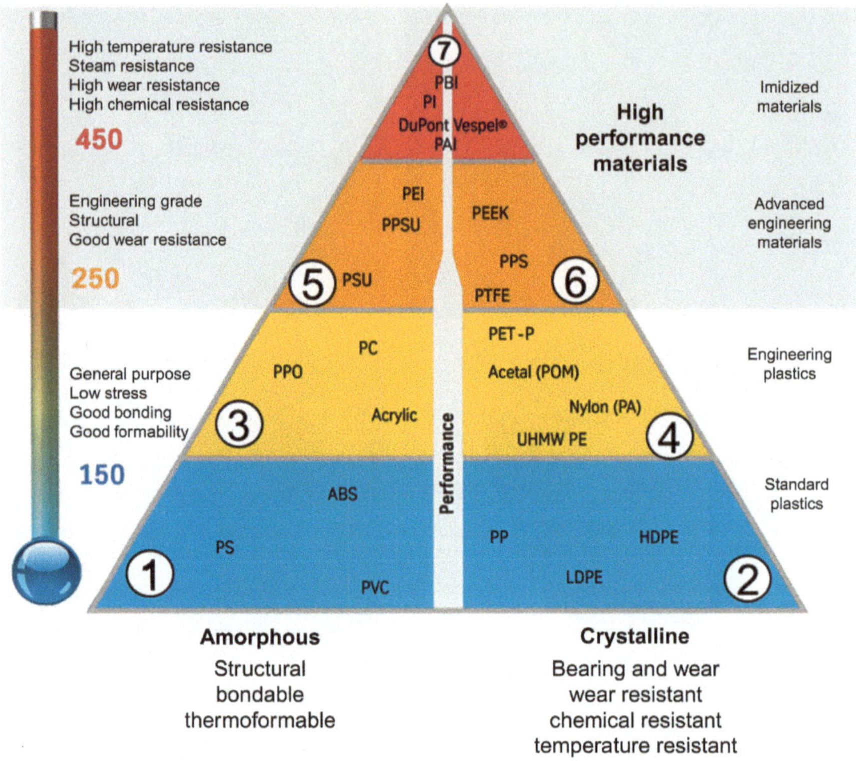

Fig. 3 Thermoplastic polymer pyramid showing seven different regions and the corresponding grouping of polymers with respect to temperature, crystallinity, and applications. Polymers in Regions 4 and 6 are interesting polymers (i.e., UHMWPE, PEEK, PPS, and PTFE) for water-lubrication applications [5]. *(Source: https://www.onlineplastics.com/).*

composites for demanding applications such as hydropower bearings can be found in Regions 4 and 6. Within demanding applications in polymer tribology, polymers with high stiffness, strength, and toughness are of interest. Therefore, parameters such as crystallinity, molecular weight, and the processing (melting) temperature of the polymer are important. It is known that the amorphous regions in the polymer allow water absorption, leading to the swelling of the polymer, which causes a shorter service life. Therefore, higher crystallinity is preferred for a polymer matrix in water-lubricated applications. Similarly, polymers such as UHMWPE, PTFE, PPS, and PEEK are reported to have better composite processability (i.e., compression molding) than other polymers in the pyramid, which is again determined by the melting point of the polymers. Hence, most of the available research on using thermoplastic polymers and their composites for hydropower sliding bearing applications will be limited to the polymers in Regions 4 and 6 in Fig. 3.

2.2 Previous investigations of thermoplastics for hydropower applications

Numerous research works have been carried out on the tribological behavior of polymer/metal tribo-pairs in the presence of water as the lubricant [6–10]. In one of our earlier works, 11 polymeric materials were tested for water-lubricated bearing applications [5]. These polymers were chosen based on performance, cost, and usage [5]. The 11 polymers can be grouped as follows: 1. *Ultrahigh-performance polymers*: polyether ether ketone (PEEK); 2. *High-performance polymers*: polytetrafluoroethylene (PTFE) and polyvinylidene fluoride (PVDF); 3. *Engineering polymers*: ultrahigh molecular weight polyethylene (UHMWPE), polyoxymethylene (POM), polymethyl methacrylate (PMMA), polyamide 6 (PA6), polyamide 66 (PA66), polycarbonate (PC), and polyethylene terephthalate (PET); 4. *Common polymer*: polypropylene (PP). A commercially available bearing material for water lubrication, lignum vitae, was also tested for performance comparison. The tribological test conditions of contact pressure, sliding speed, and test duration were 5 MPa, 0.13 m/s, and 20 h, respectively, and the lubricant was distilled water. The choice of test condition was made to mimic the conditions of a hydropower bearing application during the running time [5]. Table 2 summarizes the mechanical and physical properties of some common engineering polymers used in this study [5]. Fig. 4A and B shows summaries of the coefficient of friction (COF) and the specific wear rate (SWR), respectively [5]. PTFE showed the lowest COF of all, about ~0.06, followed by UHMWPE at ~0.12, while the traditional bearing material, lignum vitae, had a COF of ~0.25, which was four- and two-fold higher than PTFE and UHMWPE, respectively. In this study, a significant reduction in COF was achieved by using modern polymers. The wear rates show completely different trends than that of COF. For instance, PTFE showed the lowest coefficient of friction but a higher wear rate (~10^{-4} $mm^3/N m$). The lowest wear rate (or high wear resistance) was observed for UHMWPE (~0.5×10^{-6} $mm^3/N m$). Some of the other polymers such as PET and POM performed better than lignum vitae in water-lubricated conditions. However, UHMWPE was the best polymer in terms of friction and wear resistance out of the 11 selected polymers. Therefore, UHMWPE has great potential as a bearing material in water-lubricated applications. Moreover,

Table 2 Physical and mechanical properties of common engineering polymers [5].

Material	Density (g/cm^3)	Elastic modulus (GPa)	Tensile strength (MPa)	Flexural strength (MPa)	Yield strength (MPa)	Elongation at break (%)	Hardness (MPa)	Toughness (kJ/m^2)	Melting point (°C)	T_g (°C)
UHMWPE	0.93	0.8	23	40	23	> 450	65	NA	135	−160
PET	1.38	3.4	90	NA	90	50–80	170	3.5	225	76
PP	0.91	1.3–2	33	NA	30–32	700	60	NA	175	−17
POM	1.42	3	70	100	70	70–75	140	NA	175	−75
PTFE	2.15	0.4	25–36	18–20	30	400	30	NA	300–310	127
PEEK	1.30	NA	70	NA	100	50	M99	NA	334	145
PMMA	1.15	1.9	NA	NA	46	5	NA	NA	230–260	107
PVDF	1.76	0.8	40	65	55–60	25/500	110R	300	170	−35
PA6	1.15	1.5	50	40	50	200	70	> 7	220	−8
PC	1.15	2.3	65	90	60	> 80	100	NA	230	150
PA66	1.19	1.7	70	42	70	150	100	> 5	255	−6
PPS	1.35	NA	110	131	103	2.2	121R	NA	307–329	85

the results presented here for are for pristine polymers, and there is a scope for further improvement with the addition of micro- and nanofillers.

UHMWPE is a rather low-cost engineering thermoplastic that has good chemical resistance, self-lubricity, a very low COF, and good wear resistance. As far as water-based lubrication is concerned, it has great potential as a bearing lining material due to its low water absorption, low friction, and superior wear resistance [2, 5]. UHMWPE has a superior load-bearing capacity, great impact strength, abrasion resistance, and toughness due to its long linear molecular chains [11, 12]. It is mostly used in orthopedic implants, food and beverage machinery, personal armor, recreation, and general manufacturing because of its biocompatible nature [11, 12]. With these superior properties, incorporating the fillers/reinforcements into the polymer matrix can improve the mechanical and tribological properties significantly. A different scale of reinforcement (macro, micro, and nano sizes) has different effects on the final composites. The nanofillers provide a more noticeable improvement in various properties over the conventional micro- and macrofillers due to the fact that the surface-to-volume ratio becomes larger in nanofillers, which enhances the interfacial adhesion between the filler and the matrix [13, 14].

The investigation of the mechanical and tribological performances of short carbon fiber-(SCF) reinforced UHMWPE composites in dry conditions showed an increase in the yield strength and Young's modulus and an improvement in the creep behavior compared with pristine UHMWPE [15]. Furthermore, reinforced UHMWPE composites have shown a noticeable improvement in the coefficient of friction and wear rate compared to unfilled polymer in dry test conditions, with one exception where increasing the CF content above 8 wt.% showed an increase in wear [15]. The main findings from another study on the tribological behavior of CF-reinforced UHMWPE were that the water-lubricated contacts resulted in lower friction and wear for CF-reinforced UHMWPE compared to the pure UHMWPE. In this study, abrasive wear was found to be the dominant wear mechanism of CF-reinforced UHMWPE [16].

The results from studies where carbon-based nanoreinforcements were added to thermoplastic as reinforcements are presented here. The addition of nanodiamonds (NDs) to polymers resulted in significant improvement in the mechanical and tribological properties of polymers [13, 17–20]. The interfacial strength between the NDs and the polymer matrix can be altered because NDs have unique morphological and surface chemical properties as nanofillers [18]. The incorporation of 0.2% NDs into a

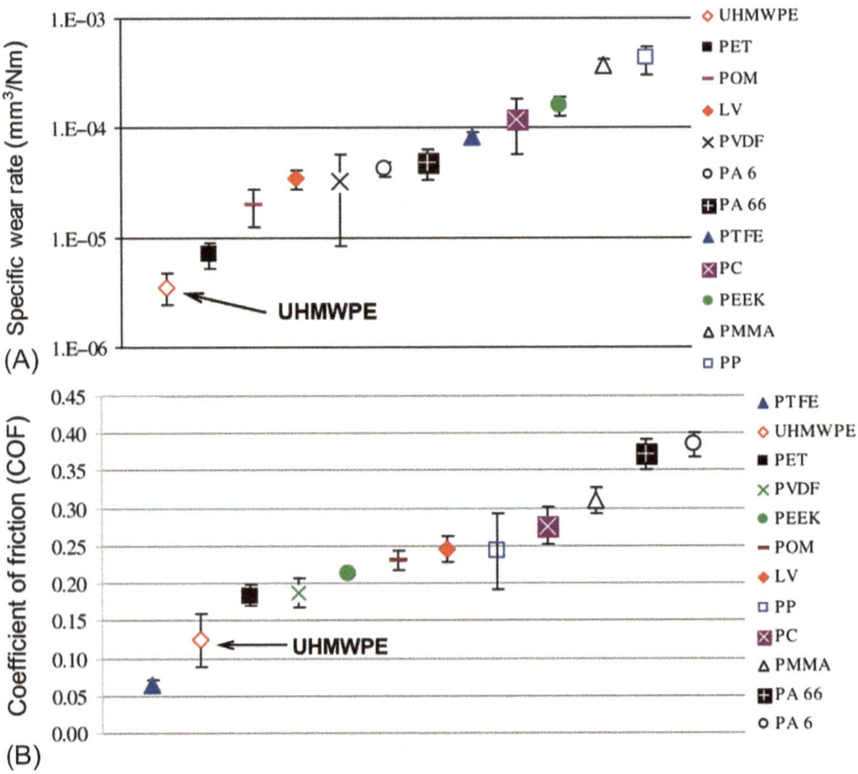

Fig. 4 The average COF and specific wear rates over the test period of 20 h in water-lubricated conditions for different polymers [5]. *UHMWPE*, ultrahigh molecular weight polyethylene; *POM*, polyoxymethylene; *PET*, polyethylene terephthalate; *PA6*, polyamide 6; *PA66*, polyamide 66; *PTFE*, polytetrafluoroethylene; *PP*, polypropylene; *PVDF*, polyvinylidene fluoride; *PEEK*, polyether ether ketone; *PMMA*, polymethyl methacrylate; *PC*, polycarbonate; *LV*, lignum vitae. *(Reproduced with permission from A. Golchin, G.F. Simmons, S. Glavatskih, B. Prakash, Tribological behaviour of polymeric materials in water-lubricated contacts, Proc. Inst. Mech. Eng. Part J: J. Eng. Tribol. 227 (2013) 811–825, https://doi.org/10.1177/1350650113476441.)*

polyurethane/epoxy (PU/EP) polymer matrix by Xia et al. [19] resulted in an increase of storage modulus, tensile strength, and wear resistance. A synergistic effect from NDs acted as a cross-linking point that restricted the movement of polymer chains, resulting in improved mechanical and wear resistance properties.

Another work that investigated the mechanical properties of epoxy composites with relatively higher contents of NDs (25%) reported a four- and three-fold increase in Young's modulus and hardness, respectively [20]. Apart from the characterization of mechanical properties, the tribological performances of the composite materials were also investigated in quite a few works. Among them, Neitzel et al. [21] observed a fourfold reduction in friction when 7.5 vol.% NDs were added to the epoxy matrix sliding against the steel counter face. In one of our previous works, the tribological characterization of a newly developed UHMWPE + ND composite in water-lubricated conditions was investigated [22]. A 72% and a 24% reduction in wear and friction coefficient were reported, respectively, compared to pristine UHMPWE. The main postulation is that the NDs may act as nanorollers in the contact, thereby shifting the contact from pure sliding to rolling-sliding configurations.

Another engineering thermoplastic is polyphenylene sulfide (PPS), a high-performance semicrystalline polymer that is extensively used in the automotive and aerospace industries due to its strength, high service temperature, chemical stability, and its low linear coefficient of thermal expansion [6, 23]. The polymer can cater to high-demanding tribological applications such as hydropower sliding bearing applications if the tribological properties are further enhanced [6]. Various reinforcements or fillers are incorporated into the PPS polymer to enhance the tribological properties of the polymer. The literature reports that the addition of SCFs into PPS results in reductions in the specific wear rate and friction coefficient [6, 15]. Apart from tribological behavior, the PPS is a relatively brittle polymer with a low fracture toughness [24]. Especially under higher-demanding tribological conditions, an increased toughness is required to prevent fracture of the bearing material. The fracture toughness of PPS could be improved by the addition of SCFs [25]. The addition of PTFE to PPS has been regarded as effective in improving the fracture toughness, resulting in better transfer film formation under sliding conditions [26]. PTFE is known as an exceptional solid lubricant polymer with good tribological properties and a higher fracture toughness than PPS. Therefore, the PPS/PTFE polymer blend is an excellent matrix to create novel multiscale composites [27].

In a recent study, both micro- and nano fillers such as PTFE, SCFs, graphene oxide (GO), and NDs were added to PPS to improve the tribological properties and fracture toughness. All the PPS multiscale composites were manufactured using the direct compression molding (DCM) technique [28]. Fig. 5 shows the summary of the coefficient of friction (COF) and the specific wear rate (SWR) of the PPS composites tested at a contact pressure and sliding speed of 1 and 5 MPa and 0.5 m/s under water lubrication, respectively. Overall, the higher contact pressure of 5 MPa showed marginally lower COF and wear rates than the lower contact pressure of

1 MPa in almost all cases. PPS + 10% SCF and PPS + 10% PTFE showed the lowest COF out of five samples while PPS + 10% SCF provided the lowest wear rate, which was about ~2 orders of magnitude less than all the samples tested here. SCF proved to be the right filler material for PPS to reduce the COF and wear considerably, irrespective of the contact pressures.

Another investigation from our own group where multiscale carbon-based fillers such as short carbon fibers (SCF), multiwalled carbon nanotubes (MWCNTs), and graphene oxide (GO) were added to PPS and multiscale carbon-based PPS composites were developed with different combinations [6]. The conditions were 5 MPa as the contact pressure, 0.13 m/s as the sliding speed, and distilled water as the lubricant, respectively. SCF provided a significant reduction in COF and wear rates out of eight different tested composites. Particularly, no appreciable synergistic effect was observed with respect to the friction and wear rates. Therefore, SCF seemed to be the right filler material for the PPS thermoplastic matrix to reduce the COF and wear considerably here as well [6].

To summarize, the addition of microfillers such as SCF and glass fibers (GF) to thermoplastic polymers resulted in improvements in composite strength, stiffness, and toughness, which subsequently increased the wear resistance and sometimes reduced the coefficient of friction. Similarly, nanofillers such as NDs, MWCNTs, and 2D materials such as graphene, molybdenum disulfide (MoS_2), and nanoparticles added to polymers improved the surface finish and heat dissipation in developed composites, which positively affected the reduction of the coefficient of friction. Therefore, multiscale polymer composites are recommended for various engineering applications by combining different dimensional-scale (i.e., nano-, micro-, and macroscales) fillers to achieve superior properties in composites (i.e., mechanical, tribological, thermal, electrical, etc.) [22, 29, 30]. It is often called the synergistic effect, where multiple filler materials were used to achieve much superior properties (i.e., mechanical, tribological, chemical, etc.) [22, 31]. Furthermore, the incorporation of fillers also the improves the processability of the polymers, besides the improvement achieved in the multiproperties of the developed thermoplastic composites [32]. Excellent adhesion between the polymer matrix and fillers and improved processability are two important factors when designing new polymer composites for tribological applications. This can be achieved through complex manufacturing processing including the following steps: tailor the target surface functionalization of the reinforcement; optimal exfoliation of the nanoreinforcement; wet-solvent casting; and chemistry matrix-reinforcement

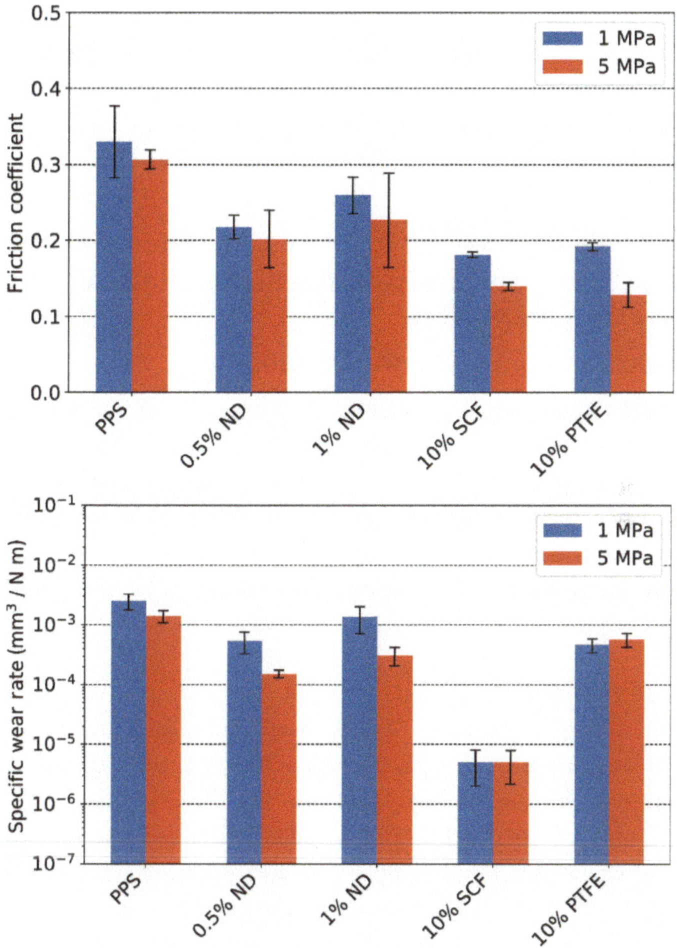

Fig. 5 Summary of the averaged COF and specific wear rates for different PPS multiscale composites in water-lubricated conditions [2]. *PPS*, polyphenylene sulfide; *0.5% ND*, PPS + 0.5 wt.% nanodiamonds (ND); *10% SCF*, PPS + 10 wt.% short carbon fibers (SCF); *10% PTFE*, PPS + 10 wt.% polytetrafluoroethylene (PTFE). *(Reproduced with permission from A. Golchin, Polymeric Materials for Bearing Applications: Tribological Studies in Lubricated Conditions (Doctoral Thesis), Luleå University of Technology (Online open access), 2015.)*

blending, combined with mechanical processing such as ball milling and consolidation through the compression molding process, also used in earlier research works [33].

Here in this chapter, detailed descriptions of processing, manufacturing, tribological, and pre- and posttribological characterizations of thermoplastic-based multiscale composites are presented. Subsequently, a detailed description on the distinctive hygrothermal aging phenomenon that is critical for water-lubricated applications is also discussed. Finally, the friction and wear trends of selected thermoplastic-based multiscale polymer composites developed by incorporating various micro- and nanofillers for water-lubricated high-load and low-speed hydropower bearing applications are also presented.

3 Manufacturing and characterization techniques of thermoplastic composites

3.1 Composite manufacturing techniques

Molding is the most commonly used technique for processing and manufacturing thermoplastic polymers and their composites. Compression, transfer, blow, injection, and extrusion molding are some of the commonly used techniques for manufacturing. For each one of these techniques, finely pulverized plastic particles, either in granular or pallet form, are forced to flow and fill, then assume the shape of a mold cavity by elevated temperature and pressure.

3.1.1 Injection molding

Injection molding is the most widely used technique for the processing of thermoplastic polymers. Injection molding is similar to die-casting in metals, where molten metal is poured into the die and the component assumes the shape of the die. The finely made pelletized polymer material is pushed into a heating chamber by a ram and cylinder arrangement. Then, the thermoplastic material melts to form a viscous liquid, and subsequently, the molted polymer is forced into the mold cavity by a ram motion through a nozzle inlet. Finally, solidification of the molted polymer occurs under the same pressure, then the sample is ejected and the same thing is repeated again and again. The main advantage of this technique is that the rate of production is extremely high and suitable for large-scale mass production, leading to very low manufacturing costs. The main limitation is that the polymers of a higher molecular weight (i.e., UHMWPE) cannot be molded [2, 33, 34].

3.1.2 Compression molding

Compression molding is a very effective technique for thermoplastic composite manufacturing, where appropriate amounts of either polymer materials or composite mixtures (polymers with fillers) are filled between male and female mold members. Then, the mold is closed and both mold members are heated and pressed against each other. Due to the heating, the polymer melts and conforms to the shape of the mold, and a solidified final product is released. The other parameters such as temperature, pressure, pressing cycles, and time duration must be chosen based on the base polymer material and its properties. This process is very time-consuming, expensive, and unsuitable for large-scale mass production. Polymers with a higher molecular weight and melt viscosity can be processed using compression molding, unlike extrusion or injection molding. Compression molding is often used in our research as well [33–35].

3.1.3 Casting

Similar to metals, thermoplastic polymer materials can be cast into desired shapes. A molten plastic material or the polymer is dissolved in a solvent, then poured into a mold and allowed to solidify. If it is molten polymer, solidification occurs upon cooling, whereas in the polymer solvent solution case, the solvent is evaporated by drying in ambient or heated conditions. Casting is a commercially economical, viable, and low-cost technique. The main drawback of this technique is the need for additional heating, which results in polymer oxidation during manufacturing. It also involves additional processing steps such as simultaneous stirring of the solution [34, 35].

3.1.4 Ball milling

Melt mixing is a commonly used technique to achieve a homogeneous composition of thermoplastic polymers and fillers. However, it requires additional processing steps in some cases where the processing temperature is significantly higher than the oxidation temperature of the polymer, leading to the inclusion of solvent in the process to avoid damage to the polymer. The subsequent postsolvent extraction process is tedious [34]. Ball milling is a common technique for grinding minerals, ceramics, and paints into fine powder as well as a technique for mechanical alloying. Ball milling is a better alternative to melt mixing due to the fact that the constituents are mixed in a solid state. Fig. 6A shows a diagram explaining the ball milling process and the milling of an individual grain between the balls. The material to be milled is loaded into a grinding jar along with the milling balls, that is, ceramic and stainless steel balls. The relative speed difference between the grinding jar and balls results in milling. As shown in Fig. 6A (right), the material is milled by compressive forces when the balls collide against each other. Fig. 6B and C shows SEM images of the UHMWPE powder before

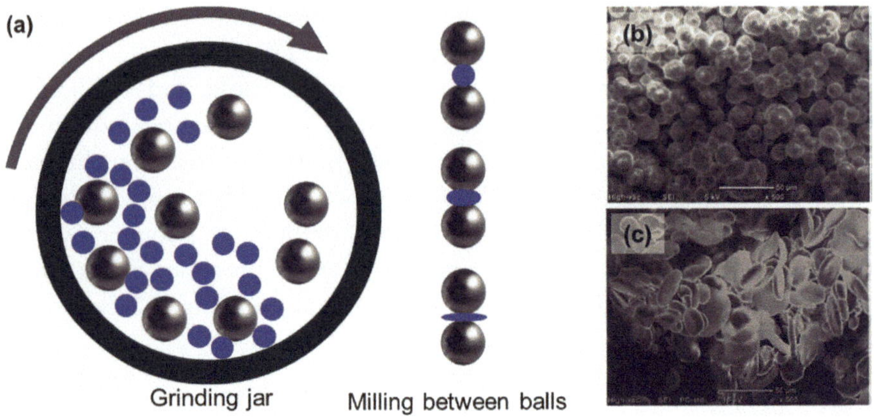

Fig. 6 (A) Schematics depicting the ball milling process and the milling of a single grain of thermoplastic polymer. SEM images of pure UHMWPE powder as an example, (B) premilling and (C) postmilling [33].

and after milling, respectively, as an example to show the morphological transformation experienced by the polymer particles. The morphology (aspect ratio) of the particles changed significantly. The optimum ball milling parameters such as ball-to-powder mass ratio, time, rotational speed, atmosphere, and type of ball mill must be chosen according to the polymer and application [35]. At the end, the ball-milled polymers will be consolidated using injection molding or compression molding.

3.1.5 Composite manufacturing: Sample preparation

Fig. 7 shows the step-by-step schematic of a sample manufacturing process used in our lab that is a combination of a few of the above-mentioned processing methods [33–35]. It involves the following major steps: (1) the characterization of the nanoreinforcement, whether it is processed in the lab or purchased (AFM, Raman, TEM, and XPS are some possible good tools for this purpose), to understand the purity level and evaluate the existing surface functionalization (if there is any); (2) suitable surface functionalization to improve the thermoplastic matrix-reinforcement interface; (3) exfoliation of the nanofillers in the solvent; (4) solvent mixing/blend of exfoliated nanoparticles with polymer powder using a homogenizer; (5) depending on the type of polymer and nanofiller, sometimes physical/mechanical blending using ball milling is the final step of the solvent blending process; (6) oven drying to remove excess solvent; and finally (7) sintering of the polymer composite powder either through direct compression molding (DCM) or injection molding. This is the basic procedure followed for sample preparation [33–35].

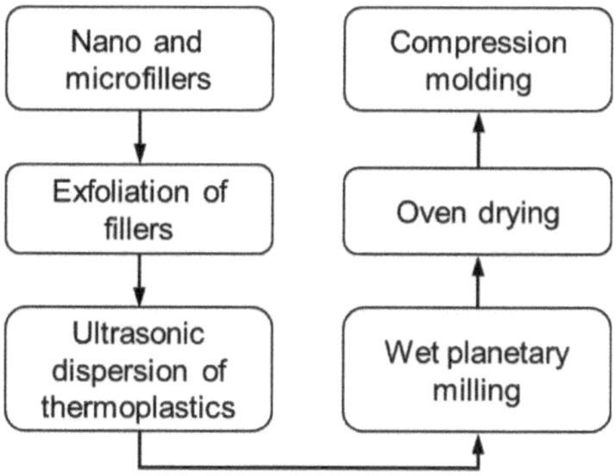

Fig. 7 A simplified step-by-step procedure used to manufacture a thermoplastic-based multiscale composite. *(Data taken from H.S. Vadivel, Tribological Behaviour of Hybrid Carbon Filled UHMWPE Composites in Water (Maters Thesis), Luleå University of Technology, 2016 (Online open access).)*

3.2 Tribological characterization

Sliding tests are performed to obtain the coefficient of friction (COF) and specific wear rates (SWR), which are essential to understand the tribological behavior of any material pair at any sliding and lubrication condition. The polymer specimens will be made to slide/rub against the counter face material in many different configurations. Some of them are pin-on-disc, pin-on-ring, pin-on-flat, pin-on-drum, ball-on-disc, ball-on-flat, cross cylinders, three- and four-ball wear test, twin disc, block-on-block, etc., which are chosen based on the testing application and objectives. Also, the relative motion between the pairs includes rotary, reciprocating, rolling, and a combination of these. Pin-on-disk is the commonly used tribo-testing technique for thermoplastic polymers and polymers in general.

Macroscale tribological friction tests were performed using a pin-on-disc tribometer (it was often used in our studies) to investigate the tribology behavior of thermoplastic-based polymer composites under water-lubricated conditions. The schematic of the tribometer setup is shown in Fig. 8. These pins will be loaded against the counter face materials such as steel, iron, Inconel, copper, or another polymer. The counter face disc will be mounted onto the spindle of the motor, which rotates along with the counter face. If the test is to be done under certain lubricants such as water, oil, or other solvents, there is an additional reservoir setup that will be mounted over the spindle, and the counter face material is placed under the liquid, as shown in Fig. 8. The probe of the linear LVDT is placed onto the dead weights over the polymer pin, and thereby the height of the pin is monitored and

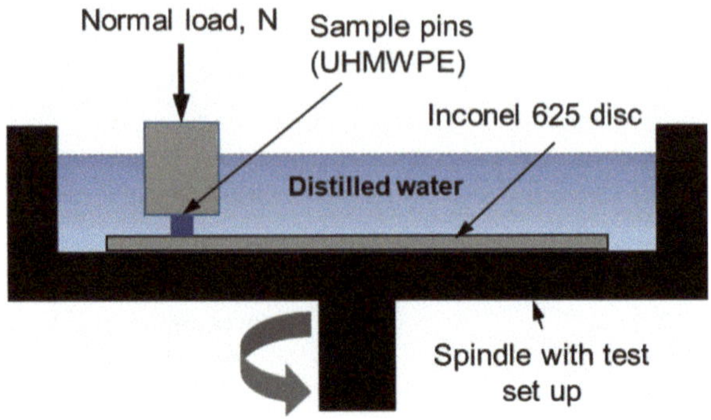

Fig. 8 Schematic diagram depicting the tribological test setup. Sample pins are made of thermoplastic multiscale polymer composites [2].

recorded simultaneously. The in situ recording of the height of the pin is essential for wear rate estimation. Typically, the COF and wear rates are obtained from sliding tests and both are essential to understand the tribological behavior of any material pair at any sliding and lubrication condition.

The coefficient of friction (COF): The steady-state coefficient of friction (COF) was calculated by averaging from the point where the steady-state friction behavior was first observed until the end of the test. Two periods such as running in and steady state were observed for every single sliding test. The time required to achieve the steady state by the tribological setup may vary depending on various aspects. The running-in period is not a representative friction behavior between the material pair as the friction values are not consistent over a period of time. Therefore, the criterion for the friction coefficient must be chosen based on one's own test observations, or ASTM/other society testing standards.

The specific wear rate: The specific wear rate (W_s) of the polymer pins was estimated using Eq. (1), where V, N, and S are the wear volume, normal load, and sliding distance, respectively [36]. The wear volume of the polymer pins was calculated by recording the real-time normal displacement (height) of the polymer pins caused by wear, using a linear variable differential transformer (LVDT). The height difference of the pins before and after the test along with the area provide the wear volume.

$$W_s = \frac{V}{N \cdot S} \tag{1}$$

Our test setup and conditions: Usually, 90 mm diameter Inconel 625 discs were used as a counter surface due to their exceptional anticorrosive nature. Prior to the sliding test, the surface roughness of the discs was increased to $R_a \sim 0.3\,\mu m$ by rubbing against sandpaper, thereby accelerating the testing. Most of the engineered metal surfaces in hydropower plants are in the roughness range of 0.2–0.3 μm. The tribological test conditions such as normal load, contact pressure, sliding velocity, and test duration were 88 N, 5 MPa, 0.13 m/s, and 20 h, respectively.

These conditions are roughly similar to the operating conditions of bearings in hydropower plants, except for the test duration (lifetime). Also, it was observed that almost all the composites attained steady-state behavior about 13 h into the test. Hence, the COF values of the last 5 h of the 20 h test were averaged to obtain the COF of any particular case. The last 5 h of COF data were found to be sufficient and reliable for all the cases tested here. Rotational speeds are varied to maintain the linear speed at 0.13 m/s. The sliding tests were carried out inside the distilled water environment. Each test must be repeated a few times and representative data should be reported. The polymer pins and Inconel discs were thoroughly cleaned with ethanol using an ultrasonic bath prior to the sliding tests.

3.3 Pre- and posttribological characterization

Pre- and posttribological characterizations such as morphological, physical, thermal, and surface chemical characterizations are necessary to evaluate the friction-induced physical and chemical changes experienced by the polymers. Pretribological characterizations were done to ascertain the state of the polymers (i.e., the physical, chemical, and mechanical properties) before the sliding test. Posttribological characterizations were done to investigate the mechanisms such as tribo-induced physical and chemical changes undergone by the polymer and their contributions to the observed tribological behavior. Also, other aspects such as material transfer, surface damage, type of wear, material compatibility, etc., will be revealed and linked to the observed tribological behavior. Overall, it is highly essential to characterize the samples before and after tribological testing for a scientific interpretation of the results.

3.3.1 Scanning electron microscopy (SEM)

A scanning electron microscope is a useful tool to study the morphology of polymer composites, both at the powder stage to study the dispersion of the reinforcement within the matrix and also after consolidation (pins-surface) before and after the sliding tests. SEM is also highly used in the characterization of the wear track on the counter face surface, the polymer transfer, and the wear debris inside the track after the test. Energy-dispersive spectroscopy

(EDS) elemental mapping was performed using two detectors coupled to the SEM (Oxford Instruments) to characterize the different elements present in the wear track and transfer film. SEM imaging and EDS analysis together will produce very useful data to understand the micro- and nanostructure materials before tribotesting and provide very useful information on wear mechanisms and tribofilm formation after the tribological tests. The samples usually need to be coated with a thin platinum/gold layer before observation; this is deposited by ion sputtering to prevent surface electron charge damage.

3.3.2 Differential scanning calorimetry (DSC)

Differential scanning calorimetry (DSC) is a useful characterization method to investigate the crystallinity and melting temperature of the thermoplastic multiscale composites. The degree of crystallinity was calculated using Eq. (2):

$$\text{Crystallinity} = \frac{\Delta H}{\Delta H_{100}} \times 100 \tag{2}$$

where ΔH and ΔH_{100} are the enthalpy of the polymer and the enthalpy of fusion for a 100% crystalline polymer, respectively [37]. The polymer composites are produced using both chemical and mechanical processing and manipulation. Therefore, it is highly important to check the degree of crystallinity of the polymer matrix before and after polymer composite manufacturing to assure that during the processing steps, the polymer matrix has not been destroyed or damaged. The crystalline part of the polymer matrix is important for the wear resistance properties and COF.

3.3.3 Thermal gravimetric analysis (TGA)

TGA is a very useful method to evaluate the thermal stability/oxidation degradation of the polymer before and after manufacturing of the new composites and the service lifetime studies of the bearing materials through hygrothermal aging. Hygrothermal aging is a degradation process that deteriorates the properties of the polymer due to water absorption after being in the water for longer durations. The polymer also swells due to water intake and exceeds the tolerances and dimensions. The thermal stability was evaluated according to the weight loss of the samples. The results are presented according to certain temperatures for which the weight loss (%) of the polymer samples was calculated using the remaining weight at the end of the test. TGA analysis was usually carried out in an inert atmosphere with a defined temperature range and heating rate; this depends on the thermoplastic polymer composite sample and its filler.

3.3.4 Fourier-transform infrared spectroscopy

FTIR or similar chemical characterization techniques are used to reveal the likely chemical changes that occurred in the composites caused by hygrothermal aging. Significant chemical changes are expected to happen during the aging of the polymer. The technique consists of measuring the infrared absorption spectrum generated by the vibration of molecules, which is characteristic for each compound. All the thermoplastic polymer samples were subjected to characterization before and after aging in water for a certain duration.

4 Hygrothermal aging of thermoplastics

The thermoplastic-based polymers and their composites may experience various reversible and nonreversible changes depending on environmental conditions such as humidity, temperature, and exposure time during their service lifetime [38, 39]. One of them is hygrothermal aging or, in other words, the water absorption behavior [39, 40]. The water absorption behavior of polymers is a very essential factor while designing the water-lubricated journal bearings. Fig. 9 shows a diagram of thermoplastic polymers before and after aging in the water for a certain time.

Most thermoplastics used for water lubrication are semicrystalline polymers, which means that there are amorphous regions as well. These amorphous regions act as sites for the water to permeate and absorb into the polymer, leading to swelling and changes in the geometry of the polymer [40]. This change might result in a potential failure of the bearing by decreasing the clearance between the shaft and bearing surfaces. Furthermore, the mechanical, tribological, chemical, and physical properties of the polymer composites are highly affected by the absorbed water in the polymer [41]. The literature reports suggest that the molecular chain mobility is amplified with water content, contributing to an increased moisture effect. Then, the absorbed moisture acts as a plasticizer, which decreases the relaxation time [42]. Similarly, this moisture-induced plastic deformation might lower the glass transition temperature (T_g). Particularly, elevated temperatures accelerate the diffusion of water molecules across the polymer [39]. According to the literature, the swelling and plasticization of the polymers are two possible adverse effects caused by water absorption. Swelling is associated with a differential strain caused by an expansion force applied by the liquid while extending polymeric chains. Plasticization prompts plastic deformation along with decreasing the glass transition temperature (T_g) [43]. These two harmful mechanisms are assumed to create capillarity in polymeric composites.

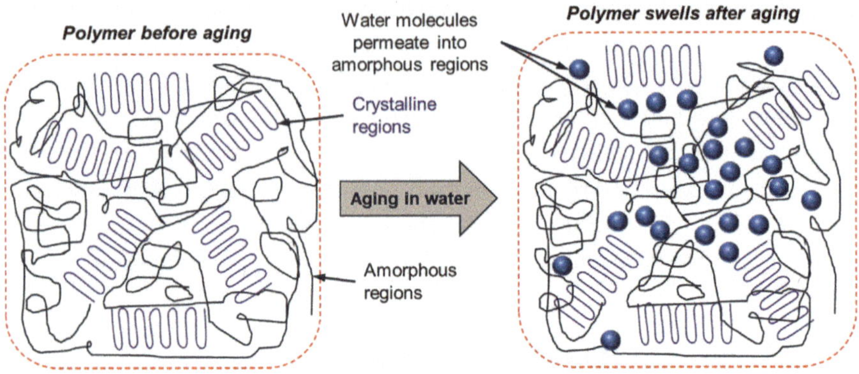

Fig. 9 Diagram depicting the thermoplastic polymer before and after hygrothermal aging [39, 40].

Fig. 10 shows the weight percentage water absorption of a few selected commonly used thermoplastic polymers for an exposure period of 4 months. Most of the polymers had water absorption levels less than 0.2% by weight. However, some polymers such as PA66, PA6, PET, PMMA, and POM displayed larger levels of water absorption. Notably PA6 and PA66 had levels of 5% and 8%, respectively, which were the highest of all the polymers by a significant margin [2]. The potential thermoplastic polymers for hydropower bearing applications such as UHMWPE, PEEK, and PTFE showed significantly lower levels of water absorption. Another important

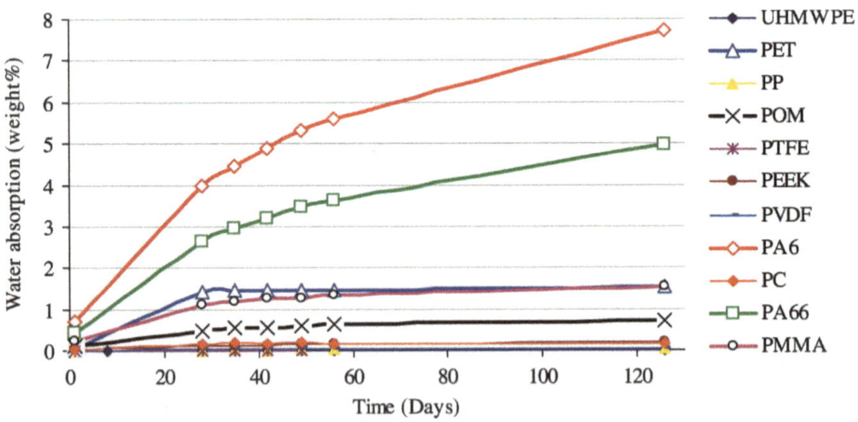

Fig. 10 Water absorption behavior of common thermoplastic polymers after an exposure period of 120 days [2]. *(Reproduced with permission from A. Golchin, Polymeric Materials for Bearing Applications: Tribological Studies in Lubricated Conditions (Doctoral Thesis), Luleå University of Technology (Online open access), 2015.)*

finding is that all polymers reached saturation levels of water absorption after 8 weeks while PA6 and PA66 continued to absorb water beyond this period, which can be seen clearly in Fig. 10. A possible explanation is that the amide groups in the molecular chains of these polymers led to water absorption by forming hydrogen bonds with water molecules owing to their high polarity [2]. Several others also investigated and postulated a similar interaction between the water molecules and polyamide chains, that is, hydrogen bonding and self-association [6, 21].

4.1 Hygrothermal aging calculations

The amount of absorbed water of thermoplastic polymeric material is estimated using gravimetric analysis. The water intake into the polymer is calculated by weighing the sample at regular time intervals. The gravimetric method is conducted using a digital balance with two decimal accuracy. The polymer specimens were immersed in water (usually DI water) for different time durations at different temperatures (that is, 20–80°C range) and the samples were wiped with tissue paper for few minutes before weighing. The uptake moisture of the specimen at a given time, t, is defined by M_t, according to Eq. (3):

$$M_t = \frac{W_t - W_0}{W_0} \times 100\,(\%) \tag{3}$$

where W_0 and W_t are the initial weight of the nonaged specimen and the weight of the wet specimen at time (t), respectively. Detailed explanations on various moisture absorption models can be found elsewhere in the literature [44].

4.2 Hygrothermal aging of PTFE composites: An example

Fig. 7 shows the water absorption characteristics of bulk and unfilled thermoplastic polymers. However, it is important to assess the susceptibility of polymer composites to water absorption and compare with pure polymers. Polytetrafluoroethylene (PTFE) and its composites were subjected to water absorption experiments performed at room temperature (23°C and 80°C) for a period of 132 days, as shown in Fig. 11. The composites were unfilled PTFE, bronze PTFE (PB); 40% bronze, black glass fiber PTFE (GF); 20% glass fiber and MoS_2 + black glass fiber PTFE (PGFM); 20% glass fiber and 5% MoS_2 [39]. These were called PTFE, PB, GF, and PGFM, respectively. Fig. 11 shows that the unfilled PTFE remained resistant to water permeation and exhibited the least weight gain among all samples at both room temperature and a high temperature of 80°C. In the case of room

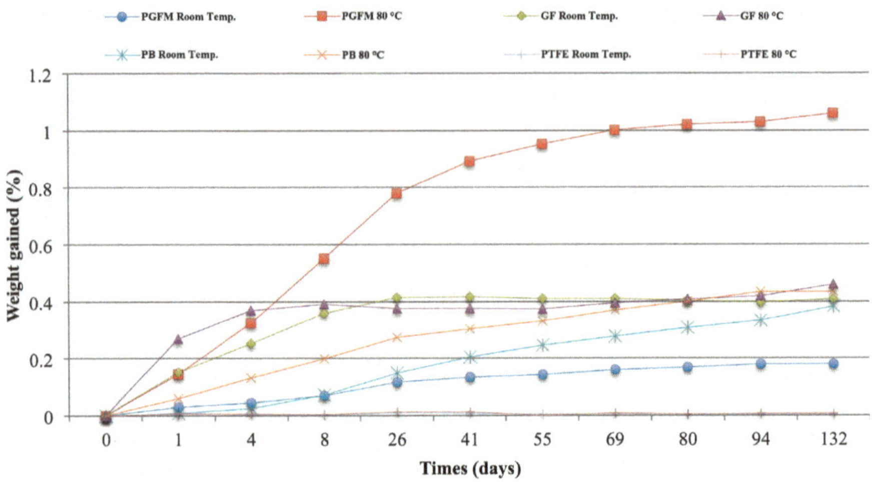

Fig. 11 Water absorption behavior of PTFE and its composites after an exposure period of 132 days [40]. *(Reproduced with permission from M.R. Homayoun, A. Golchin, N. Emami, Effect of hygrothermal ageing on tribological behaviour of PTFE-based composites, Lubricants 6 (2018) 103, https://doi.org/10.3390/lubricants6040103.)*

temperature, the glass fiber-filled PTFE (GF) had the biggest weight gain from water absorption. Also, it conceded the highest diffusivity at both temperatures (RT and 80°C). The maximum weight gain due to water uptake was displayed by GF + MoS_2-filled PTFE (PGFM) at 80°C; this was the highest gain witnessed of all the samples [40].

Concisely, the polymer (PTFE) composites were a lot more susceptible to hygrothermal aging than the unfilled polymers. The aging temperature influenced the water absorption kinetics of the polymers and their composites. A saturation state was reached for almost all the samples within the time frame studied; this trend was common across the majority of thermoplastics and conditions. Furthermore, hygrothermal aging always affects the microstructure robustness of the polymers and their composites, contributing to defects and a reduction in strength [40].

5 The friction and wear trends of thermoplastic-based multiscale composites in water-lubricated conditions: Recent investigations

The friction and wear trends of the two novel thermoplastic-based multiscale composites (UHMWPE and PTFE) at water-lubricated conditions are reported here. The addition of different types and sizes of reinforcements such as micro- (i.e., SCF, GF) and nanofillers (i.e., ND, GO) to these thermoplastic-based polymers resulted in superior properties (i.e., mechanical, tribological) through a synergistic effect obtained via multiscale composite mode. Therefore, multiscale polymeric-based composites are preferred to improve the performances of components in various engineering applications.

5.1 Tribology of UHMWPE multiscale composites

5.1.1 Friction and wear trends

Prior to tribological tests, the prepared sample pins were thoroughly characterized using morphological, surface-chemical, physical, and mechanical characterizations to ensure the integrity of the manufactured composites. These characterizations are not reported here but can be found in the literature elsewhere [2, 33, 35]. Fig. 12A and B summarizes the coefficient of friction (COF) and the specific wear rate (SWR) results of the UHMWPE multiscale composites sliding against the Inconel counter surface in water-lubricated conditions, respectively. The test conditions such as normal load, contact pressure, sliding speed, test duration, temperature, and lubricant are 88 N, 5 MPa, 0.13 m/s, 20 h, room temperature, and DI water, respectively. The unfilled UHMWPE has a COF of ~0.12 while the composites have COF values in the range of 0.97–0.15 [33, 45]. The ND-containing composites such as ND + SCF and GO + ND + SCF provided a significant reduction in friction, irrespective of other fillers present in the composite [33]. Therefore, NDs appear to have a significant influence on the friction of UHMWPE over other fillers used here (GO, SCF).

The wear rate data is very essential because COF alone may not provide a complete understanding of the tribological behavior of the composite. In many occasions, the decrease in wear contributes to a decrease in friction. However, this is not always the case and no proportional relationship between COF and wear rate is always observed. Unfilled pure UHMWPE has $1.4 \times 10^{-6}\,mm^3/N\,m$. All the composites have more or less similar wear rates to that of unfilled UHMWPE. A 1% GO composite exhibited the lowest wear rate ($1.1 \times 10^{-6}\,mm^3/N\,m$) of all the composites tested [33]. Summarizing the sliding test results, GO + ND + SCF is the optimized multiscale composite that provided a significant reduction in friction and the highest wear resistance of all the composites tested in this work. The mechanisms behind the superior tribological properties of the GO + ND + SCF composite will be investigated in the subsequent discussion section.

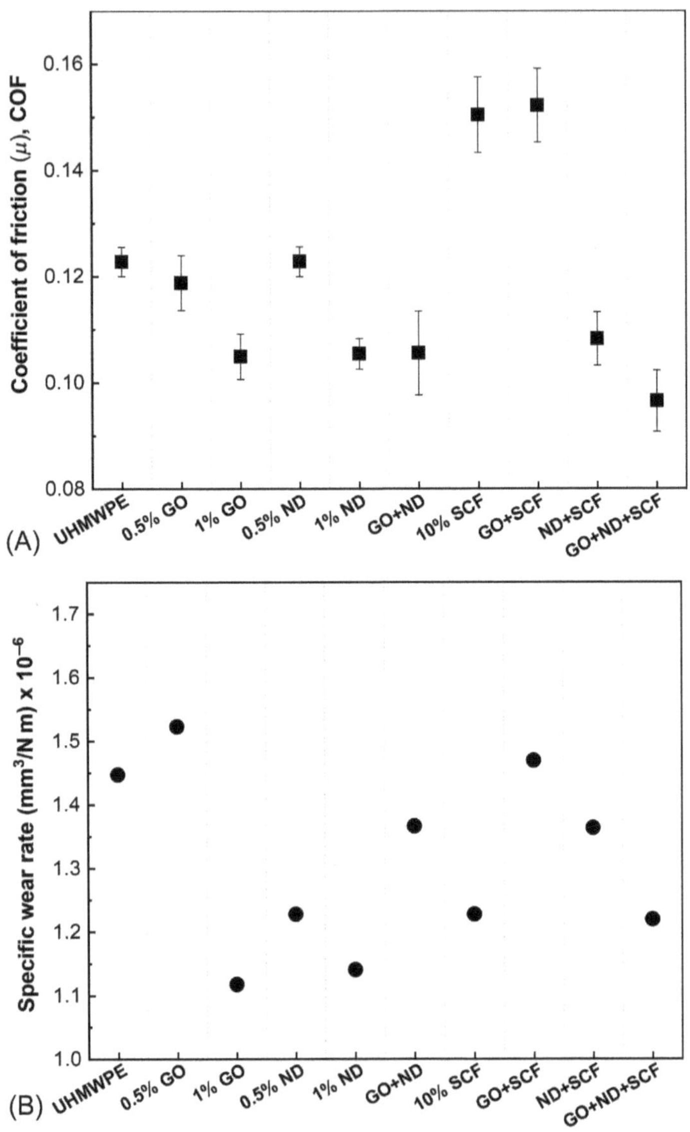

Fig. 12 Summaries of (A) friction coefficients and (B) specific wear rates of unfilled UHMWPE and nine different UHMWPE composites tested in water lubricated conditions [33].

5.1.2 Scientific discussion of results

Given the above findings, the underlying mechanisms of the tribological behavior of the UHMWPE composites can be investigated further by plotting the tribological results as a function of other bulk and surface properties of composites. Crystallinity and hardness values were plotted against the obtained coefficient of friction and wear rates for further data interpretation.

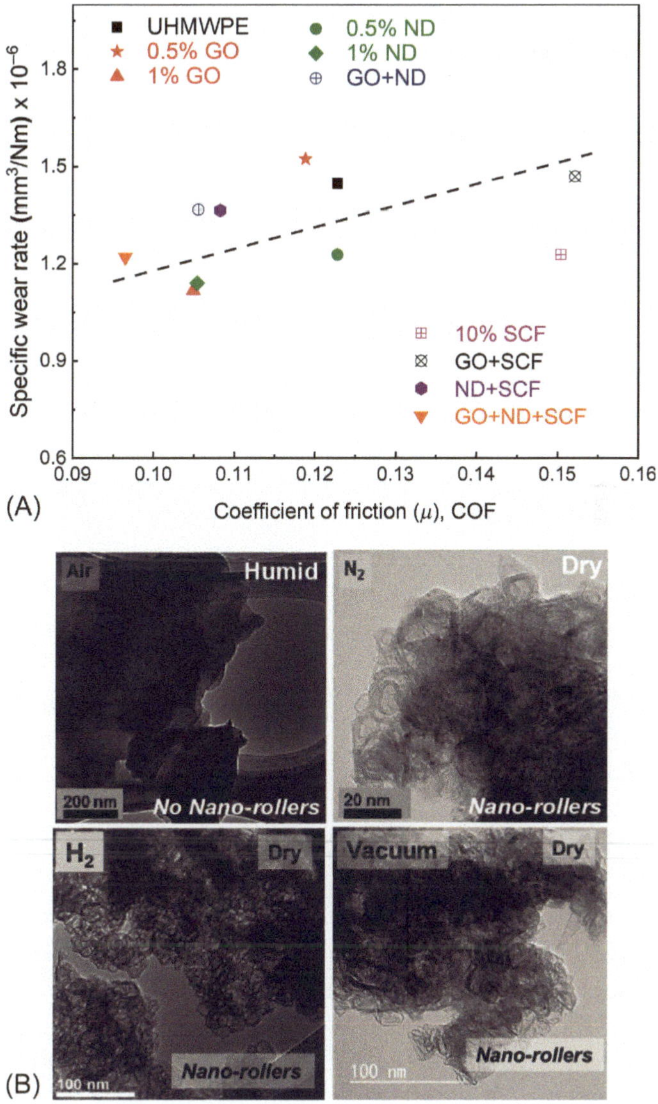

Fig. 13 (A) Specific wear rate (SWR) is plotted against the coefficient of friction (COF) for pure UHMWPE and its composites. (B) Transmission electron microscopy (TEM) images of GO wear debris showing the formation of nanorollers in dry atmospheres, which resulted in superlubricity.

However, no useful trends were observed and therefore, this is not reported here. Fig. 13A shows a correlation between the COF and SWR when COF is plotted against SWR. There appears to be a linear relationship between the COF and wear rate, as indicated by the dotted line in Fig. 13A. All data points are more or less closer to the trend line except the 10% SCF and 0.5

GO composites. Pure UHMWPE has a COF of ~0.12 and has a higher wear rate than most of the composites tested, appearing as a reference point at the center of the graph [33].

Overall, GO and/or ND inclusion to UHMWPE decreased the COF while SCF inclusion increased the COF and reduced the wear rate marginally. Moreover, the incorporation of SCF to UHMWPE increased the friction significantly higher than pristine UHMWPE, except when ND was added among other fillers. GO+SCF and 10% SCF composites showed the highest COFs among all samples while ND+SCF and GO+ND+SCF composites showed a COF much lower than pure UHMWPE; these were, in fact, the lowest of all. This indicates that the ND as a filler has a more significant influence on friction than GO and SCF [33].

NDs plays the role of lubricant effectively in UHMWPE composites, irrespective of the presence of other fillers. In a recent work, superlubricity was achieved between a DLC-coated ball and SiO_2 by the addition of ND to the solution-cast graphene sheets [46]. Subsequently, the ND particles enhanced the formation of graphene nanoscrolls in a dry atmosphere. Therefore, the observed macroscale superlubricity was mainly associated with a reduction in the interfacial contact area by more than 65% and incommensurability between the DLC and graphene scrolls [46]. Also in this work, the NDs were postulated to act as nanoscale ball bearings between the two contacting surfaces, thereby reducing the contact area, friction, and wear. Also, the asperity contact between the surfaces was prevented by the nanoscrolls, which converted the sliding into rolling or partial rolling friction [46–49].

SCF has been found to increase the friction in the literature as well [31]. The increase in COF can be attributed to the SCF in the pin surface sliding and abrading against the counter face surface. The higher hardness of SCFs on the pin surface increases the interfacial shear strength of the contact, thereby increasing the COF. However, SCFs improve the mechanical properties of the composite, which enhances the wear resistance of the composite. Hence, using SCF is a clear trade-off between the mechanical and tribological properties of the composite, as the lubrication effect is stifled [33].

The GO+ND+SCF multiscale composite has emerged as the best composite in terms of COF and SWR among all the composites tested here. As postulated earlier, NDs and SCFs have bigger influences on the friction and wear rate, respectively. Possibly, the presence of the third filler, GO, might have resulted in a synergistic effect where NDs and SCFs delivered their roles in the composite without any interference from other fillers. GO might have provided the sites (functional groups on the surface) for better bonding of SCFs, which reduced the wear rate, while NDs had better bonding with the UHMWPE matrix and behaved as an effective lubricant. This ND+UHMWPE mix might have covered the surface of the SCF, preventing it from direct contact with the counter face [50]. Therefore, a synergistic effect is achieved in both areas of friction and wear. However, the synergistic effect needs further explanation and investigation.

5.2 Tribology of PTFE multiscale composites

5.2.1 Friction and wear trends

Fig. 14 summarizes the COF and SWR results of the unfilled PTFE and three PTFE multiscale composites sliding against the Inconel counter surface in both dry and water-lubricated conditions before and after aging, respectively. The polymer pins were subjected to hygrothermal aging for a period of 60 and 132 days at room temperature and 80°C, respectively, and then used for sliding tests. The pin specimens were unfilled PTFE, bronze PTFE (PB); 40% bronze, black glass fiber PTFE (GF); 20% glass fiber and MoS_2 + black glass fiber PTFE (PGFM); 20% glass fiber and 5% MoS_2 [40]. These were called PTFE, PB, GF, and PGFM, respectively. The test conditions such as normal load, contact pressure, sliding speed, test duration, temperature, and lubricant, were 88 N, 5 MPa, 0.13 m/s, 1 h (sliding distance: 468 m), room temperature, and DI water, respectively. Both the COF and SWR have four graphs each. Each graph has two datasets (water-lubricated and dry) and each dataset has five data points (nonaged to aged (132 days)—80°C) [51].

Unfilled PTFE and its composites showed more or less the same friction coefficient (COF) values before and after aging with different days at different temperatures. This suggests that the aging had a very marginal effect on the friction of PTFE and composites. Unfilled PTFE (all aging and nonaging) had a COF about ~0.13 and 0.19 in water-lubricated and dry conditions, respectively. PGFM had a COF about ~0.21 and 0.22 in water-lubricated and dry conditions, respectively, suggesting the least effect due to the testing environments on friction. PGF had a COF about ~0.33 and 0.15 in water-lubricated and dry conditions, respectively. Finally, the PB had a COF around ~0.26 and 0.22 in water-lubricated and dry conditions, respectively; this is the only case where the aged and nonaged samples showed some variation compared to the other three cases where the effect of aging on friction was almost negligible [51].

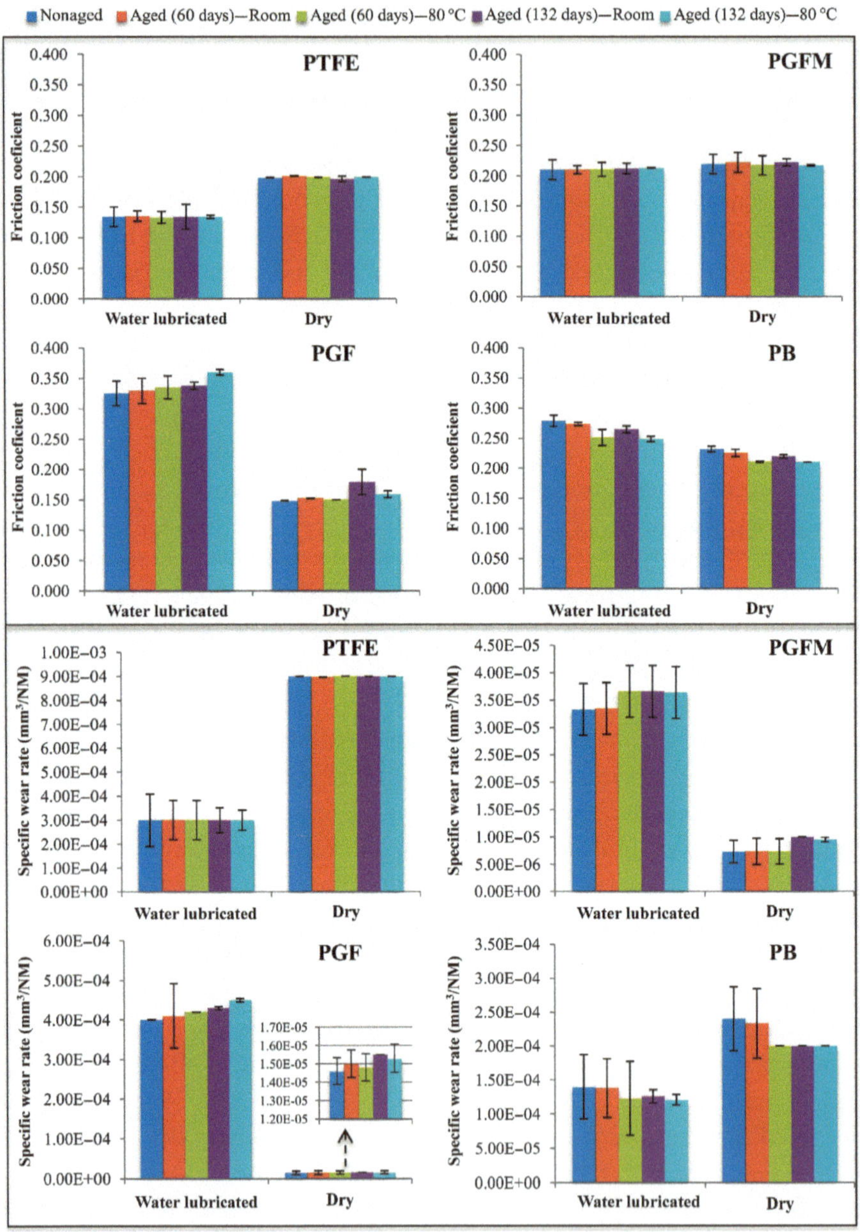

Fig. 14 Summaries of friction coefficients and specific wear rates of unfilled PTFE and three different PTFE composites tested in dry water lubricated conditions before and after aging for two different periods of time (60 and 132 days) at two different temperatures (room temperature and 80°C) [51].

In the case of wear rates, the unfilled PTFE and PB showed ~3 and 1 times lower wear rates under the water-lubricated condition than the dry condition, respectively. In contrast, PGF and PGFM showed a one-order (i.e., 10^{-5}–10^{-6}) lower wear rate under the dry condition than the water-lubricated condition. This indicates that water lubrication may not provide low friction and wear for all polymer composite cases. Furthermore, the influence of hygrothermal aging on the friction and wear of PTFE and its composites was also very marginal, except for a few cases [51].

5.2.2 Scientific discussion of results

In summary, the lowest friction coefficient in the dry condition was observed for PGF, followed by unfilled PTFE, PGFM, and PB. In the water-lubricated condition, the lowest friction coefficients were observed in the following order: unfilled PTFE, PGFM, PB, and PGF [51]. Additionally, the lowest wear rate in the dry condition was observed for PGFM, followed by PGF, PB, and unfilled PTFE. In the water-lubricated condition, PGFM showed the least wear rate, followed by PB, PTFE, and PGF. Unfilled PTFE had the lowest friction and PGFM had the lowest wear rate, respectively, across all testing conditions [51].

The heat generated at the interface during sliding and also the absence of lubricant can increase the friction and wear. The higher COF and wear rate of PTFE in the dry condition may be attributed to heat generation and the absence of water and the opposite is true for the water-lubricated condition. The lowest COF of PGF in the dry condition is due to the formation of an effective discontinuous transfer film on the counter face, composed of an enriched glass fiber-filled PTFE transfer layer. No such transfer films were formed in the presence of water. PGFM also followed a similar trend as PGF because of the presence of the glass fiber. On the contrary, the friction and wear of the PB (PTFE + 40% bronze) appears to be controlled by an oxide layer formed at the interface that acts as a solid lubricant [51]. Except for unfilled PTFE, the presence of water appears to be interrupting the transfer layer as well as the formation and development in all three composites.

To investigate the wear behavior further, differential scanning calorimetry (DSC) characterization was performed for PTFE and its composites before and after aging at room temperature (RT) and 80°C for time periods of 60 and 120 days. The melting temperature and relative crystallinity values obtained through DSC analysis are summarized in Table 3. The crystallinity values are important to understand the tribological behavior in dry

Table 3 Melting temperature and crystallinity of PTFE and its composites [51].

Polymer samples	Melting temperature (°C)	Relative crystallinity (%)				
		Nonaged samples	60 days, RT	60 days, 80°C	132 days, RT	132 days, 80°C
Unfilled PTFE	329	30.46	30.42	30.43	30.47	30.41
PB	330	27.48	25.98	23.46	23.53	23.22
PGF	329	37.34	37.23	36.80	36.43	35.93
PGFM	328	38.97	38.56	38.51	37.88	37.49

conditions, as there is a clear correlation emerging between the degree of crystallinity and the specific wear rate values of samples tested in dry conditions [51]. PGFM with the highest crystallinity showed the lowest wear rate, followed by PGF, PTFE, and PB in that order. Also, the difference in wear rates between the samples became larger with the difference in crystallinity between them. For instance, PGFM and PB had ~15% difference in crystallinity (highest to lowest), which resulted in two orders of difference in the wear rates (i.e., 10^{-6}–10^{-4}) [51]. This good agreement between the wear rates and crystallinity is due to the fact that the heat generated by friction cannot be dissipated and there is no presence of additional lubricants such as water. These factors make this system simple and make the prediction of wear rates possible.

However, no distinctive trend was observed between the wear rates and the crystallinity values of the polymer samples tested under the water-lubricated condition as opposed to the dry case. The PGFM with the highest crystallinity showed the lowest wear rate, followed by PB, PTFE, and PGF showing more or less similar values of wear rates irrespective of their crystallinity values. The presence of water in the PTFE and PB cases helped to reduce friction and wear by dissipating the heat generated due to friction and also acted as a coolant lubricant. In the case of PB, the water promoted the formation of oxide layers at the interface, which reduced the friction and wear [51]. However, in the PGF and PGFM cases, the water prevented the formation of an effective transfer film on the counter surface as opposed to the dry condition, where glass fiber-rich PTFE transfer films were formed [51]. Though sufficient results and discussions are presented on the effect of hygrothermal aging on the friction and wear of PTFE composites, still, future work is necessary to further investigate the friction and wear trends of PTFE composites.

5.3 Higher contact pressure sliding tests

The friction and wear trends reported in Sections 5.1 and 5.2 were tested at a low contact pressure of 5 MPa. However, the typical contact pressure of hydropower sliding bearing would reach up to 30 MPa [2]. Hence, it is necessary to conduct the sliding tests at higher contact pressure conditions.

The tribological results of selected thermoplastic polymer composites tested at 20 MPa are included here for better understanding. A specially designed, environmentally adaptive lubricant (EAL), which was composed of 50% water and 50% glycerin, was used as a lubricant to enhance the tribological performance of the polymers. Fig. 15A and B summarizes the COF and SWR results of selected multiscale composites sliding against the steel counter surface in dry and EAL (50% water + 50% glycerin)-lubricated conditions, respectively. The other test conditions such as sliding speed, test duration, and temperature were 20 mm/s, 20 h, and room temperature, respectively. The selected thermoplastic composites were classified into three groups. Group 1 has Thorplas blue and Orkot TXM, which are the commercial polymer composites exclusively designed for hydropower sliding bearing applications.

Group 2 has PPS blue (15% PTFE, 15% glass fiber), PPS Svart (10% PTFE, 10% Gr, 10% carbon fiber), PAI duarton (pure PAI), and PAI torlon (12% Gr, 8% PTFE), which are the commercial polymer composites exclusively designed for automotive bearing applications. Group 3 has PPS 1, PPS 2, and PPS 3 (all three have certain amounts of PTFE, SCF, GO, and CNT; the exact percentage cannot be disclosed), which are the in-house made (LTU made) polymer composites exclusively designed for hydropower bearing applications based on our knowledge and expertise. The COFs of all composites were reduced around 50% from dry to EAL-lubricated conditions. Group 1 (Thorplas and Orkot) performed very well in terms of friction and wear, as expected [52]. Interestingly, the Group 3 (LTU made) composites (PPS 1, 2, and 3) showed similar COFs and wear rate values to Group 1 when compared with Group 2, which had the highest COFs and wear rates in both dry and EAL conditions. However, the poor performance of UHMWPE in dry and EAL conditions was highly unexpected, suggesting a need for further investigation [33, 34]. Scientific interpretation of the reported results has yet to be performed because these results are a part of ongoing research. Future work will address further investigations of the tribological mechanisms at higher contact pressures.

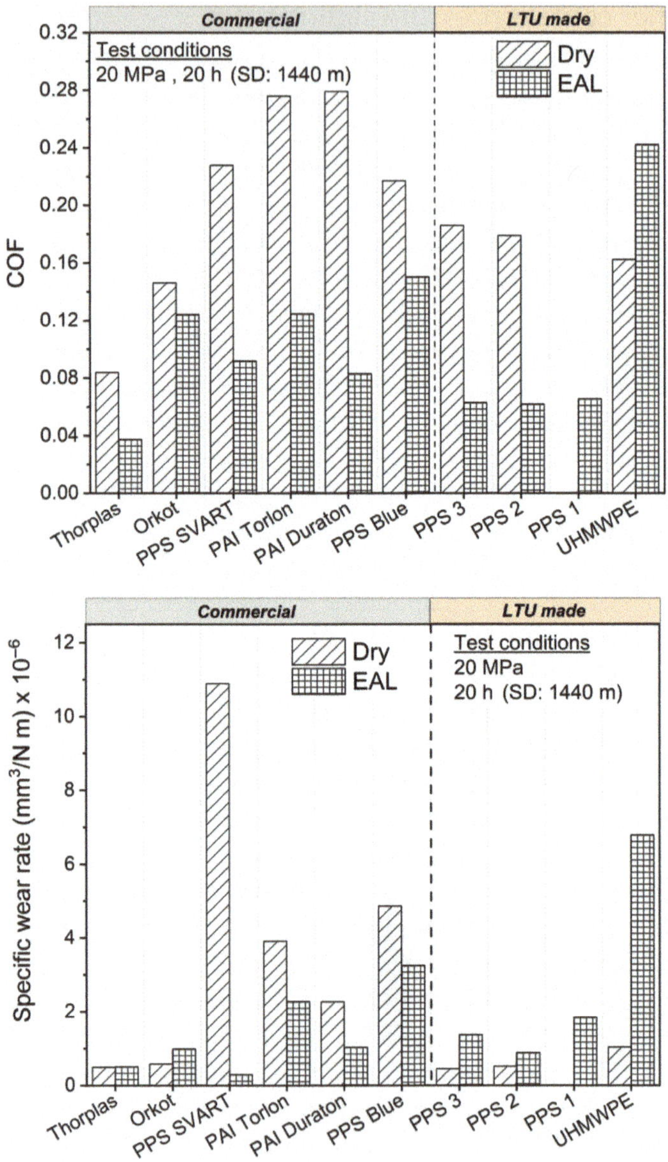

Fig. 15 Summaries of (A) friction coefficients and (B) specific wear rates of selected thermoplastic composites at a 20 MPa contact pressure under dry and EAL lubricated conditions.

The friction and wear trends of three thermoplastic-based multiscale composites (i.e., UHMWPE, PTFE, and PPS) were investigated for hydropower bearing applications, and various tribological mechanisms in dry and water-lubricated conditions were proposed. Though sufficient results and discussions are presented on the effect of hygrothermal aging on the friction and wear of PTFE composites, future work remains necessary to further investigate the friction and wear trends of PTFE composites. Additionally, there is a greater necessity to explore and investigate the potential of other thermoplastics and their composites for water-lubricated bearing applications.

6 Conclusion

In conclusion, the hydropower plant sliding bearing is an extreme application due to its extreme contact pressure during operation (~30 MPa), the longevity of operation (~40 years), and the necessity to adapt oil-free or water lubrication from a sustainability perspective. Therefore, the use of novel multiscale thermoplastic polymer composites is essential to combat the friction and wear issues along with water-based lubrication. The amorphous regions in the polymer allow water absorption, leading to swelling of the polymer, which deteriorates its properties. Therefore, higher crystallinity is preferred for water-lubricated applications. The UHMWPE, PTFE, PPS, and PEEK are most suitable for water lubrication due to their semicrystalline nature and the better composite process ability (i.e., compression molding) than other amorphous and semicrystalline polymers.

In this chapter, we provided examples of novel multiscale thermoplastic polymer composites developed by the addition of macro- and microreinforcements (i.e., SCF, GF, CF), carbon-based nanofillers (i.e., ND, CNTs), and 2D materials (i.e., GO, MoS_2) to polymers that have resulted in superior properties (i.e., mechanical, tribological) through a synergistic effect obtained via the multiscale mode. The typical detailed multiscale composite manufacturing and characterization techniques for thermoplastic-based polymers are reported, along with the procedures used in our earlier works. While testing the polymers for water-lubricated journal bearing applications, the water absorption behavior or otherwise and the hygrothermal aging of polymers must be taken into consideration because the mechanical, tribological, chemical, and physical properties of polymer composites are highly affected by the absorbed water in the polymer. These amorphous regions act as sites for the water to permeate and absorb into the polymer, leading to swelling and a change in the microstructure and geometry of the polymer, thereby altering all properties.

Finally, the recent investigations of the friction and wear trends of two thermoplastics, UHMWPE and PTFE, and their multiscale composites under water-lubricated conditions were discussed and explored from a hydropower bearing perspective. The UHMWPE multiscale composite with GO, ND, and SCF provided a significant reduction in friction and wear compared to pure UHMWPE in water-lubricated conditions. ND and SCF are responsible for friction and wear reduction, respectively, resulting in a synergetic effect from the micro- and nanofillers. The NDs are postulated to act as nanoscale ball bearings between the two contacting surfaces, thereby reducing the contact area, friction, and wear. Similarly, SCF addition improved the mechanical properties (i.e., hardness) of UHMWPE, which enhanced the wear resistance of the composite.

Furthermore, a detailed investigation showed that the effect of hygrothermal aging on the friction and wear of the PTFE and its composites is very marginal, with minor exceptions. The unfilled PTFE had the lowest friction and PTFE reinforced with GF and MoS_2 had the lowest wear rate, respectively, across all samples and testing conditions. Also, there was a very distinctive trend observed between the degree of crystallinity and the specific wear rate values of samples tested in dry conditions. No such trend was observed between the wear rates and crystallinity values of the polymer samples tested in water-lubricated conditions. Moreover, higher COF and wear rates in the dry condition were attributed to the heat generation by friction and the absence of water while in the water-lubricated condition, the heat dissipation and lubrication by water did help to achieve lower friction and wear. Other factors also contributed marginally to the observed tribological behavior. However, the exact mechanisms occurring inside the synergistic effect of thermoplastic multiscale composites require further detailed investigation with more sophisticated analyzing techniques.

Hereinafter, sustainable or green tribology can only be realized by developing a new generation of thermoplastic polymer composites and environmentally adaptive lubricants (EAL) as well as concern for the nature and behavioral change. In the materials aspect, thermoplastics enable us to develop a new generation of composites with greater strength, lower cost, and less harm to the environment. Similarly, a new generation of EALs has to be formulated to provide the same or higher lubrication performance than that of fossil fuel-based oils. Notably, leaking of this lubricant will not be harmful to the environment at all. Finally, the environment itself is so irreplaceable and it is our responsibility to preserve it by adapting some behavior changes and cleaner techniques. Hence, the tribological behavior of other thermoplastic-based polymers in water or EAL-lubricated conditions must be investigated for a better understanding of the relevant issues and future solutions. By focusing on al1 three aspects, sustainable tribology can be achieved in hydropower bearing applications.

References

[1] M. Wasilczuk, Friction and lubrication of large tilting-pad thrust bearings, Lubricants 3 (2015) 164–180, https://doi.org/10.3390/lubricants3020164.
[2] A. Golchin, Polymeric Materials for Bearing Applications:Tribological Studies in Lubricated Conditions (Doctoral Thesis), Luleå University of Technology, 2015 (Online open access).
[3] S. Åstrand, Environmental Effects of Turbine Oil Spills from Hydro Power Plants to Rivers, (Online). Available from: http://www.wprogram.nu/filer/exjobb/Stina_%C3%85strand.pdf, 2008.
[4] G. Stachowiak, A. Batchelor, Engineering Tribology. Elsevier, Amsterdam, 2013. https://doi.org/10.1016/B978-0-7506-7836-0.X5000-7.
[5] A. Golchin, G.F. Simmons, S. Glavatskih, B. Prakash, Tribological behaviour of polymeric materials in water-lubricated contacts, Proc. Inst. Mech. Eng. Part J: J. Eng. Tribol. 227 (2013) 811–825, https://doi.org/10.1177/1350650113476441.
[6] A. Golchin, K. Friedrich, A. Noll, B. Prakash, Tribological behavior of carbon-filled PPS composites in water lubricated contacts, Wear 328–329 (2015) 456–463, https://doi.org/10.1016/j.wear.2015.03.012 (5 in this paper).
[7] A. Golchin, K. Friedrich, A. Noll, B. Prakash, Influence of counter surface topography on the tribological behavior of carbon-filled PPS composites in water, Tribol. Int. 88 (2015) 209–217.
[8] A. Golchin, A. Wikner, N. Emami, An investigation into tribological behaviour of multi-walled carbon nanotube/graphene oxide reinforced UHMWPE in water lubricated contacts, Tribol. Int. 95 (2016) 156–161.
[9] C. Dong, L. Shi, L. Li, X. Bai, C. Yuan, Y. Tian, Stick-slip behaviours of water lubrication polymer materials under low speed conditions, Tribol. Int. 106 (2017) 55–61.
[10] W. Litwin, C. Dymarski, Experimental research on water-lubricated marine stern tube bearings in conditions of improper lubrication and cooling causing rapid bush wear, Tribol. Int. 95 (2016) 449–455.
[11] M. Chanda, Plastics Technology Handbook, Taylor and Francis group, Florida, 2006.
[12] S. Kurtz, UHMWPE Biomaterials Handbook, third ed., Academic Press, 2009.
[13] V.N. Mochalin, Y. Gogotsi, Nanodiamond-polymer composites, Diamond Relat. Mater. 58 (2015) 161–171.
[14] M. Ullah, A. Kausar, M. Siddiq, M. Subhan, M.A. Zia, Reinforcing effects of modified nanodiamonds on the physical properties of polymer-based nanocomposites: a review, Polym. Plast. Technol. Eng. 54 (2015) 861–879.
[15] D.I. Chukov, A.A. Stepashkin, A.V. Maksimkin, V.V. Tcherdyntsev, S.D. Kaloshkin, K.V. Kuskov, V.I. Bugakov, Investigation of structure, mechanical and tribological properties of short carbon fiber reinforced UHMWPE-matrix composites, Compos. Part B: Eng. 76 (2015) 79–88, https://doi.org/10.1016/j.compositesb.2015.02.019.
[16] X. Dangsheng, Friction and wear properties of UHMWPE composites reinforced with carbon fiber, Mater. Lett. 59 (2–3) (2005) 175–179.

[17] M. Bao, C. Zhang, D. Lahiri, A. Agarwal, The tribological behavior of plasmasprayed Al-Si composite coatings reinforced with nanodiamond, JOM 64 (2012) 702–708.
[18] E.P. Koumoulos, P. Jagadale, A. Lorenzi, A. Tagliaferro, C.A. Charitidis, Evaluation of surface properties of epoxy-nanodiamonds composites, Compos. Part B: Eng. 80 (2015) 27–36.
[19] S. Xia, Y. Liu, L. Zhang, D. Wang, W. Zou, J. Peng, S. Cao, Tribological mechanism improving the wear resistance of polyurethane/epoxy interpenetrating polymer network via nanodiamond hybridization, J. Appl. Polym. Sci. 131 (2014) 1–9.
[20] I. Neitzel, V. Mochalin, I. Knoke, G.R. Palmese, Y. Gogotsi, Mechanical properties of epoxy composites with high contents of nanodiamond, Compos. Sci. Technol. 71 (2011) 710–716.
[21] I. Neitzel, V. Mochalin, J.A. Bares, R.W. Carpick, A. Erdemir, Y. Gogotsi, Tribological properties of nanodiamond-epoxy composites, Tribol. Lett. 47 (2012) 195–202.
[22] A. Golchin, A. Villain, N. Emami, Tribological behaviour of nanodiamond reinforced UHMWPE in water-lubricated contacts, Tribol. Int. 110 (2017) 195–200.
[23] Y. Yamamoto, M. Hashimoto, Friction and wear of water lubricated PEEK and PPS sliding contacts: part 2. Composites with carbon or glass fibre, Wear 257 (2004) 181–189, https://doi.org/10.1016/j.wear.2003.12.004.
[24] K. Friedrich, Microstructural efficiency and fracture toughness of short fiber/thermoplastic matrix composites, Compos. Sci. Technol. 22 (1985) 43–74.
[25] G.M. Lin, J.K. Lai, Fracture mechanism in short fibre reinforced thermoplastic resin composites, J. Mater. Sci. 28 (1993) 5240–5246.
[26] W. Luo, Q. Liu, Y. Li, S. Zhou, H. Zou, M. Liang, Enhanced mechanical and tribological properties in polyphenylene sulfide/polytetrafluoroethylene composites reinforced by short carbon fiber, Compos. Part B: Eng. 91 (2016) 579–588.
[27] S. Biswas, K. Vijayan, Friction and wear of PTFE—a review, Wear 158 (1992) 193–211.
[28] A. Jain, J. Somberg, N. Emami, Development and characterization of multi-scale carbon reinforced PPS composites for tribological applications, Lubricants 7 (2019) 34, https://doi.org/10.3390/lubricants7040034.
[29] S.V. Panin, L.A. Kornienko, T. Nguen Suan, L.R. Ivanova, M.A. Korchagin, S.V. Shil'ko, Y.M. Pleskachevskii, Wear resistance of composites based on hybrid UHMWPE–PTFE matrix: mechanical and tribotechnical properties of the matrix, J. Frict. Wear 36 (2015) 249–256, https://doi.org/10.3103/S1068366615030113.
[30] M.M. Thwe, K. Liao, Durability of bamboo-glass fiber reinforced polymer matrix hybrid composites, Compos. Sci. Technol. 63 (2003) 375–387, https://doi.org/10.1016/S0266-3538(02)00225-7.
[31] H.S. Vadivel, A. Golchin, N. Emami, Tribological behaviour of carbon filled hybrid UHMWPE composites in water, Tribol. Int. 124 (2018) 169–177, https://doi.org/10.1016/j.triboint.2018.04.001.
[32] R.L. Orndorff, New UHMWPE/rubber bearing alloy. J. Tribol. 122 (1) (1999) 367–373, https://doi.org/10.1115/1.555361.
[33] H.S. Vadivel, Tribological Behaviour of Hybrid Carbon Filled UHMWPE Composites in Water (Maters Thesis), Luleå University of Technology, 2016 (Online open access).
[34] E. Enqvist, Carbon Nanofiller Reinforced UHMWPE for Orthopaedic Applications (Doctoral Thesis), Lulea University of Technology, 2013 (Online open access).
[35] S.S. Moreno, Nanodiamond/Ultra-High Molecular Weight Polyethylene Composites for Bearing Applications (Doctoral Thesis), Lulea University of Technology, 2014 (Online open access).
[36] P. Saravanan, R. Selyanchyn, H. Tanaka, S. Fujikawa, S.M. Lyth, J. Sugimura, Ultra-low friction between polymers and graphene oxide multilayers in nitrogen atmosphere,

mediated by stable transfer film formation, Carbon N.Y. 122 (2017) 395–403, https://doi.org/10.1016/j.carbon.2017.06.090.
[37] S. Peng, D.X. Yan, J.F. Gao, Z.M. Li, Temperature-resistivity behaviour of CNTS/UHMWPE composites with a two-dimensional conductive network, Polym. Plast. Technol. Eng. 48 (4) (2009) 478–481.
[38] R. Taktak, N. Guermazi, J. Derbeli, N. Haddar, Effect of hygrothermal aging on the mechanical properties and ductile fracture of polyamide 6: experimental and numerical approaches, Eng. Fract. Mech. 148 (2015) 122–133.
[39] M. Eftekhari, A. Fatemi, Tensile behavior of thermoplastic composites including temperature, moisture, and hygrothermal effects, Polym. Test. 51 (2016) 151–164.
[40] M.R. Homayoun, A. Golchin, N. Emami, Effect of hygrothermal ageing on tribological behaviour of PTFE-based composites, Lubricants 6 (2018) 103, https://doi.org/10.3390/lubricants6040103.
[41] B.C. Ray, Temperature effect during humid ageing on interfaces of glass and carbon fibers reinforced epoxy composites, J. Colloid Interface Sci. 298 (2006) 111–117.
[42] T. Bárány, E. Földes, T. Czigány, Effect of thermal and hygrothermal aging on the plane stress fracture toughness of poly(ethylene terephthalate) sheets, Express Polym. Lett. 1 (2007) 180–187.
[43] H.J. Barraza, L. Aktas, Y.K. Hamidi, J. Long, E.A. O'Rear, M.C. Altan, Moisture absorption and wet-adhesion properties of resin transfer molded (RTM) composites containing elastomer-coated glass fibers, J. Adhes. Sci. Technol. 17 (2003) 217–242.
[44] C.-H. Shen, G.S. Springer, Moisture absorption and desorption of composite materials, J. Compos. Mater. 10 (1976) 2–20.
[45] J.L. Tipper, N. Emami, S. Suner, R. Joffe, Ultra high molecular weight polyethylene/graphene oxide nanocomposites: thermal, mechanical and wettability characterisation, Compos. Part B: Eng. 78 (1) (2015) 185–191.
[46] D. Berman, S.A. Deshmukh, S.K.R.S. Sankaranarayanan, A. Erdemir, A.V. Sumant, Macroscale superlubricity enabled by graphene nanoscroll formation, Science 348 (2015) 1118–1122, https://doi.org/10.1126/science.1262024.
[47] D. Berman, B. Narayanan, M.J. Cherukara, S.K.R.S. Sankaranarayanan, A. Erdemir, A. Zinovev, A.V. Sumant, Operando tribochemical formation of onion-like-carbon leads to macroscale superlubricity, Nat. Commun. 9 (2018), https://doi.org/10.1038/s41467-018-03549.
[48] P. Saravanan, R. Selyanchyn, H. Tanaka, D. Darekar, A. Staykov, S. Fujikawa, S.M. Lyth, J. Sugimura, Macroscale superlubricity of multilayer polyethylenimine/graphene oxide coatings in different gas environments. ACS Appl. Mater. Interfaces 8 (2016) 27179–27187, https://doi.org/10.1021/acsami.6b06779.
[49] Y. Gogotsi, V.N. Mochalin, O. Shenderova, The properties and applications of nanodiamonds, Nat. Nanotechnol. 7 (1) (2012) 11–23.
[50] Z. Jiang, L.A. Gyurova, A.K. Schlarb, K. Friedrich, Z. Zhang, Study on friction and wear behavior of polyphenylene sulfide composites reinforced by short carbon fibers and sub-micro TiO_2 particles, Compos. Sci. Technol. 68 (2008) 734–742, https://doi.org/10.1016/j.compscitech.2007.09.022.
[51] M.R. Homayoun, Self-Lubricated Hybrid Polymeric Composites for Water Lubricated Bearing Applications: Effect of Hygrothermal Aging on Wear Mechanism and Friction Behavior (Maters Thesis), Lulea University of Technology, 2017 (Online open access).
[52] M. Rodiouchkina, K. Berglund, J. Mouzon, F. Forsberg, F.U. Shah, I. Rodushkin, R. Larsson, Material characterization and influence of sliding speed and pressure on friction and wear behavior of self-lubricating bearing materials for hydropower applications, Lubricants 6 (2018) 1–30, https://doi.org/10.3390/lubricants6020039.

CHAPTER 14

Tribological characterization of different mesh-sized natural barite-based copper-free brake friction composites

R. Vijay and D. Lenin Singaravelu
Department of Production Engineering, National Institute of Technology, Tiruchirappalli, Tamil Nadu, India

1 Introduction

A brake is a mechanical member used to alter kinetic energy to heat energy through friction. During braking, a lot of heat is generated at the braking interface [1]. To sustain that temperature without performance deterioration, a cocktail of ingredients is mandatory. The prime characteristic of friction composites is that they provide better friction with less wear [2]. To produce such desired characteristics, a cocktail of 10–15 ingredients is used. These are broadly categorized under four main headings: fibers, binders, fillers, and frictional modifiers [3]. Fillers and frictional modifiers are further divided into inert, functional and abrasive, and lubricant, respectively [4]. Asbestos was used in friction composites due to its advantages such as low cost, high thermal stability, and good strength, but its carcinogenic nature in the working environment causes lung diseases and cancer. Thus, it was banned in 1989 by the US Environmental Protection Agency [5]. In the current global context, the use of copper is reduced due to its adverse effect on ecology. The wear debris produced by braking contains copper that is washed away by rain and reaches water bodies, leading to imbalances in the aquatic food chain. Thus, it was decided to reduce copper usage by 0.5% by 2025 in friction material formulations by enacting stringent laws. Thus, copper-free brake friction composites are being developed by researchers using various new ingredients [6]. An inert filler means being stable without altering anything. But with inert fillers, the optimal size and purity must be used or the performance properties of the brake friction composites are altered [7]. In brake friction composites, inert fillers are used to make up the

Tribology of Polymer Composites
https://doi.org/10.1016/B978-0-12-819767-7.00014-1

volume of the composition and reduce the manufacturing cost. There are several inert fillers used in formulations available commercially, including barites, calcium carbonate, and fluorite. Among them, barites are used widely as inert fillers due to their higher density, minimum water solubility, and higher heat resistance [8]. Additionally, they are cheap compared to other ingredients used in formulations. It was also found that if the barites are used in an optimal percentage with metals in brake formulations, the wear rate is reduced. Many research works in the literature use barites as a filler to compensate for the varying ingredients in the brake friction material formulation to seek the performance results of the ingredients used. But the barites as such have some vital impacts on the performance of the brake friction composites, which are explored from the following literature. Headson [9] utilized barites in the friction material composition, but the filler role of barites in phenolic plastics was known from the work of Baekeland in 1909 [10]. Barites with other materials in the brake friction material formulations were experimentally investigated by Handa and Kato [11]. It was proved that barites, when used in an optimal percentage ranging from 10% to 20%, showed better tribological performance compared to other weight percentages. Kim et al. [12] experimentally showed that barites affect the μ levels badly at higher speeds and higher concentrations, leading to stick-slip during sliding, which induced noise generation. It was proposed that noise generation be avoided by choosing the barites at optimal levels. McCormick [13] observed the interactions of barites with the matrix in brake friction composites. It was shown that phenolic resins provided a carbon source that reacted with barites at elevated temperatures. It was also suggested that barium sulfate was reduced to a sulfate, providing additional support for inert filler reactivity at the interface during braking. Ma et al. [14] used X-ray diffraction analysis to show that barites formed a tribofilm on the mating disc at operating temperatures of 300°C and above. The formation of film was seen by heavy debris filling the mechanism where barites filled the voids formed interstitially, leading to wear resistance. Handa and Kato [11] proved a similar phenomenon. Tiwari et al. [15] compared and analyzed barium sulfate cenosphere-based brake friction composites as per ECE-90 regulations. It was inferred from experimental results that the μ levels seemed to be better for barium sulfate-based brake friction composites compared to others. Menapace et al. [16] experimentally analyzed the dry sliding friction characteristics of copper-free friction materials containing 3 wt.% of barites using a pin-on-disc tribometer. It was seen that barites improved the tribological behavior by producing stable tribofilms. Thus, it was clear that barites improved the performance properties of brake friction materials. It was seen from the above results that barytes enhanced the tribological performance by forming stable tribofilms, but there is no study regarding the influence of the particle size of barites on the friction and wear characteristics of brake friction composites. Thus, the current study deals with the study of the influence of three different mesh-sized natural barites on the tribological performance characteristics of brake friction composites and analyzes the worn surface using scanning electron microscopy (SEM).

2 Materials and methods

2.1 Materials

Barites of natural form were used in the current study. The barites were ground into three different sizes, namely 150, 250, and 350 BSS (British Sieve Size), that were commercially produced by mechanical milling and used in the brake friction material formulations. Based on the size, there was a slight variation in the purity levels based on the mining region. The technical specifications of the three different mesh-sized barites are given in Table 1 and the SEM images are seen in Fig. 1A–C.

2.2 Development of brake friction composites

Brake friction composites were developed in the form of brake pads to suit the Indian passenger car vehicle scenario. The formulation of the developed brake pads possessed 15 constant ingredients: fibers with additives (14 wt.%) aramid pulp, hydrated lime, rock wool fiber; binders (15 wt.%) with straight phenolic resin, crumb rubber, NBR; the friction modifiers (18 wt.%) silicon dioxide (abrasive) and synthetic graphite, MoS_2 (lubricant); and the fillers (functional) (38 wt.%) of exfoliated vermiculite, magnesium oxide, tin powder, kaolin clay, boron-graphite modified friction dust, and mica. The varying ingredients were the three different mesh-sized natural barites with 15 wt.%. The brake pads so developed were designated as BC1, BC2, and BC3 based on the mesh sizes 150, 250, and 350 BSS, respectively.

The development methodology for the manufacture of brake pads included mixing, curing, postcuring, and finishing; it is given in Fig. 2 with all manufacturing parameters. Prior to the commencement of manufacturing (curing), backplate preparation was done by cleaning the sand-blasted mild steel plate using acetone. Then, anabond was applied over the surface-cleaned backplate, followed by drying at 60°C for 45 min in a hot air oven [17, 18]. The developed brake pads are shown in Fig. 3.

Table 1 Technical specifications of three different mesh-sized natural barites as received from the supplier.

S. no.	Properties	Unit	Specification (as per industrial internal standards)	Natural barites 150 BSS	Natural barites 250 BSS	Natural barites 350 BSS
				95% Purity grade selected		
1	Appearance	Visual	White in color Free flowing powder	White in color Free flowing powder	White in color Free flowing powder	White in color Free flowing powder
2	Ash content	%	98.0 min	98.4	98.7	99
3	Moisture content @ 110°C	%	1.0 max	0.12	0.15	0.16
4	Particle size: 100 g. Sample by brushing method up to 250 BSS and wet sieving process above 250 BSS					
a	+ 100 BSS	%		NIL	NIL	NIL
b	− 100 + 150 BSS	%	95 min	97	1	0.5
c	− 150 + 250 BSS	%	95 min	1	98	2
d	− 250 + 350 BSS	%	95 min	1	0.5	97.5
e	− 350 (PAN)	%		1	0.5	TRACES
5	Density	g/cc	4.00–4.30	4.1	4.18	4.23
6	Purity as $BaSO_4$ (by acid digestion method)	%	90 min	95.2	95.4	95.7
7	Impurity as SiO_2 (by acid digestion method)	%	8 max	3.2	2.7	2.3
8	Impurity others (Fe_2O_3)	%	5 max	1.6	1.9	2.0

Fig. 1 SEM images of barites with mesh size (A) 150 BSS; (B) 250 BSS; and (C) 350 BSS.

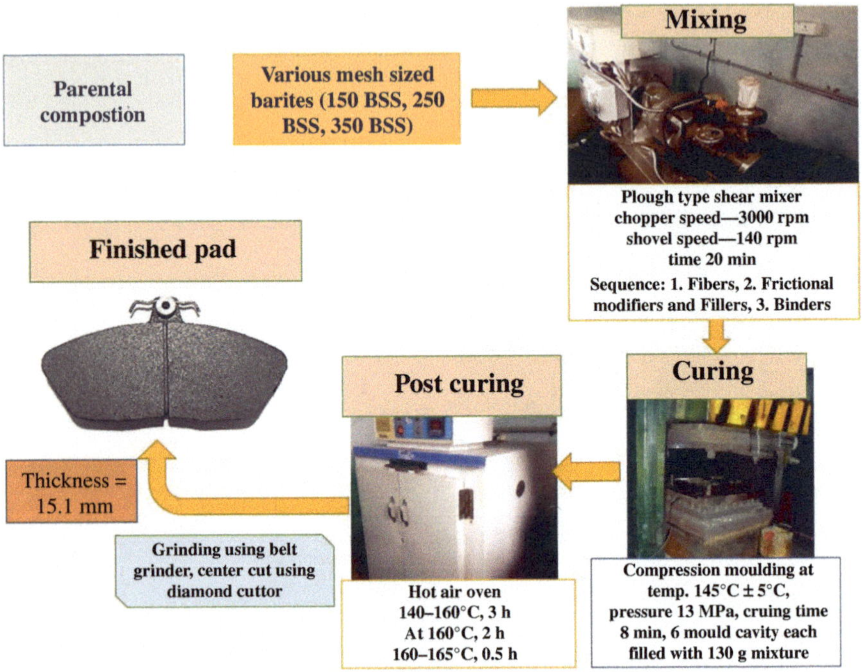

Fig. 2 Brake pad development methodology.

Fig. 3 Developed brake pads.

2.3 Characterization of the developed brake friction composites

A density apparatus that uses the Archimedes principle was used to find the density. The hardness was measured by Rockwell hardness using the *K* scale (steel ball indenter with ϕ 3.125 with 1500 N load). Resins that were not appropriately cured were found using the Soxhlet extraction apparatus by acetone extraction. Uncured resins were found using the acetone extraction test by using a Soxhlet extraction apparatus. The mass loss due to the ignition of the brake pads was found by considering 5–10 g of the developed sample in a silica crucible heated to 800°C and maintained at that temperature for 2 h in a muffle furnace. The difference in mass loss was noted and reported. The pad swell on heating was found by considering a sample of size of 10 mm × 10 mm × 4 mm that was maintained in a hot air oven for 40 min at 200 ± 3°C. The brake pad samples were cut to a 50 mm × 25 mm size and immersed in water in ambient conditions for 30 min. For both tests, the difference in thickness was noted. The above tests were performed as per IS2742 Part 3. The hot shear strength was measured by heating the pad to 200°C and maintaining at that temperature for 30 min while cold shear strength was found by testing the brake pads under ambient conditions. The above shear tests were performed as per ISO 6312. The image analysis technique used by industries was followed to find the adhesion. Porosity was found by placing brake pad samples of 625 mm^2 in a desiccator for 24 h and then soaking it in a preheated SAE 90-grade oil bath at 90°C ± 1°C for 8 h. After the required time, the heater was switched off, and the cooling of the samples was performed at room temperature. It was carried out according to JIS D 4418 [19]. To check the consistency, three samples were tested and the average values reported. A quick assessment of the tribological characteristics was done using the Chase test following IS 2742 Part 4 for a brake pad sample size of 625 mm^2 that was made to slide against a 280 mm diameter cast iron drum [20, 21].

The fade, recovery, and effectiveness characteristics of the developed brake pads were analyzed using a full-scale inertia brake dynamometer as per the JASO C-406 schedule. The parameters were an inertia of 50 $kg\,ms^2$, a rolling radius of 0.293 m, an effective radius of 0.103 m, a pad thickness with a backplate of 15.1 mm, and a mating surface of cast iron. The calibration of the data acquisition system used to record the data was done through process variable equipment, namely pressure sensors, temperature sensors, speed sensors (encoders), and torque measurement (load cell) as per the National Traceability Standards of NABL-certified laboratories. For consistency in results, two samples were tested. If the error was more than 8%, then a third sample was tested. The establishment of 90% contact between the pad and disc was done by a bedding test at a 50 km/h speed with 3.0 m/s^2 deceleration with 200 applications having an initial brake temperature of 100°C. Three effectiveness studies were performed. In the first effectiveness study, two different speeds (50 and 100 km/h) with two different decelerations (3.0 and 6.0 m/s^2) with an initial temperature of 80°C were considered. In the second effectiveness study, three different speeds (50, 100, and 130 km/h) with two different decelerations (3.0 and 6.0 m/s^2) with an initial temperature of less than 80°C were considered. Finally, in the third effectiveness study, three different speeds (50, 100, and 130 km/h) with two different decelerations (3.0 and 6.0 m/s^2) with an initial temperature of less than 100°C were considered. For all effectiveness studies, the number of applications was repeated until the measurements for four or more points were equal within the deceleration range. For the above test, the air blower was kept off.

Two decelerations (3.0 and 6.0 m/s^2) were carried out at a 100 km/h vehicle speed for the fade test with a braking interval of 35 s for 15 applications in each deceleration with an initial brake temperature of 80°C under blower off conditions. In the case of recovery, two decelerations (3.0 and 6.0 m/s^2) were repeated at 120 s intervals for a 50 km/h speed, with the air blower on and with an initial brake temperature of less than 80°C. The μ was measured and the average values reported. The fade and recovery rates were calculated as per the formulas that are given in Eqs. (1), (2). The pad and rotor wear were reported based on the after test measurement using a digital micrometer [18, 19].

$$\text{Fade rate} = \left(\left(\mu_{\text{fmax}} - \mu_{\text{fmin}}\right) / \mu_{\text{fmax}}\right) * 100\,(\%) \tag{1}$$

$$\text{Recovery rate} = \left(\mu_{\text{rmax}} / \mu_{\text{rmin}}\right) * 100\,(\%) \tag{2}$$

Where μ_{rmax} and μ_{rmin} are the maximum and minimum coefficients of friction measured during the recovery cycle, and μ_{fmax} and μ_{fmin} are the maximum and minimum coefficients of friction measured during the fade cycle.

The K-type (button type) thermocouple was used to measure temperature. For measurement, drilling was performed through the spigot hole and a thermocouple was placed just beneath the pad surface. The final and initial temperatures were noted. Another through-hole was made for contact with the disc to measure the braking interface temperature. In the current study, the pad temperature was measured and reported. In some instances, as prescribed by the customer or standard, a temperature rise in the disc would be reported. A scanning electron microscope (Tescan VEGA 3LMU SEM) was used to analyze the worn surfaces of the tribological-tested brake pad samples.

3 Results and discussion

3.1 Physical, chemical, thermal, and mechanical characteristics of the developed brake pads

The test results of the various characteristics are given in Table 2. The density of BC1 is lower than that of the other two brake pads. This was mainly due to the higher mesh size having lower-density barites preventing molecular attraction, which was predominant in lower mesh-size materials [3]. If it is in the nanoscale, preparation methods are needed, but in this work, no such preparation was done. Hardness refers to resistance to indentation. In the current study, BC3 showed a higher hardness mainly due to the smaller-sized particles that filled the space between the ingredients, leading to such results [21]. BC3 composites showed a higher extract value that resembled the more uncured resin, mainly due to the reduced pores enabling adequate curing of the pads. This is in tandem with the literature [22–24]. The obtained values were in accordance with the industrial standard range [25]. The resistance of heat degradation can be confirmed from the lower weight loss in the pads. BC1 composites showed less deterioration in heat and had better resistance due to the presence of higher mesh-size barites. This is because of the excellent thermal resistance due to the reduced penetration of heat to the core of the particle that is inducing degradation compared to the other two mesh-sized barites. BP1 composites showed a higher porosity due to the higher mesh-size barites in its formulation, leaving adequate space between the ingredients and resulting in good porosity [3]. Another aspect is that the lower the density and hardness, the higher the porosity [26]. This similar behavior is also seen in the current study. It is also essential to maintain optimal porosity as per the industrial standards to uphold tribological performance, as explained in forthcoming sections. Cold and hot shear strengths are higher for the BC1 pads due to better curing of the composites, which forms a firm interlock with the shot-blasted backplate and leads to such higher values. The obtained values are in accordance with the standard values [27]. The adhesion found using image analysis as shown in Fig. 4 had a higher value in BC1 mainly due to the better bonding of the

Table 2 Various characteristics of the developed brake pads.

S. No.	Properties	Unit	BC1	BC2	BC3
1	Density	g/cc	2.43	2.47	2.54
2	Hardness	No unit	81	92	99
3	Acetone extraction	%	1.58	1.71	1.83
4	Loss of ignition	%	19	24	27
5	Porosity	%	7.5	5.1	3.9
6	Cold shear strength	kg/cm^2	49	45	41
7	Adhesion	%	100	95	85
8	Hot shear strength	kg/cm^2	29	25	22
9	Adhesion	%	100	90	80
10	Normal friction	μ	0.45	0.46	0.5
11	Hot friction	μ	0.420	0.430	0.44
12	Weight loss	%	3.2	3.9	4.6
13	Thickness loss	%	2.9	3.7	5.2

Fig. 4 Adhesion-tested brake pads after cold shear strength.

friction layer with the backplate, thus preventing higher peel off compared to other pads. This behavior is in tandem with the literature findings [18]. Chase test results of the three pads showed that BC3 had more friction coefficient due to a smaller particle size, producing less tribofilm as explained in detail in the forthcoming sections. This increased friction value increased the wear rate (both weight and thickness loss), causing more polymeric ingredient degradation [2, 21].

3.2 Tribological characterization of the developed brake pads using a full-scale inertia brake dynamometer

3.2.1 Effectiveness studies

Changes in μ with speed and pressure play a vital role in braking. The μ variation must be minimized so that the driver feels comfort throughout the service life of the brake friction composites [2]. The μ for various effectiveness levels is given in Fig. 5A–C. In general, all μ values decreased with an increase in pressure and speed, mainly due to the following reasons. The increase in pressure leads to an increase in the real contact area, leading to more heating and degradation of polymeric ingredients. These get transferred to the counter surface, then are back-transferred to the contact interface, leading to reduced μ results [28]. The increase in speed breaks the contact asperities by thermal and mechanical overloading, leading to increased frictional heating and plateau exposure. This changes the μ values abnormally and the values are also not decreased linearly [8]. The μ values obtained in the current study are in a range of 0.3–0.5, which is within industrial standards [29]. It is necessary to have lower undulations. In the preliminary stage, the μ decreased with an increase in speed for two different pressures. Then, it stabilized in the case of BC1 composites. It is a general postulate that lower mesh-size materials enhance friction. Similar behavior was seen in BC3 composites, as in the initial stages there was an increase in μ levels. At elevated temperatures, there was more reduction in μ levels, leading to high wear. This behavior was mainly due to a reduction in the level of separation caused by the irregular and thin tribofilms formed during braking [3]. This was less compared to higher mesh-size barite-based brake pads. It is inferred from the literature that at high temperature, the barites formed a film at higher operating temperatures (that is, 300°C and above) by reacting with the sulfide with the carbons of the phenolic resin. It formed a film by filling the interstitial voids with the heavy debris of high-density barites. This also protects from wear [11, 14]. It was also stated that the film-forming ability was influenced by the mesh size. Upon sliding, the barite particles get fractured and are reduced to smaller particles, filling and forming a film over the surface. It is important to note that the fracture of the larger barite particles fetches more tiny particles that fill effectively. But upon fracture, lower mesh-size barite particles produce more small particles that do not fill and form films effectively. This leads to a deterioration in the μ performance [30]. Another possible reason for stabilized μ levels is that better primary plateaus that are predominant and protected by the sulfide film are formed in the case of BC1. In BC2 and BC3, the films

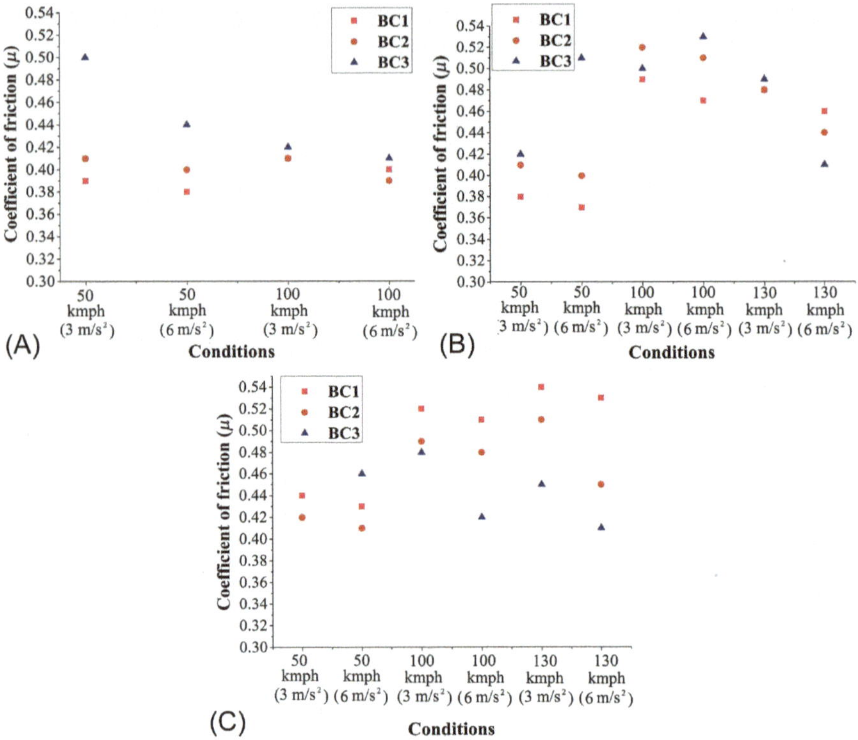

Fig. 5 Effectiveness studies (A) first; (B) second; and (C) third for the developed brake pads.

formed at the interface do not protect the plateaus formed, thereby leading to the destruction and formation of more secondary patches formed by the mechanical, chemical, and surface energy interactions caused by the least thermal stable ingredients [31]. Similar behavior was seen by Blau and McLaughlin in the near fade regions leading to increased local temperature that was much higher than the bulk temperature of the material, causing pad material softening and strength decrease [32]. The increase in speed increases the interface heat, thereby reducing μ levels as explained from the frictional heat theory. Then, the third bodies formed by the disintegration of the hard particles and broken fibers damage the protection films and primary plateaus, thereby leading to variations in the μ results [22].

3.2.2 Fade and recovery characteristics

The degradation of friction properties at high temperature, that is, severe braking conditions, is called fade and the corresponding μ is called μ_{fade}. As per industrial standards, an ideal brake pad must have μ_{fade} in a range of 0.3–0.5; a similar range is seen in the current study [2]. Generally, μ_{fade} values are less than recovery μ values, mainly due to the higher temperature rise causing thermal decomposition and glass transition of the phenolic resin that weakens the bonding characteristics of the ingredients and the matrix. Also, it creates a pyrolysis reaction, that is, the release of gas due to the degradation of polymeric and organic constituents. The released gas gets trapped in the sliding interface, thereby exerting an opposite force on the applied load and reducing the μ [17]. The μ level should always be consistent for effective braking. In BC1, the difference in μ values is comparatively less than in the other two pads. It is high, as found from the fade rate. The μ_{fade} for the highest to lower range is given below.

$$\mu_{\text{fade}} = \text{BC1}\,(0.43) > \text{BC2}\,(0.40) > \text{BC3}\,(0.37)$$

The fade rate showed μ deterioration at elevated conditions. The industrial range for the fade rate is 0%–20%, and the current study range is within this [29]. The lowest fade rate of 12% is seen in BC1 pads, that is, having the highest fade resistance compared to the other brake pads developed, as shown in Fig. 6. The fade rate followed the highest to lowest sequence as given below.

Fade rate: BC3 > BC2 > BC1

This lowest fade rate, that is, the higher fade resistance in BC1, is mainly due to the formation of a stable tribofilm caused by higher mesh-sized

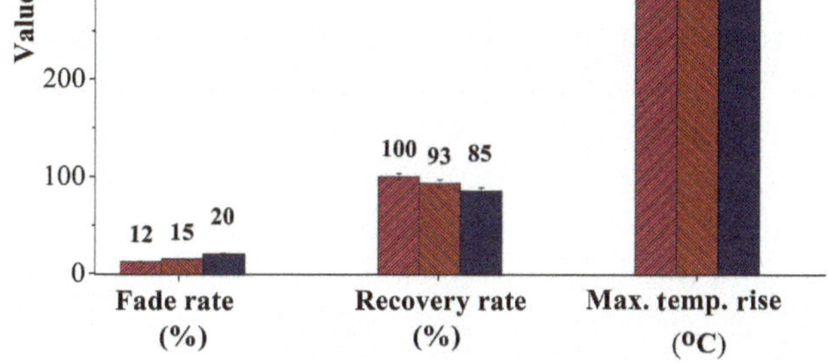

Fig. 6 Fade, recovery rates, and maximum temperature rise of the dyno-tested pads.

barites that prevent the degradation of the ingredients from the pad surface, thus sustaining the μ levels. In lower mesh-sized barites, film formation is poor and unable to provide good separation of the pad surface from the mating surface, thereby sustaining friction without leading to much pyrolysis [17].

The phenomenon of bringing the brake pad to room temperature by reducing the temperature rise at the interface using an air blower is termed recovery. The viscous film that was formed due to a higher temperature rise becomes solid due to cooling. This film gets embedded with the hard particles, forming a third body abrasion. The $\mu_{recovery}$ is higher than μ_{fade} due to good plateau formation [22]. The film formation is sustained without causing much degradation of polymeric the least stable ingredients in the friction interface. BC1 composites showed more $\mu_{recovery}$ values compared to the other brake pads due to better film formation and the silica present in the chemical composition of barites, which increases μ levels to a certain level by becoming embedded in the polymeric back transfer patches upon cooling. Similar behavior was seen in various mesh-sized MoS_2 brake pads [3]. The $\mu_{recovery}$ of the dyno-tested brake pads is given below in the higher to lower sequence. The obtained values are in the industrial range of 0.3–0.5.

$$\mu_{recovery}: \text{BC1}\left(0.45\right) > \text{BC2}\left(0.42\right) > \text{BC3}\left(0.40\right)$$

Another key deciding factor for tribological performance is the recovery rate. This is dependent on the ingredients used and the composite's behavior. The industrial range of the recovery rate is 75%–100% [18]. It is advisable that a higher recovery rate always ensures good tribological performance. The developed pads in the current study showed similar behavior. BC1 pads showed a higher recovery rate compared to the other brake pads due to the higher mesh size of the barites. The prime contributor for recovery rate is porosity. A higher optimal porosity enables rapid cooling of the braking interface, leading to the reduced degradation of polymeric ingredients [21]. The μ should be consistent throughout testing so that it can suit working conditions. It is also important to note that variations in the μ should be less for a better working life. BC1 brake pads showed such behavior due to good tribofilm formation to sustain μ. The trend of the recovery rate in the order of the highest to the lowest is given below.

$$\text{Recovery rate: BC1} > \text{BC2} > \text{BC3}$$

It is always advisable to have a lower temperature rise for brakes to be effective. In general, the temperature rise is mainly dependent on fade cycles. BC1 composites showed a lower temperature rise compared to the other two brake pads due to good tribofilm formation caused by the higher mesh size of the barites and good heat dissipation caused by higher porosity in the pads. This collectively reduces the temperature rise, leading to counter friendliness [18]. The temperature rise in the brake pads is given below in the highest to lowest order.

Temperature rise: BC3 > BC2 > BC1

3.2.3 Wear of the dyno-tested brake pads

Wear in the brake pads is a complex mechanism due to abrasion, adhesion, delamination, pitting, etc., which are influenced by temperature and working conditions. In general, a higher μ causes more wear [2]. This postulate is true in the current study. The μ of the BC3 and BC2 brake pads showed a higher value at the start and then deteriorated due to an increase in temperature, causing polymeric ingredient degradation and unstable film formations. The film formation is dependent on the following mechanism: material transfer to the interface leading to micron-sized wear debris. In another mechanism, two bodies are formed independently on both the mating and pad surfaces, which are dependent on the composition and temperature rise at the interface. Film formation helps to prevent the direct contact of the friction surface with the mating surface, leading to stabilized friction and less wear rate [18]. BC1 composites that possessed higher mesh-sized barites helped to produce good tribofilms that helped to prevent the plastic deformation of the resin matrix upon application of load during sliding. The third bodies formed during braking also induce wear rate, as third bodies are formed by the chip-off of the ingredients present in the formulation at a minute level [33]. This also causes plough on both the surface of the pad and on the rotor. The rotor debris upon oxidation forms iron oxide, which gets entrapped in the braking interface and leads to abrasive wear [28]. Another aspect is that the higher the hardness, the lower the wear rate. But this postulate fails in friction composites in most of the cases, as porosity is necessary to reduce the temperature rise at the interface through heat dissipation [34].

The reduced disc wear helps reduce noise, judder, and vibration in braking. Thus, it is mandatory for an ideal pad to have an optimal μ with less wear rate [18]. Similar behavior was seen in BC1 brake pads. The values of

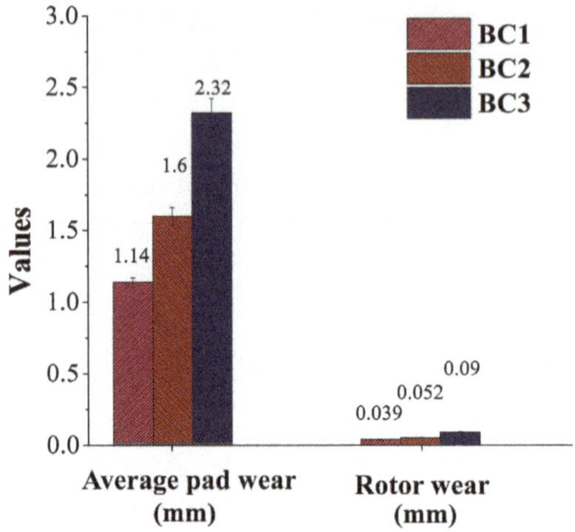

Fig. 7 Pad and rotor wear of the dyno-tested pads.

rotor and pad wear for the developed brake pads are seen in Fig. 7. Pad and rotor wear in the highest to lowest order are given below

Pad and rotor wear: BC1 > BC2 > BC3

3.3 Empirical relationship between mesh size, material, and temperature rise on μ_{fade} for the developed brake pads

The influence of mesh size, material, and temperature rise on μ are analyzed by considering other evaluation conditions like pressure and speed as constant. This is carried out using a qualitative correlation similar to the wear analysis equation given by Rhee [35]. Thus, μ is expressed in terms of Eq. (3) obtained from the literature [36].

$$\mu = KT^{a}\,\phi^{b} \tag{3}$$

where T, ϕ, and K are temperature rise, particle size, and material constant, respectively. Then log is applied on both sides of Eq. (3) and given in Eq. (4).

$$\log(\mu) = \log(K) + a\log(T) + b\log(\phi) \tag{4}$$

Fade dependence is found by considering μ_{fade}. A typical equation is shown in Eq. (5) by substituting the μ_{fade} and other values of BC1 in Eq. (4).

$$\log(0.41) = \log(K) + a\log(380) + b\log(105) \tag{5}$$

Upon substituting and solving all the values in Eq. (2) for all composites, we get

$$K = 4.4785 \times 10^{-6}, a = 4.6463 \text{ and } b = 1.6392 \text{ for fade performance}$$

It is clear that temperature plays a crucial role in determining μ_{fade} and mesh size which helps to sustain μ values. The material coefficient seems to be less influential. The recovery μ cannot be found using the above formula, as the temperature rise is not measured during that test. Thus, it is clear that temperature rise and mesh size play a synergetic role in determining μ_{fade}.

3.4 Worn surface characterization of the dyno-tested brake pads

Worn surface images of the dyno-tested brake pads are given in Fig. 8A–C. Plateau formation governs the tribological performance of the brake pads in general. The higher the contact plateaus, the better the tribological performance, as it aids the load-bearing capacity. It is also essential that contact

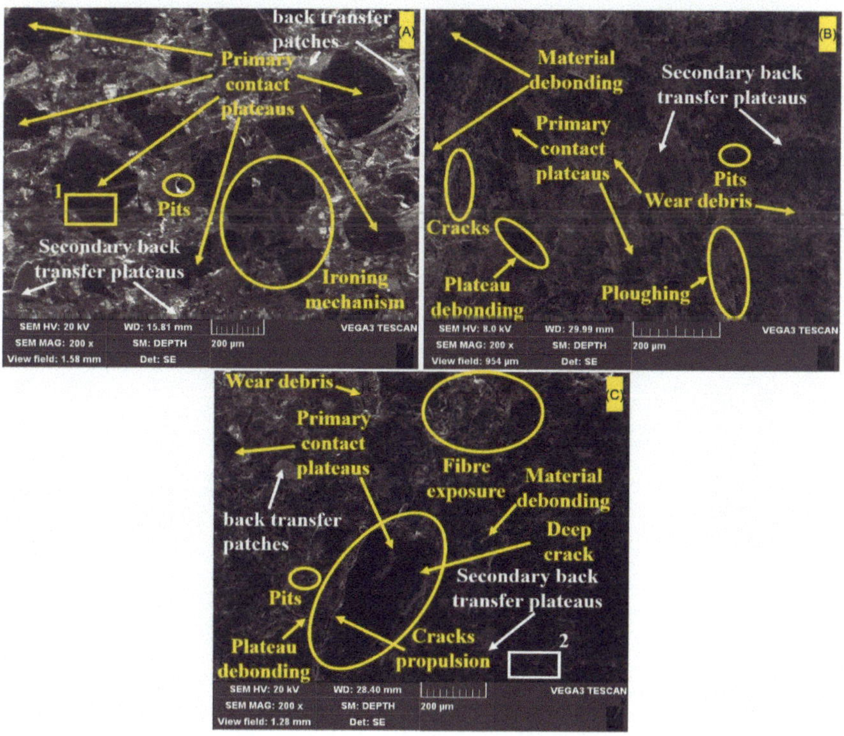

Fig. 8 Worn surface SEM images of (A) BC1; (B) BC2; and (C) BC3.

plateaus formed should not be destroyed soon. Primary plateaus comprise highly stable materials and fibers combined to prevent wear debris from further movement [2]. The secondary plateaus are formed from the least stable ingredients such as rubber and aramid, which are degraded upon application of pressure; this increases heat at the interface, causing reduced performance [20]. Large-sized primary contact plateaus that enhance friction levels are seen in BC1 composites.

Further, back-transfer patches are also comparatively less due to tribo-film formation over the surface, as confirmed from EDAX measured from spectrum "1" as shown in Fig. 9A. The Ba and S are in that region form a protective film coating from plateaus destruction due to abrasion by third

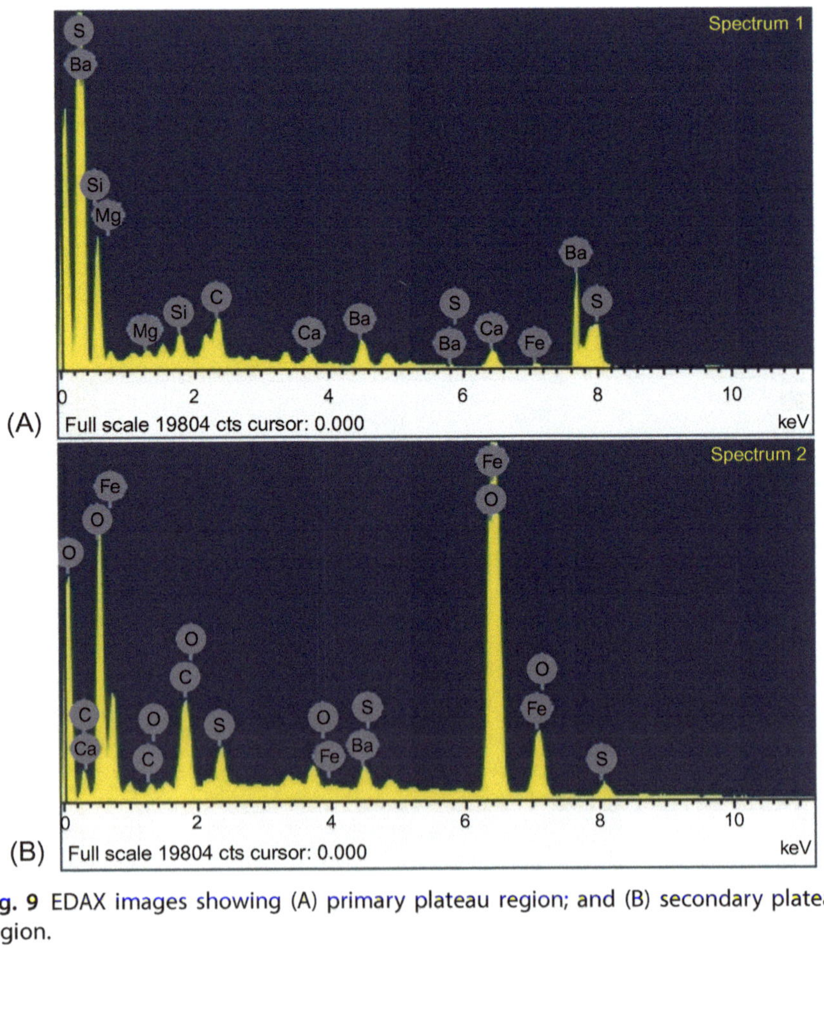

Fig. 9 EDAX images showing (A) primary plateau region; and (B) secondary plateau region.

bodies or ploughing by the mating surface. It is visualized that an ironing mechanism has taken place in the primary plateau region, indicating mild wear, which is in tandem with the literature [37,38]. This higher contact plateau formation maintains μ consistency without causing many undulations. This helps to reduce the wear rate. In the case of the BC2 composites, there seems to be a little higher back-transfer of polymeric ingredients forming back-transfer patches, leading to poor wear resistance and μ reduction compared to BC1, as shown in Fig. 8B [2]. Finally in BC3, there are more nucleated primary plateaus and secondary plateaus with a higher back-transfer of polymeric ingredients showing more shiny patches, as shown in Fig. 8C. This proves that BC3 has poor wear resistance.

It is severe due to the least tribofilm formation resulting in a higher wear rate and additional patches of polymeric ingredients due to back-transfer from the mating surface, as inferred from the EDAX studies measured for spectrum "2" as shown in Fig. 9B [21]. The iron on oxidation from the mating surface upon debonding gets trapped in the friction interface as shown in the rotor wear in BC3 these hard particles damage primary plateaus. Crack formation and propulsion as seen in the BC3 composite proves that a higher temperature rise leads to such poor behavior. Debonding of plateaus due to increasing temperature is also seen in Fig. 8C. Thus, overall the BC1 pads showed improved plateau formation, less back-transfer, and debris with good tribofilm formation, resulting in better tribological performance.

4 Conclusions

Three different mesh-sized barites were used with other parental ingredients in brake pad development as per standard practice. The developed pads were tested for various properties as per industrial standards. Based on the test results, the following inferences were drawn:

- Barites with 150 BSS mesh-sized brake pads showed lower density, hardness, acetone extraction, and loss on ignition with good shear strength and porosity compared to the other two brake pads. The adhesion levels for brake pads with a backplate were also higher in such cases.
- Barites with 150 BSS mesh-sized brake pads showed consistency in μ levels for effectiveness, fade, and recovery conditions. They also showed a higher recovery rate and lesser fade rate due to proper tribofilm formation and porosity, which reduced the temperature rise.
- The wear rate of barites with 150 BSS mesh-sized brake pads was less due to better film formation preventing more back-transfer.

- An empirical relationship showed that μ_{fade} was primarily dependent on the synergy of the temperature rise and the mesh size of the barites.
- Barites with 150 BSS mesh-sized dyno-tested brake pads showed more primary plateaus with lesser secondary back-transfer patches due to good film formation, which protected it from destruction as confirmed from EDAX.

Thus, BC1 brake pads comprising barites with a 150 BSS mesh size as a varying ingredient are a positive solution in the current global automobile braking scenario.

References

[1] D. Chan, G.W. Stachowiak, Review of automotive brake friction materials, Proc. Inst. Mech. Eng. Part D: J. Automob. Eng. 218 (9) (2004) 953–966.
[2] R. Vijay, D. Lenin Singaravelu, R. Jayaganthan, Development and characterization of stainless steel fiber-based copper-free brake liner formulation: a positive solution for steel fiber replacement, Friction (2019) 1–25.
[3] R. Vijay, D. Lenin Singaravelu, P. Filip, Influence of molybdenum disulfide particle size on friction and wear characteristics of non-asbestos-based copper-free brake friction composites, Surf. Rev. Lett. (2019) 1950085.
[4] I. Justin Antonyraj, R. Vijay, D. Lenin Singaravelu, Influence of WS_2/SnS_2 on the tribological performance of copper-free brake pads, Ind. Lubr. Tribol. 71 (3) (2019) 398–405.
[5] D. Lenin Singaravelu, R. Vijay, M. Rahul, Influence of Crab Shell on Tribological Characterization of Eco-Friendly Products Based Non Asbestos Brake Friction Materials, 2015. (No. 2015-01-2676). SAE Technical Paper.
[6] D. Lenin Singaravelu, R. Ragh, R. Vijay, S. Manoharan, M. Kchaou, Development and performance evaluation of eco-friendly crab shell powder based brake pads for automotive applications, Int. J. Automot. Mech. Eng. 16 (2) (2019) 6502–6523.
[7] S. Manoharan, A.I. Shihab, A.S.A. Alemdar, L. Ganesh Babu, R. Vijay, D. Lenin Singaravelu, Influence of recycled basalt-aramid fibres integration on the mechanical and thermal properties of brake friction composites, Mater. Res. Express 6 (11) (2019). 115310.
[8] S. Manoharan, R. Vijay, D. Lenin Singaravelu, S. Krishnaraj, B. Suresha, Tribological characterization of recycled basalt-aramid fiber reinforced hybrid friction composites using grey-based Taguchi approach, Mater. Res. Express 6 (6) (2019), 065301.
[9] F. Headson, Lining or Friction Member for Brake Bands, Brake Shoes, or Other Articles, 1923. United States Patent No. 1468634, Filed May 9, 1921, and Issued September 25, 1923.
[10] L. Baekeland, Composition Containing Phenolic Condensation Products, 1918. United States Patent No. 1259472, Filed December 2, 1909, and Issued March 19, 1918.
[11] Y. Handa, T. Kato, Effects of Cu powder, $BaSO_4$ and cashew dust on the wear and friction characteristics of automotive brake pads, Tribol. Trans. 39 (2) (1996) 346–353.
[12] S.J. Kim, J.Y. Lee, H. Jang, Effect of humidity on friction characteristics of automotive friction materials, KSTLE Int. J. 2 (2) (2001) 150–153.
[13] B. McCormick, The interactions of phenolic polymers in friction material, in: Braking, Professional Engineering Publishing, London, 2002.
[14] Y. Ma, G.S. Martynková, M. Valášková, V. Matějka, Y. Lu, Effects of $ZrSiO_4$ in non-metallic brake friction materials on friction performance, Tribol. Int. 41 (3) (2008) 166–174.

[15] A. Tiwari, H.S. Jaggi, R.K. Kachhap, B.K. Satapathy, S.N. Maiti, B.S. Tomar, Comparative performance assessment of cenosphere and barium sulphate based friction composites, Wear 309 (1–2) (2014) 259–268.
[16] C. Menapace, M. Leonardi, V. Matějka, S. Gialanella, G. Straffelini, Dry sliding behavior and friction layer formation in copper-free barite containing friction materials, Wear 398 (2018) 191–200.
[17] S. Manoharan, R. Vijay, D. Lenin Singaravelu, M. Kchaou, Experimental investigation on the tribo-thermal properties of brake friction materials containing various forms of graphite: a comparative study, Arab. J. Sci. Eng. 44 (2) (2019) 1459–1473.
[18] D. Lenin Singaravelu, R. Vijay, P. Filip, Influence of various cashew friction dusts on the fade and recovery characteristics of non-asbestos copper free brake friction composites, Wear 426 (2019) 1129–1141.
[19] V. Thiyagarajan, K. Kalaichelvan, R. Vijay, D. Lenin Singaravelu, Influence of thermal conductivity and thermal stability on the fade and recovery characteristics of non-asbestos semi-metallic disc brake pad, J. Braz. Soc. Mech. Sci. Eng. 38 (4) (2016) 1207–1219.
[20] S. Manoharan, R. Vijay, M. Kchaou, Investigation on tribological and corrosion characteristics of oxide-coated steel and mild steel fiber-based brake friction composites, Ind. Lubr. Tribol. 71 (3) (2019) 341–347.
[21] G. Saikrishnan, L.S. Jayakumari, R. Vijay, D. Lenin Singaravelu, Influence of iron-aluminum alloy on the tribological performance of non-asbestos brake friction materials—a solution for copper replacement, Ind. Lubr. Tribol. 72 (1) (2020) 66–78.
[22] M. Kumar, J. Bijwe, Optimized selection of metallic fillers for best combination of performance properties of friction materials: a comprehensive study, Wear 303 (1–2) (2013) 569–583.
[23] T. Singh, A. Patnaik, Performance assessment of lapinus–aramid based brake pad hybrid phenolic composites in friction braking, Arch. Civ. Mech. Eng. 15 (1) (2015) 151–161.
[24] T. Singh, A. Tiwari, A. Patnaik, R. Chauhan, S. Ali, Influence of wollastonite shape and amount on tribo-performance of non-asbestos organic brake friction composites, Wear 386 (2017) 157–164.
[25] B.S. Rajan, M.A.S. Balaji, S.S. Saravanakumar, Effect of chemical treatment and fiber loading on physico-mechanical properties of Prosopis juliflora fiber reinforced hybrid friction composite, Mater. Res. Express 6 (3) (2018), 035302.
[26] T.R. Jaafar, M.S. Selamat, R. Kasiran, Selection of best formulation for semi-metallic brake friction materials development, Powder Metall. (2012) 1–30.
[27] B.S. Rajan, M.A.S. Balaji, A.B. Mohamed Aslam Noorani, Effect of silane surface treatment on the physico-mechanical properties of shell powder reinforced epoxy modified phenolic friction composite, Mater. Res. Express 6 (6) (2019), 065315.
[28] M. Kumar, J. Bijwe, NAO friction materials with various metal powders: tribological evaluation on full-scale inertia dynamometer, Wear 269 (11–12) (2010) 826–837.
[29] B.S. Rajan, M.A. Saibalaji, S.R. Mohideen, Tribological performance evaluation of epoxy modified phenolic FC reinforced with chemically modified *Prosopis juliflora* bark fiber, Mater. Res. Express 6 (7) (2019), 075313.
[30] W. Österle, M. Griepentrog, T. Gross, I. Urban, Chemical and microstructural changes induced by friction and wear of brakes, Wear 251 (1–12) (2001) 1469–1476.
[31] X. Hu, On the size effect of molybdenum disulfide particles on tribological performance, Ind. Lubr. Tribol. 57 (6) (2005) 255–259.
[32] P.J. Blau, J.C. McLaughlin, Effects of water films and sliding speed on the frictional behavior of truck disc brake materials, Tribol. Int. 36 (10) (2003) 709–715.
[33] I. Mutlu, C. Oner, F. Findik, Boric acid effect in phenolic composites on tribological properties in brake linings, Mater. Des. 28 (2) (2007) 480–487.
[34] M. Kchaou, A. Sellami, R. Elleuch, H. Singh, Friction characteristics of a brake friction material under different braking conditions, Mater. Des. (1980-2015) 52 (2013) 533–540.

[35] S.K. Rhee, Friction properties of a phenolic resin filled with iron and graphite—sensitivity to load, speed and temperature, Wear 28 (2) (1974) 277–281.
[36] D.K. Kolluri, A.K. Ghosh, J. Bijwe, Performance evaluation of composite friction materials: influence of nature and particle size of graphite, J. Reinf. Plast. Compos. 29 (18) (2010) 2842–2854.
[37] A.K. Ilanko, S. Vijayaraghavan, Wear behavior of asbestos-free eco-friendly composites for automobile brake materials, Friction 4 (2) (2016) 144–152.
[38] S. Manoharan, G. Saikrishnan, L. Ganesh Babu, R. Vijay, D. Lenin Singaravelu, Synergistic effect of red mud-iron sulfide particles on fade-recovery characteristics of non-asbestos organic brake friction composites, Mater. Res. Express 6 (10) (2019) 105311.

CHAPTER 15

Tribological studies and modal analysis on biocomposite/PVC core sandwich panels

I. Siva[a], M. Chithirai Pon Selvan[b], Krishasamy Senthilkumar[c], D. Sivakumar[d], P. Arulraj[e], and Sandro C. Amico[f]
[a]Centre for Composite Materials, International Research Centre, Kalasalingam Academy of Research and Education, Krishnankoil, Tamil Nadu, India
[b]School of Science and Engineering, Curtin University Dubai, Dubai, United Arab Emirates
[c]Center of Innovation in Design and Engineering for Manufacturing (CoI-DEM), King Mongkut's University of Technology North Bangkok, Bangsue, Bangkok, Thailand
[d]Department of Mechanical Engineering, Technical University of Malaysia Melaka, Melaka, Malaysia
[e]Department of Mechanical Engineering, Einstein College of Engineering, Tirunelveli, Tamil Nadu, India
[f]Laboratory of Polymeric Materials, Federal University of Rio Grande do Sul, Porto Alegre, RS, Brazil

1 Introduction

Sandwich composites are comprised of multilayered structures that include thin two-face sheets (also called skins) on the extreme outer layers and are enclosed by a thick but low density core. The face sheets can be used as natural fibers for the development of ecofriendly composites. However, the inherent stiffness is not better than the artificial/synthetic fibers such as carbon and glass. One well-known method to enhance the stiffness of the homogeneous composite/monolithic laminate is to fabricate it as a sandwich composite. Nowadays, the sandwich structures are widely used in many fields such as aerospace, aircraft, automobile, civil structural works, etc., because they have a high stiffness-to-mass ratio. The performance of the sandwich composites is tailor made by altering the core materials and skins in the composite, varying the thickness of the skins and the core material used, etc. [1, 2]. The general advantages of core materials are good stiffness, shear strength, good acoustic and thermal insulation, good damage tolerance, better vibration properties such as good damping, and good compatibility with thermoset resins.

Many researchers have studied various works on the sandwich panels [3–5]. Garay et al. [6] fabricated a sandwich composite structure by using polymeric resin. The core is made up of polyethylene terephthalate and a polyvinyl chloride core. In this research work, the authors studied the effect of cores and skins. The sandwich composites are subjected to mechanical testing such as flexural, flatwise, and edgewise compression as well as

Tribology of Polymer Composites
https://doi.org/10.1016/B978-0-12-819767-7.00015-3

Shore-D hardness tests. Mancuso et al. [7] fabricated a sandwich composite by using flax fibers as the skin material and agglomerated cork as the core. The sandwich composites were fabricated by using the hand layup, resin transfer molding, and infusion techniques. In this work, the fabricated composites were subjected to flexural and tensile behavior. Abdi et al. [8] fabricated a foam core sandwich (FCS) and a polymer-pin reinforced foam core sandwich (PRFCS) by using the vacuum infusion process. The FCS and PRFCS sandwich composites were made by using chopped strand mat glass as the face sheets and polyurethane foam as the core material with a polyester matrix. The sandwich composites were subjected to flexural and flatwise compression loadings. It was found that the introduction of the polymer-pin reinforced foam enhanced the flatwise compression and the flexural properties of the composites increased significantly. Menta et al. [9] used a modified vacuum-assisted resin transfer molding process to fabricate a honeycomb sandwich structure with commercially available film adhesives. The sandwich panels were subjected to flatwise tension, flatwise/edgewise compression, and three-point bending tests to investigate the mechanical performance of the sandwich composites. Hoto et al. [10] addressed the flexural and water absorption properties of sandwich composites. These sandwich composites were made up of cork panel, basalt, and flax fibers. The authors reported that the entire composite possessed good energy-absorbing capability in bending tests. Further, the moisture uptake was reduced considerably in the sandwich composites. Yaman et al. [11] fabricated sandwich structures using balsa wood as the core while glass fibers were placed as the skin of the structures. The authors tried the natural-based materials as the core in the sandwich structures instead of trying conventional materials. In this study, the dynamic properties of the structures were studied experimentally and numerically. The properties were influenced by the number of layers, the layer thickness, and the fiber orientations. It was found that the core thickness was optimized and the dynamic properties were enhanced. The work addressed the application of sandwich materials to fabricate wind turbine blades rather than using conventional materials [12]. The authors also discussed the potential advantages and challenges of sandwich structures for the load-carrying capacity of parts used in wind turbine blades. They also suggested that some of the challenges need to be addressed in extremely loaded wind turbine blades by using the sandwich structures. A new type of sandwich structure was produced by using a new core known as a fiber network sandwich material. Shahdin et al. [13] studied the mechanical and vibrational properties of the developed new sandwich structures. The study reported that the sandwich structure possessed low compression and bending strength but produced higher vibration test results. So, the authors suggested that these kinds of sandwich structures can be used for vibroacoustic applications such as internal paneling of a helicopter, where the structural strengths are secondary. Reis et al. [14] used a cork-based material as the core to produce sandwich structures. The cork-based material possesses isolated thermal and acoustic properties compared to the currently used materials in sandwich structures. The objective of this work was to investigate the mechanical behavior (three-point bending and shear tests) of sandwich composites made up of carbon as face sheets and different cork agglomerates as cores.

The tribological properties of materials are mainly influenced by hardness and density. Increase in the hardness of materials significantly enhance the wear resistance. On the other hand, modern materials are interested in reduced density perhaps to increase the strength-to-weight ratio. Hence, the wear resistance may be reduced due to their ultralightweight molecular structure. Stacking and hybridizing such lightweight structures may result in enhanced wear resistance.

Though many researchers have reported on sandwich composites, very few works have been reported on hybrids in sandwich composites and almost none on the tribological properties. In this context, the present study describes the effect of hybridization on the tribological and vibrational properties of sandwich composites made of woven sisal, banana, and glass polyester composites. The vibrational properties are studied by using the impulse hammer technique.

2 Materials and methods

Banana and sisal woven mats were supplied by Shiva Exports, Tirunelveli, TN, India. The E-Glass fabric of 300 GSM was supplied by Saint-Gobain Vetrotex Fiberglass, India. Unsaturated isothalic polyester resins with reagents were supplied by Huntsman, Chennai. A foam core of 80 GSM was supplied by Vortex, Chennai. Sandwich composites were made using three different fibers such as woven sisal (S), woven banana (B), and woven glass fiber (G) with a polyvinyl chloride (PVC) core. The sandwich structures contain two external faces with polyvinyl chloride (C) as the core sheet. Nine different layering patterns were fabricated: SCS, BCB, GCG, SCG, BCG, BGCGB, BGCBG, SGCGS, and SGCSG. Table 1 displays the notations, codes, and corresponding descriptions followed in the sandwich production.

Table 1 Notations, codes, and their descriptions.

Notation	Code	Descriptions
1	SCS	Sisal/PVC/sisal
2	BCB	Banana/PVC/banana
3	GCG	Glass/PVC/glass
4	SCG	Sisal/PVC/glass
5	BCG	Banana/PVC/glass
6	BGCGB	Banana/glass/PVC/glass/banana
7	BGCBG	Banana/glass/PVC/banana/glass
8	SGCGS	Sisal/glass/PVC/glass/sisal
9	SGCSG	Sisal/glass/PVC/sisal/glass

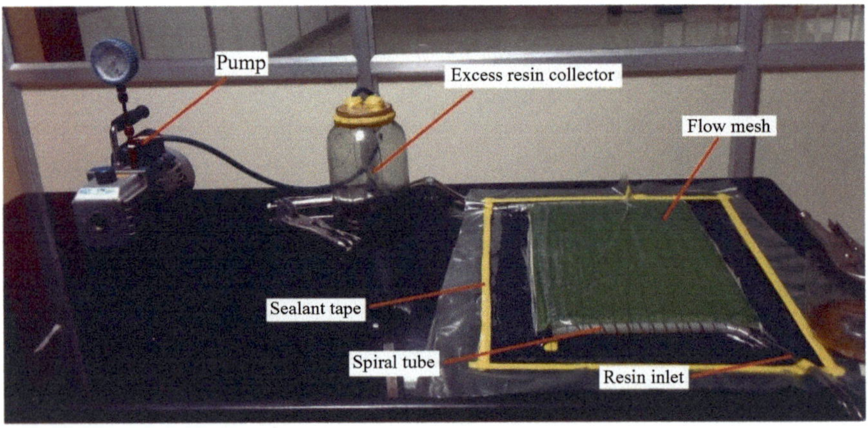

Fig. 1 Manufacturing set up of the sandwich composite.

The sandwich composites were fabricated using a vacuum-assisted infusion technique. Initially, the surface of the produced composite area was cleaned and waxed for ease of release. According to the layering sequences, the skins (fibers) and core (polyvinyl chloride) were layered over together on the cleaned surface of the surface plate. Fig. 1 shows the complete set up of the vacuum-infusion technique.

After the skins and core were arranged over the cleaned surfaces, the peel ply, flow mesh, and vacuum bag were positioned with the help of sealant tape. A full vacuum was ensured before the final pass of resin into the flow mesh. Enough time and a gradual increase in inlet pressure were given to encapsulate the entire preforms. Upon curing, the sandwich composite was removed from the base plate and support layers were nailed out.

2.1 Model analysis

Modal analysis is one of the methods used to evaluate the dynamic characteristics of mechanical components. Moreover, this technique can explain a structure in terms of its natural characteristics. The natural characteristics are natural frequency, damping, and mode shapes. To perform the modal analysis, there are two well-known methods: the impact hammer excitation method and the vibration shaker method. In terms of the theoretical point of view, the measured frequency response functions (FRF) may be collected from the impact hammer excitation and vibration shaker tests. But, due to some of the practical aspects, usually differences in values occur. In this investigation, the impact hammer test was used to study the modal analysis of sandwich composites. Fig. 2 shows the line diagram of the free vibration analysis test.

The test comprises an impact hammer with a sharp hardened tip (Kistler model 9722A500), and further, an accelerometer has been fixed at the free end of the cantilever type of sandwich beam (200×20×3 mm). This set up has been used to find the natural frequencies by hitting the impulse hammer in the equally spaced sandwich composite. From the accelerometer, a displacement signal is recorded in a personal computer by using a data acquisition system (DAS) (DEWE43, Dewteron Corp., Austria) and an ICP conditioner (MSIBRACC). In this set up, two adaptors have been used to capture the output signal—one for the accelerometer signal and the other for the responses measured by using the hammer after hitting the laminate.

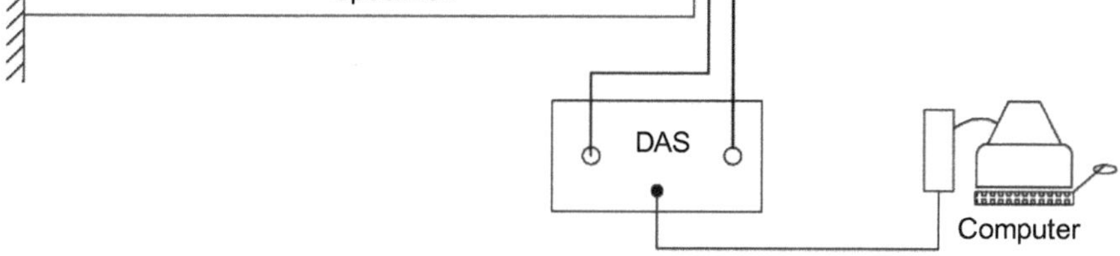

Fig. 2 Experimental impact hammer excitation method.

2.2 Damping factor

To understand the dynamic characteristics of fiber-reinforced composites, studying damping is important. The damping mechanism of natural fiber composites is entirely different from the conventional materials. Further, the energy dissipation mechanism changes depending upon the interphase and the viscoelastic nature between the fiber matrix and the visco-plastic characteristics. Generally, studying the damping mechanism of the natural fiber composites is difficult owing to its chemical constituents, yet it gives better damping due to the inherent porous nature. To find the damping coefficients, the half-power band width method has been used to study the woven sisal, woven banana, and woven glass fiber-reinforced composites by using the FRF curves obtained from the FFT analyzer. Based on Eq. (1), the damping has been calculated.

$$\zeta = \frac{\Delta\omega}{2\omega n} \tag{1}$$

where ζ—damping coefficient, $\Delta\omega$—bandwidth, and ωn—natural frequency.

2.3 Wear and frictional studies

Wear and friction analyses were conducted on a pin-on-disc instrument as per ASTM G-99, (Magnum Engineers, Bangalore). Under ambient conditions, the abrading was performed for varying velocity and applied load. The ambient temperature was roughly 23°C and the relative humidity was $50 \pm 5\%$. Sandwich composites were cut into $10 \times 10 \times 3$ mm blocks and they encountered a hardened alloy steel disc of 62 HRC and 0.54 μm rougher. A diagram of the counter surface contact and sample is shown in Fig. 3. Experiments were run on the specified samples. Based on the vibrational analysis, with which damping was found to be higher, those samples underwent tribological analysis. Abrading was done for varying normal loads such as 10, 20, and 30 N. Commonly, all the runs were done for a 1200 m distance and velocities were taken as 1.5 and 2.5 m/s. Friction was measured in steps of a 300 m run. To enable complete contact between the specimen and counter, the sandwich samples were polished with a 600 grade SiC sheet. The surfaces of both the samples and the disc were cleaned with a soft paper soaked in acetone and thoroughly dried before the test. The pin assembly was initially weighed to an accuracy of 0.0001 g in an electronic balance (SHIMADZU AUX 220). The difference between the initial and final weights was the measure of the sliding wear loss. The wear

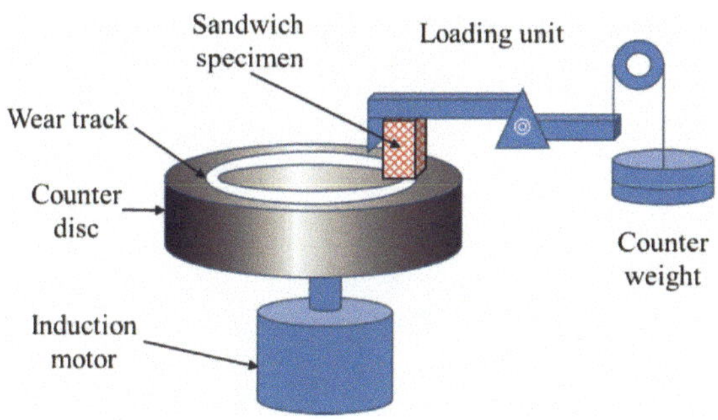

Fig. 3 Pin-on-disc wear analysis unit.

was measured by the loss in weight, which was then converted into wear volume using the measured density data. The specific wear rate (K_s) was calculated from the equation:

$$K_s = \frac{\Delta V}{L \times d} \frac{\mathrm{m}^3}{\mathrm{N\,m}} \tag{2}$$

where ΔV is the volume loss in m^3, L is the load in Newton, and d is the abrading distance in meters.

3 Result and discussion

3.1 Mechanical characterizations

The tensile strength of the fabricated sandwich composites was measured as per the ASTM C297 in an INSTRON machine. The variations in tensile strength as a function of layered and hybrid sandwich composites are shown in Fig. 4. The tensile strength of the sandwich composites varies between 171 and 216 MPa and does not show a clear trend regarding the optimum layering pattern of sandwich composites.

From Fig. 4, a linear increase in tensile strength was observed as a result of the addition of glass fibers. The percentage improvements from sisal/PVC/sisal to sisal/PVC/glass and banana/PVC/banana to banana/PVC/glass were observed as 10.32% and 9.72%, respectively. It can be noticed that the tensile strength was higher when stronger banana fiber composites were used as the face sheets because the banana fibers are the main load-bearing elements in the tensile strength. The same kinds of results have been published by Mariatti et al. [15]; the authors fabricated bilayer, trilayer, and intimate mix types of hybrid composites using banana and sisal fibers. Under the trilayer combinations, banana/sisal/banana and sisal/banana/sisal hybrid

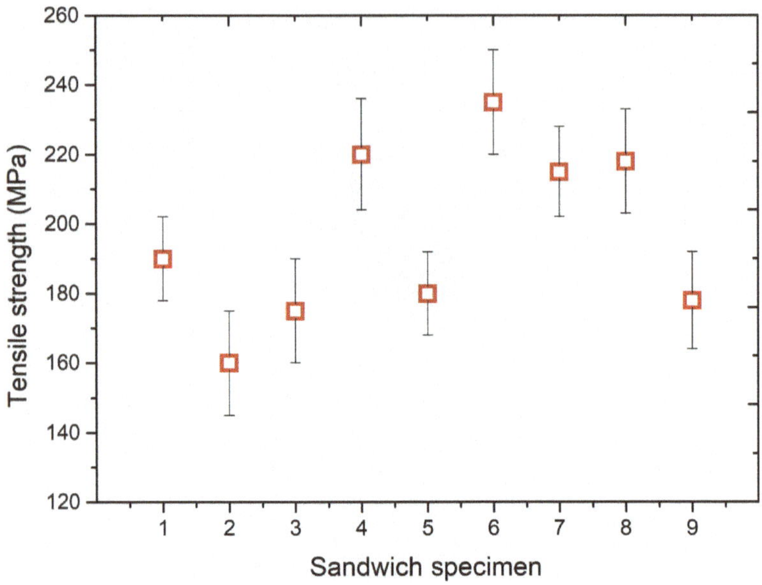

Fig. 4 Variation of tensile strength as a function of hybridization.

composites were fabricated. Among the different types of combinations, the banana/sisal/banana hybrid composite showed the highest tensile strength.

Further, it was noticed that although the banana/glass/PVC/glass/banana and the banana/glass/PVC/banana/glass were built with the same number of layers of banana and glass, the tensile strength varied. Also, the same kind of trend was observed between the sisal/glass/PVC/glass/sisal and the sisal/glass/PVC/sisal/glass type of sandwich composite. From these results, it was observed that the asymmetrical ones particularly were more efficient than the symmetrical ones in the five-layered type of sandwich composite. Among all the types of sandwich composites, the banana/glass/PVC/glass/banana showed the highest tensile strength among the studied combinations.

Flexural strength was experimented with as per the ASTM C 393 in an INSTRON machine. Fig. 5 shows the sandwich composite before and after bending failure. A transverse crack at the loading point can be noticed across the thickness. It was observed during experimentation that upon maximum loading, the brittle skin composites fractured at first and the crack then initiated at the foam followed by the full fracture.

The variation of flexural strength with different layered sandwich composites is shown in Fig. 6. It can be noticed that five-layered combinations

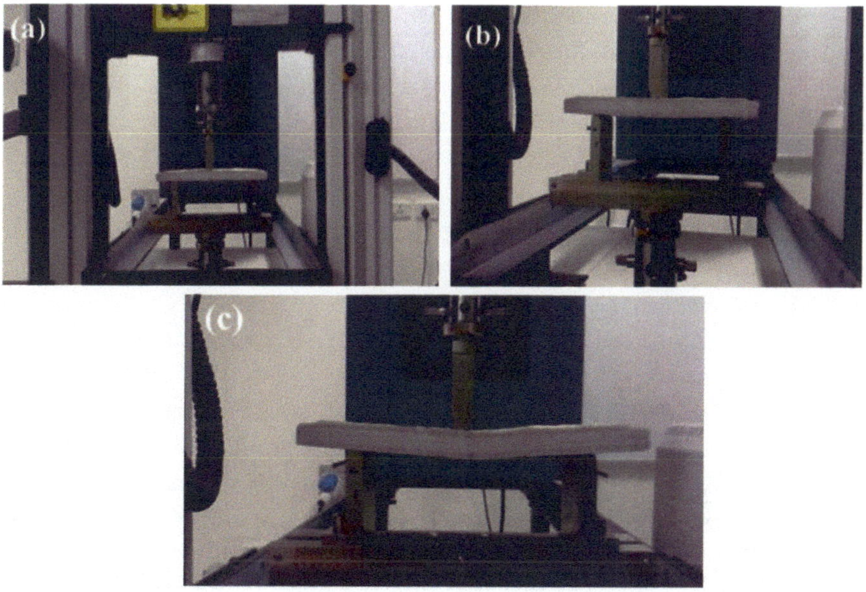

Fig. 5 Sandwich specimen (A, B) before flexural test and (C) fractured specimen.

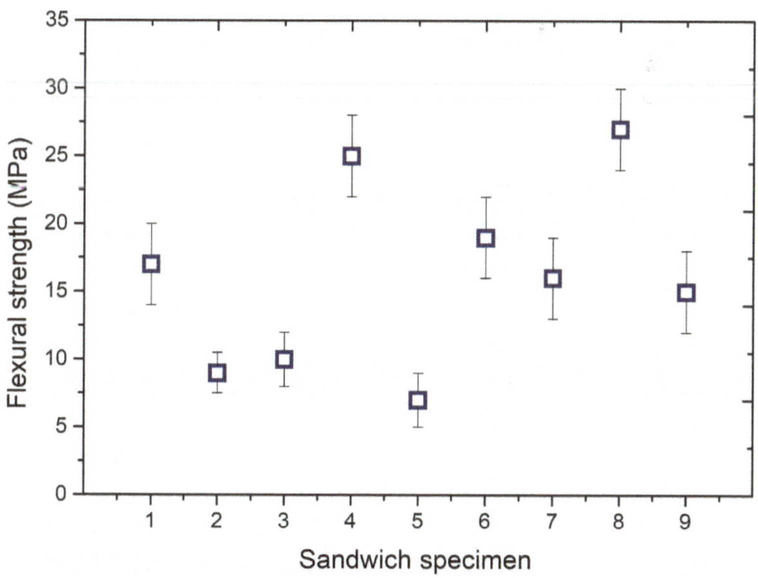

Fig. 6 Variation in flexural strength as a function of hybridization.

of sandwich composites could withstand higher strength (sisal/glass/PVC/glass/sisal and banana/glass/PVC/glass/banana) than trilayered sandwich composites of both sisal and banana fiber sandwich composites. As Mariatti et al. [15] pointed out, higher fiber loading results in stronger composites because more fibers could share the applied forces. In general, the flexural strength of the studied composites is an increasing trend upon the incorporation of the glass fiber content. The percentage improvements from sisal/PVC/sisal to sisal/glass/PVC/glass/sisal and sisal/PVC/glass are 62.63% and 55.23%, respectively.

Further, it is noticed that the value of the flexural strength reduces when the sisal woven mat is placed next to the PVC foam. This effect has been noticed in the combination of both sisal/PVC/sisal and sisal/glass/PVC/sisal/glass. In case of banana fiber sandwich composites, the banana/glass/PVC/Glass/Banana combination provides the higher flexural strength than the other type of combinations such as banana/glass/PVC/banana/glass, banana/PVC/banana and banana/PVC/glass.

Moreover, it is noticed that the both the hybrid sandwich composites banana/glass/PVC/glass/banana and banana/glass/PVC/banana/glass possess a similar number of banana and glass layers in sandwich composites, whereas the flexural strength varies. The percentage improvement from banana/glass/PVC/banana/glass to banana/glass/PVC/glass/banana is observed as 24.98%. Likewise, the percentage improvement from sisal/glass/PVC/sisal/glass to sisal/glass/PVC/glass/sisal is observed as 69.31%. It can be justified from the fact that the flexural strength is controlled by the layering pattern of the composites.

3.2 Vibrational properties

Experimental modal analysis was carried out for all the above sandwich composites. Nine different types of sandwich composites were fabricated by using vacuum infusion technique: sisal/PVC/sisal, banana/PVC/banana, glass/PVC/glass, sisal/PVC/glass, banana/PVC/glass, banana/glass/PVC/glass/banana, banana/glass/PVC/banana/glass, sisal/glass/PVC/glass/sisal, and sisal/glass/PVC/Sisal/glass. To investigate the effect of the layering pattern, the natural frequency and modal loss factor associated with the first three lowest frequency bending modes were found. Fig. 7A and B shows the vibrational response time domain and the typical FRF (frequency response function) for a banana/PVC/glass hybrid sandwich composite with the first three modes of the natural frequency from Dewesoft software.

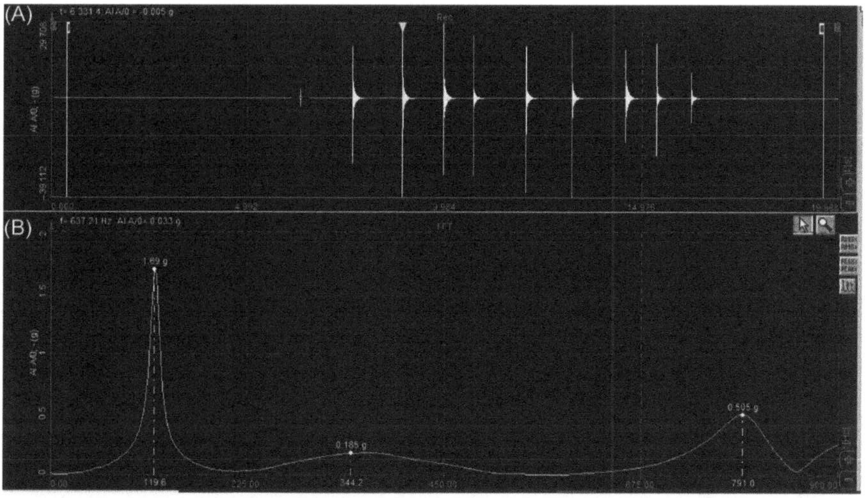

Fig. 7 Vibrational response time domain (A), frequency response function of banana/PVC/glass sandwich composite (B).

Table 2 Natural frequency and model damping variation as a function of hybridization.

Specimen code	Natural frequency (Hz)			Damping		
	Mode-I	Mode-II	Mode-III	Mode-I	Mode-II	Mode-III
1	85.45	163.57	334.47	**0.2700**	0.1410	0.0690
2	75.68	490.7	1008.3	0.0631	0.0097	0.00473
3	124.5	363.8	747.1	0.1591	0.0544	0.0265
4	87.89	644.5	1123	0.0166	0.0022	0.0013
5	119.6	344.2	791	0.1576	0.0547	0.0238
6	95.21	612.8	642.1	**0.2200**	0.0350	0.0335
7	136.7	800.8	1601.6	**0.2160**	0.0369	0.0184
8	136.7	1035.2	2177.7	0.1920	0.0253	0.0120
9	117.2	771.5	1240.2	0.1510	0.0220	0.0142

Table 2 shows the fundamental frequency and damping of the produced sandwich composites. From Table 2, one can observe that the addition of glass fiber increases the natural frequency in sisal/PVC/glass and banana/PVC/glass compared to the sisal/PVC/sisal and banana/PVC/banana sandwich composites. The percentage improvements from sisal/PVC/sisal to sisal/PVC/glass and banana/PVC/banana to banana/PVC/glass are observed as 2.77% and 36.72%, respectively. When the sandwich composites were made up of five layers, the natural frequency was further modified.

An asymmetric sandwich composite built with banana, glass, and PVC gave the maximum natural frequency compared to other symmetric-type combinations. But, in the case of a sisal, glass, and PVC sandwich with a five-layer combination, an inverted trend compared to the banana, glass, and PVC combined sandwich composites was noted.

It is understood from Table 2 that the damping of the studied sandwich materials varies with the modification in layering pattern. The maximum increase was observed in the sisal/PVC/sisal combinations. The values appeared in "bold" indicates the highest damping values recorded. Further, most of the five-layered sandwich composites possess higher damping than the trilayered sandwich combinations.

3.3 Wear of plain and hybrid sandwich composites

The friction and wear responses of sandwich composite were studied for customized specimens. Based on the vibrational results, the specimens for wear analysis were chosen. Specimens with higher damping were chosen because the larger machine vibration absorption and the effective load transfer from loading to the counter would simulate real-time tribological conditions. Fig. 8 shows the variation in wear resistance as a function of sliding velocity with respect to the nature of the samples.

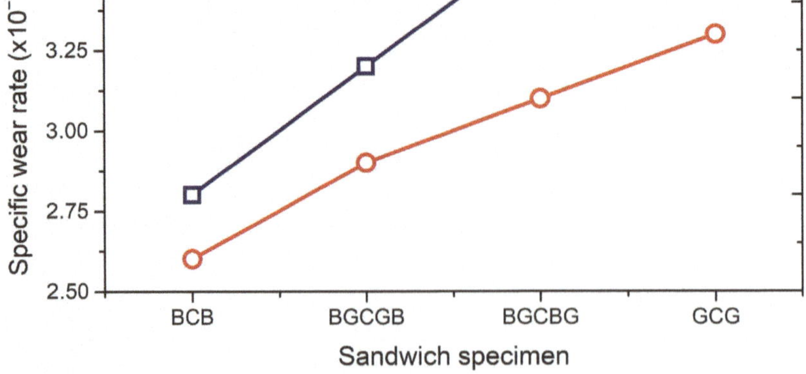

Fig. 8 Effect of sliding velocity on the specific wear.

From Table 2, it is observed that the sandwich samples made of banana/polyester skins and the hybrid possess greater damping properties. The properties of higher damping may be due to the vegetable plant nature of the fibers. Their spongy nature may promote a lot of energy absorption. Hence, the BCB, BGCBG, and BGCGB samples alone were further studied. As a reference, the performance of the glass fiber sandwich also underwent wear analysis. As noted from Fig. 8, the natural fiber composite skin protects the structure well in abrading. The inherent fibrous nature of the banana offered stronger resistance against abrading. Meanwhile, the incorporation of the E-Glass fibers into the sandwich system significantly stimulated mass loss during sliding. This could happen due to the incompatibility of the banana fibers with glass fibers. The hybrid entanglements may be insufficient to hold the creamy layer as bound together. Hence the loss of the brittle polymer may carry additional fibrous materials during abrasion.

Electron microscopy (Fig. 9) of the worn samples revealed that the predominant removal mechanism would be abrasion. At onset, the preliminary polymer layers were removed by the adhesion action, further the subsurface fibers were then worn by abrasive actions. Fig. 9A shows the worn BCB sandwich composite; one can notice the patch removal of huge material from the surface. Such abrasion is attributed to the surface fatigue failures due to the high compression. Further, the synergy of adhesion and abrasion resulted in such huge surface fatigue and the patch removal in massive quantities. The worn BGCGB is displayed in Fig. 9B; in several places, the crazing is noted (*circle* marked in Fig. 9B) predominantly. Crazing is a well-known localized deformation found on worn morphology. Crazing is attributed to the presence of microcracks initiated due to the normal loading around the fibril places. The initiated crack then transforms to the brittle polymeric zone, leading to material failure.

Further, the friction exerted between the sandwich specimen and the counter is also accounted for in order to understand the lubrication of the studied sandwich composites; this is presented in Fig. 10. The friction of the banana (BCB) sandwich composite was found to be higher upon the 1200 m run. Upon the incorporation of the glass fibers in the banana sandwich, considerable surface smoothening was noticed. The removal of the polymer skin brought the banana fiber into contact with the steel counter. The entire load applied on the upper side may not be transferred to the sample-counter interface, resulting in easy erosion and debris deposition.

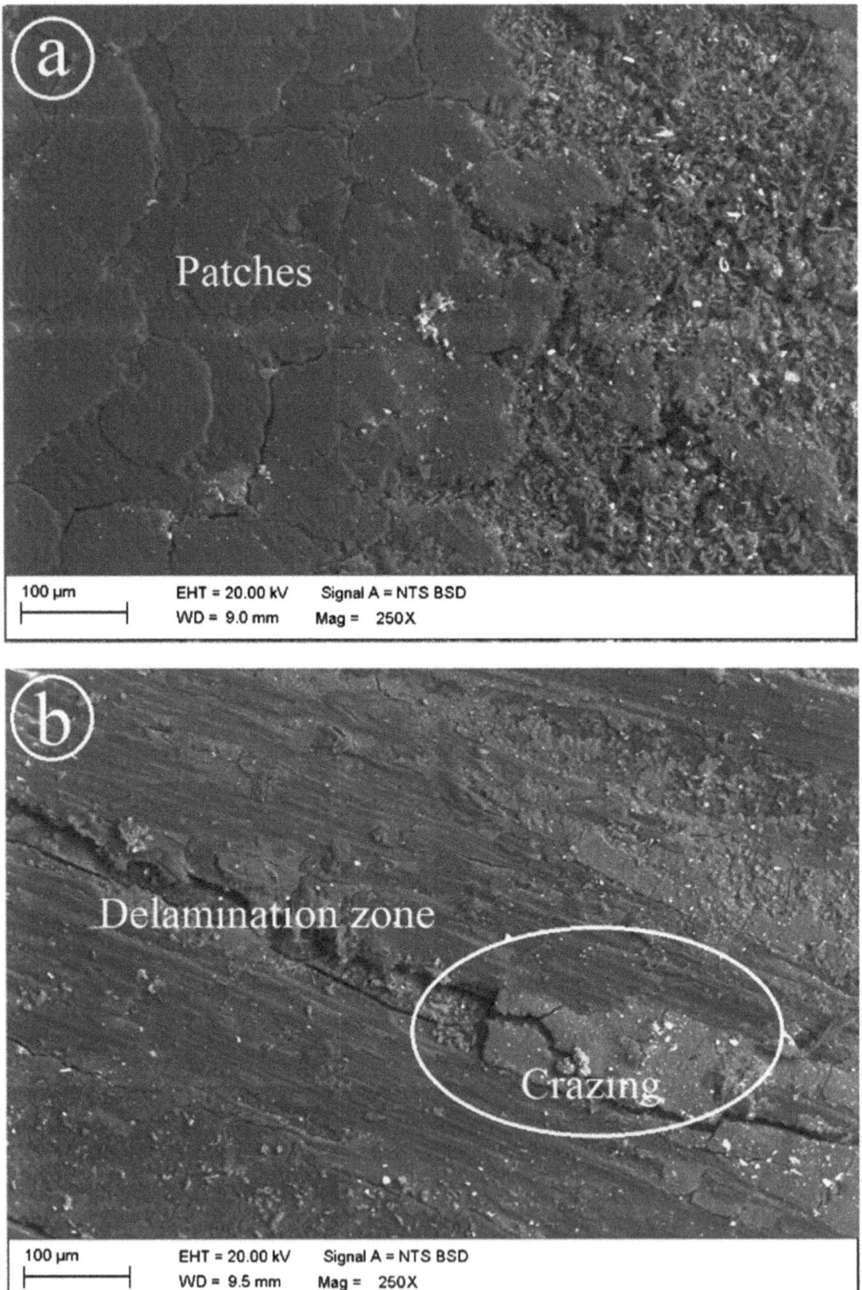

Fig. 9 Worn morphology, sandwich with patch crack due to fatigue (A), surface with crazing and major delamination (B).

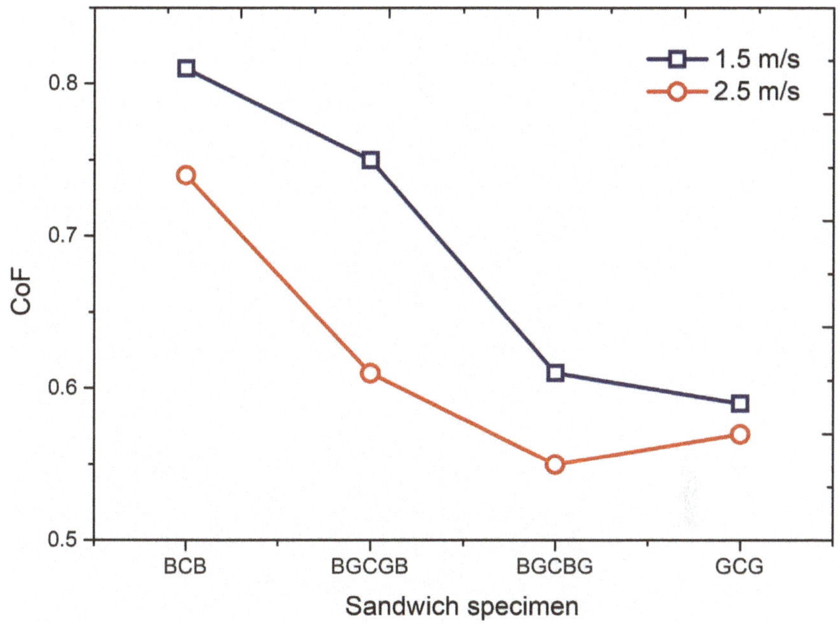

Fig. 10 Effect of sliding velocity on the coefficient of friction.

Fading may be experienced due to the deposited polymer and vegetable fiber debris. Further with the BGCBG sandwich, because the glass fiber was exposed, the surface smoothening was attained shortly, resulting in a low friction run.

Table 3 presents the wear and frictional results of the run for 20 and 30 N normal loads. A similar trend was noted on the wear and frictional behaviors of the pure and hybrid sandwich structures. Nevertheless, friction was significantly increased due to the higher loading of the pin. Perhaps the higher the compressive force applied on the sandwich specimen, the greater the friction exerted at the interface.

Fig. 11 illustrates the frictional trend of the BCB and its hybrid sandwich composites. Initially, the polymer matrix layers were worn out resulting in severe friction; the same was then reduced due to the fiber-counter disc interactions. The wipe out of polymer and fiber debris occurred due to the continuous specimen loading, which in turn resulted in polishing/surface fading. Hence, the friction became stable at the outset. Similar trends in friction were noted with hybrid and full synthetic sandwich composites.

Table 3 Friction and wear of plain and hybrid sandwich composites for various applied loads.

Sandwich specimen	20 N applied load				30 N applied load			
	Specific wear $\times 10^{-13}$ $mm^3/N\,m$		CoF		Specific wear $\times 10^{-13}$ $mm^3/N\,m$		CoF	
	1.5 m/s	2.5 m/s	1.5 m/s	2.5 m/s	1.5 m/s	2.5 m/s	1.5 m/s	2.5 m/s
BCB	3.1	2.9	0.82	0.73	3.3	3.6	0.84	0.74
BGCGB	3.5	3.6	0.78	0.75	3.8	4.2	0.81	0.71
BGCBG	3.8	3.5	0.68	0.72	4.1	4.3	0.78	0.69
GCG	4.2	4.1	0.61	0.67	4.4	4.6	0.62	0.66

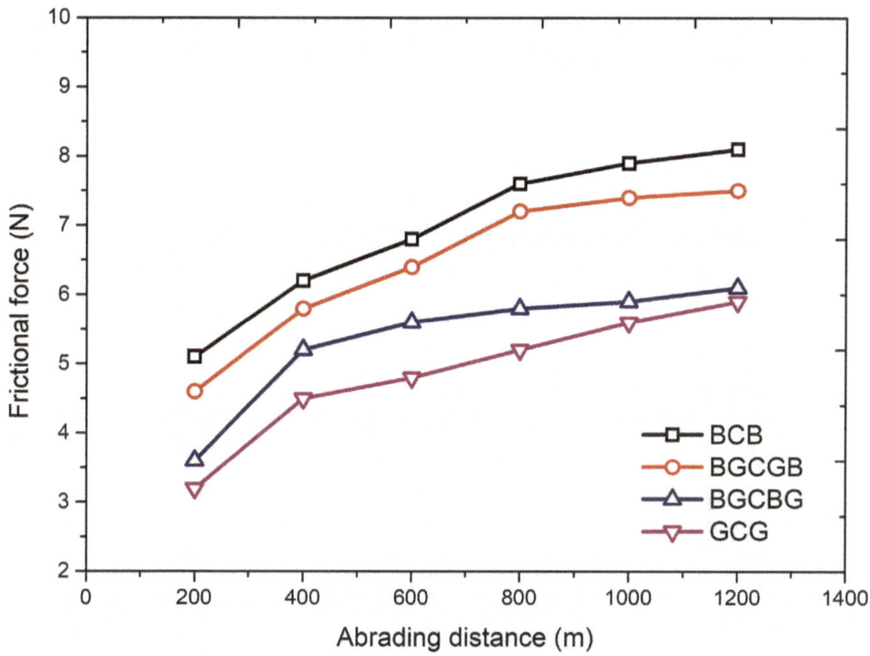

Fig. 11 Friction progress with respect to abrading distance.

4 Conclusion

Recently, most of the lightweight polymeric materials are reinforced with bio-fillers. Many times, lowering the density of materials leads to a loss of wear resistance, even though they possess greater vibration damping. Perhaps incorporating synthetic materials may sustain more when compared to plain materials. The current report discusses the tribological and vibrational performance of plain and hybridized sandwich composites. The following are a few points derived from this study:

- The tensile strength of sandwich composites varies between 171 and 216 MPa and does not show a clear trend regarding the optimum layering pattern of sandwich composites. A linear increase in tensile strength is observed as a result of the addition of glass fibers.
- Among all types of sandwich composites, the banana/glass/PVC/glass/banana has shown the highest tensile strength.
- In terms of flexural strength, the five-layered combinations of sandwich composites (sisal/glass/PVC/glass/sisal and banana/glass/PVC/glass/banana) withstand higher strength.

- Further, it is noticed that the value of the flexural strength is reduced when the sisal woven mat is placed next to the PVC foam. This effect has been noticed in the combination of both in sisal/PVC/sisal and sisal/glass/PVC/sisal/glass.
- In free vibration characteristics, it is observed that the addition of glass fiber increases the natural frequency in sisal/PVC/glass and banana/PVC/glass compared to the sisal/PVC/sisal and banana/PVC/banana sandwich composites.
- In asymmetric, a sandwich composite built by banana, glass, and PVC gives the highest natural frequency compared to the symmetric type of combination. But a sisal, glass, and PVC combined sandwich with a five-layered combination gives the reverse trend compared to the banana, glass, and PVC combined sandwich composites.
- The maximum increase is observed in sisal/PVC/sisal combinations. However, no significant change is noted in damping due to the change in variation of the layering pattern.
- The wear resistance of plain sandwich composites looks better than the hybrid and full glass fiber sandwich.
- However, from a damping point of view, the incorporation may not spoil the wear resistance significantly, irrespective of the applied load and abrading velocity.
- Crazing and surface fatigue are found as the predominant wear mechanisms. However, under higher loading, fading resulted.
- The friction increased drastically at the onset and the deposition of polymer and fiber debris layers over the new surface prevented further mass loss and stabilized the friction.

On the whole, it is concluded that hybridization promotes the structural application/serviceability of the vegetable fiber composite skin sandwich without losing the wear resistance. Especially, the banana fiber/polyester has great potential to substitute the glass fiber/polymer skins for the foam core sandwich structures.

Acknowledgments

The authors wish to thank the Centre for Composite Materials, Department of Mechanical Engineering, Kalasalingam University, for the kind permission to carry out the preparation and testing of the composites. Also, the authors wish to acknowledge the BJT/CNPq and CNPq/DST (Brazil/India) joint projects.

References

[1] V. D'Alessandro, G. Petrone, F. Franco, S. De Rosa, A review of the vibroacoustics of sandwich panels: models and experiments, J. Sandw. Struct. Mater. 15 (5) (2013) 541–582.

[2] L.A. Carlsson, G.A. Kardomateas, Characterization of the mechanical properties of face sheet and core materials, in: Structural and Failure Mechanics of Sandwich Composites. Solid Mechanics and Its Applications, vol. 121, Springer, Dordrecht, 2011, pp. 19–37.

[3] E.E. Gdoutos, I.M. Daniel, K.A. Wang, Compression facing wrinkling of composite sandwich structures, Mech. Mater. 35 (3–6) (2003) 511–522.

[4] J.L. Avery, B.V. Sankar, Compressive failure of sandwich beams with debonded face-sheets, J. Compos. Mater. 34 (14) (2000) 1176–1199.

[5] E.E. Gdoutos, I.M. Daniel, K.A. Wang, J.L. Abot, Nonlinear behavior of composite sandwich beams in three-point bending, Exp. Mech. 41 (2) (2001) 182–189.

[6] A.C. Garay, J.A. Souza, S.C. Amico, Evaluation of mechanical properties of sandwich structures with polyethylene terephthalate and polyvinyl chloride core, J. Sandw. Struct. Mater. 18 (2) (2016) 229–241.

[7] A. Mancuso, G. Pitarresi, D. Tumino, Mechanical behaviour of a green sandwich made of flax reinforced polymer facings and cork core, Procedia Eng. 109 (2015) 144–153.

[8] B. Abdi, S. Azwan, M.R. Abdullah, A. Ayob, Y. Yahya, Comparison of foam core sandwich panel and through-thickness polymer pin-reinforced foam core sandwich panel subject to indentation and flatwise compression loadings, Polym. Compos. 37 (2) (2016) 612–619.

[9] V. Menta, R. Vuppalapati, K. Chandrashekhara, D. Pfitzinger, N. Phan, Manufacturing and mechanical performance evaluation of resin-infused honeycomb composites, J. Reinf. Plast. Compos. 31 (6) (2012) 415–423.

[10] R. Hoto, G. Furundarena, J.P. Torres, E. Muñoz, J. Andrés, J.A. García, Flexural behavior and water absorption of asymmetrical sandwich composites from natural fibers and cork agglomerate core, Mater. Lett. 127 (2014) 48–52.

[11] M. Yaman, T. Önal, Investigation of dynamic properties of natural material-based sandwich composites: experimental test and numerical simulation, J. Sandw. Struct. Mater. 18 (4) (2016) 397–414.

[12] O.T. Thomsen, Sandwich materials for wind turbine blades—present and future, J. Sandw. Struct. Mater. 11 (1) (2009) 7–26.

[13] A. Shahdin, J. Morlier, Y. Gourinat, L. Mezeix, C. Bouvet, Fabrication and mechanical testing of a new sandwich structure with carbon fiber network core, J. Sandw. Struct. Mater. 12 (5) (2010) 569–589.

[14] L. Reis, A. Silva, Mechanical behavior of sandwich structures using natural cork agglomerates as core materials, J. Sandw. Struct. Mater. 11 (6) (2009) 487–500.

[15] M. Mariatti, M. Nasir, H. Ismail, Effect of stacking structure on the properties of satin laminated composites, J. Thermoplast. Compos. Mater. 16 (4) (2003) 303–315.

CHAPTER 16

Synergism of fillers on adhesive and abrasive wear of thermoplastic copolyester elastomer composites

R. Hemanth[a,b] and B. Suresha[b,c]
[a]Department of Mechanical Engineering, NIE Institute of Technology, Mysore, Karnataka, India
[b]Visvesvaraya Technological University, Belagavi, Karnataka, India
[c]Department of Mechanical Engineering, The National Institute of Engineering, Mysore, Karnataka, India

1 Introduction

The interest in fiber and particulate-fortified thermoplastic compounds is emerging rapidly in industrial and research front. Their accessibility, reduced density, ease of fabrication, minimal pricing, and rational characteristics brand these compounds a reasonable substitution for conventional materials. In the mainstream of examples, the strand incorporation in polymers has upgraded the mechanical characteristics to fulfill the particular needs. Strand-strengthened compound materials include filaments installed in or attached to a grid with discrete interfaces (frontiers) between them. In this form, both strands and matrix conserve their physical and chemical attributes, yet they produce a mix of properties that can't be compared with either of the constituents acting alone. As a rule, strands are the significant load-bearing individuals while the connecting matrix keeps them in the favored location and alignment. Furthermore, the matrix additionally transmits the load to the strands and shields them against atmospheric impedances, for example, high temperature, dampness, etc., as was noted by Agarwal and Broutman [1]. The majority of the strand-strengthened polymer compounds (FRPCs) yield a mix of mechanical properties that are comparable to various traditional materials. As a result of the aforementioned reasons, FRPCs have become a well-known gathering of materials and are generally utilized in an expansive range of areas from local to space applications. The short strands are strengthened with arbitrary direction, and the long continuous filaments are oriented in a specific axis. The utilization of short-strand fortified polymer compounds (SFRPCs) has quickly developed, mainly in the

Tribology of Polymer Composites
https://doi.org/10.1016/B978-0-12-819767-7.00016-5

automobile manufacturing sector. SFRPCs are effectively molded with the capacity of being produce multifaceted profiles. A high production rate by processing methods such as injection molding as well as a lower cost are the noteworthy benefits of SFRPCs over ceaseless long filament compounds. The orientation and perspective proportion of the fiber are two significant elements that show the microstructures of a compound with straight short strands. Studies have noted [2] that when the perspective proportion of the strands doesn't pursue a symmetric dispersion, the general elastic properties achieved by the normal perspective proportion of the filaments may contrast by more than 30% from those achieved by the strategy considering the perspective proportion dissemination. Among various kinds of filaments used as fortification in PMCs, strands brewed of glass, carbon, aramid, etc., are widely utilized. Polymers are additionally fortified with diverse grouts that are at hand naturally or mass-produced as chips, platelets, specks, etc., to improve their processability and mechanical properties as well as to decrease the material expense. Examination reports are accessible on acquiring synergistic performances by the combined reinforcement of fiber and fillers into polymer lattices.

In PMCs, various polymers such as thermoplastics, thermosets, elastomers, and thermoplastic elastomers (TPE) can be utilized as matrix materials. The TPE is one of the most encouraging polymers of late. TPEs show qualities of both rigid plastics as well as elastomers. This demeanor has started another road in polymer science. Brydson [3] has characterized them by their chemistry into the following six classes: polystyrene, polyolefin, polyetherimide, polyurethanes, polyamides, and polyesters. Salman and Muhammad [4] stated that thermoplastic elastomers have been the subject of examinations by scientists the world over during past few decades and are consciously employed in extensive applications. In the current framework, the polyester-centered TPE, that is, the thermoplastic copolyester elastomer (TCE) with the trade name Arnitel EM 740, is employed as a budding matrix substance that has been investigated once in a while.

Notwithstanding the basic applications, the parts/elements made of PMCs are exposed to various wear circumstances. Stölarski [5] has characterized wear as a dynamic loss of materials because of the relative movement between two surfaces. Wear prompts the substitution of machine parts/elements, bringing about substantial use as well as a decline in the pace of manufacturing and related burdens. Sometimes, it may bring about a loss of human life because of a calamitous disappointment of the machine elements. Budinski [6] in his work portrayed a couple of different wear modes that can be put into four major classifications: abrasion, adhesion, erosion, and surface fatigue. Päivi [7] revealed that wear issues continue to exist in industries despite broad scientific investigations. This proposed wear is an exceptionally perplexing phenomenon and not totally comprehended so far. From the accessible reports, it has been understood that among various methods of wear, abrasion and adhesion are the most noticeable and inescapable modes contrasted with the fatigue and erosion modes.

Therefore, while picking a material for tribological utilization, it is critical to comprehend its behavior under various wear modes. However, examinations on concurrent wear characterization of the material under various wear modes are rare. Thus, the assurance of wear attributes of polymer composites is a critical necessity before picking a material for tribo-applications. Table 1 depicts a literature review pertaining to the synergistic performances (mechanical and tribological) of selected crossbred polymer compounds fortified with filaments and fillers.

Not many studies are accessible in the literature on the synergistic study of strand and filler-filled thermoplastic copolyester elastomer (TCE) composites on friction and wear studies. This has motivated the authors to write their research findings in this chapter.

2 Friction and wear performance of thermoplastic copolyester compounds

Table 2 presents the constituents used in the preparation of the TCE hybrid composites. Table 3 provides the details of the constituents of the compounds. Fig. 1 presents the stages involved in the preparation of TCE hybrid composites.

The tribological execution of TCEs and their compounds was assessed by presenting different wear examinations:

1. Two-body abrasive wear (2-BAW)
    a. 2-BAW beneath a multipass (MP) situation done on pin-on-disc (POD) attrition assessment equipment (Make: DUCOM, Bangalore) as indicated by ASTM G99-05 [38]. Fig. 2 displays the diagram of the assessment set up. The 2-BAW test procedure pursued as discussed about somewhere else [39].
    b. 2-BAW beneath a single-pass (SP) situation at low speed as indicated by ASTM G133 [40] investigation standards on a specially fabricated abrasion apparatus (Make: Magnum Engineers, Bengaluru).

Table 1 Literature review of selected hybrid PMCs.

References	Material system	Wear mode	Key findings
Sathishkumar et al. [8]			Fillers inclusion enhanced the properties of PMCs
Kistaiah et al. [9]			
Kuljanin et al. [10], Weidenfeller et al. [11], Bijwe, and Indumathi [12]			Filler inclusion into FRPCs resulted in increased modulus and reduced strength and toughness
Hemanth et al. [13]			
Parikh and Gohil [14]			Discussed the significance of material parameters, chemical/physical treatments of fibers and operating parameters that effects friction and wear performance of PMCs
Lucas et al. [15]	HDPE + UHMWPE blends	Two-body abrasion	Addition of UHMWPE into HDPE improves wear resistance and mechanical properties
Friedrich [16]	PET, glass fiber, glass bead	Two-body abrasion	Wear resistance decreases with increasing size of abrasive grains
Harsha and Tewari [17]	PSU, glass fiber	Two-body abrasion	Unfilled PSU demonstrated better wear resistance
Ravikumar et al. [18]	LDPE + EVA blends, nanoclay	Two-body abrasion	LDPE/EVA blend reinforced with nanoclay demonstrated superior wear resistance to abrasion
Bijwe et al. [19]	PEI, SGF, PTFE, MoS_2, and graphite	Three-body abrasion	Unfilled PEI exhibited better abrasion resistance
Suresha et al. [20]	PU, glass fiber	Three-body abrasion	Wear volume increased with the addition in SGF content in the PU composite

Ravikumar et al. [21]	PA66/PP blend, SCF, nanoclay	Three-body abrasion	Inclusion of SCF and nanoclay into PA66/PP blend was found to be detrimental to three-body abrasion. However, presence of SCF improved strength and hardness, whereas presence of nanoclay improved hardness alone
Poomalai et al. [22]	PMMA and TPU	Three-body abrasive wear	Unfilled PMMA showed better abrasion wear resistance compared to PMMA/TPU blends. Reduction in tensile strength and modulus was observed with an increase in TPU content
Sole and Ball [23]	PP, $CaCO_3$, $BaSO_4$, and fly ash	Three-body abrasion	Inclusion of fillers to the PP reduces the wear resistance under harsh abrasion situation
Pettarin et al. [24]	HMWHDPE, MoS_2	Dry sliding and three-body abrasion	10 wt.% MoS_2 in HMWHDPE enhances the adhesive and abrasive wear performances
Zhaohong et al. [25]	POM, CF, and nano SiO_2	Dry sliding wear	POM + 3 vol.% nano SiO_2 + CF demonstrated good toughness, modulus with reduced friction coefficient and wear volume
Jian and Tao [26]	PPS, CF, and PA6 as filler	Dry sliding wear	Addition of PA6 increases bending strength and decreases wear loss and friction coefficient of CF/PPS composites

Continued

Table 1 Literature review of selected hybrid PMCs—cont'd

References	Material system	Wear mode	Key findings
Bahadur et al. [27–32]	CuO and CuS to PTFE, CuS, CuF_2, CaO, and PbS to PA11, CuO, CuS, and CuF_2 to PEEK	Dry sliding wear	Wear rate decreased substantially
Bahadur et al. [30, 31]	BaF_2, CaF_2, ZnF_2, SnF_2, ZnS, SnS, ZnO, and SnO	Dry sliding wear	Enhanced the wear rate and contradictory to the findings of [27–32]
Li and Xia [33]	PA6 and CF	Dry sliding wear	20 vol.% CF content exhibited better wear resistance. Increase in applied load resulted in increase in friction coefficient and wear volume
Li [34]	PTFE/PA6 blends	Dry sliding wear	30 vol.% PA6 in PTFE exhibited better impact, friction and wear properties
Wei [35]	PA6, PTFE, and SCF	Dry sliding wear	30 vol.% SCF in PA6/PTFE blend exhibited least wear and friction properties
Chilan et al. [36]	SBR, CF, and CNT	Dry sliding wear	SBR/CF composite with CNT exhibited lower friction coefficient and wear
Hausberger et al. [37]	TPU and graphite	Dry sliding wear	The graphite filled TPU composite showed a significant difference in the performance under vacuum, regarding to the friction and wear behavior

Table 2 Constituents used in preparing the TCE hybrid composites.

Constituent		Particulars with designation	Density ($g\,cm^{-3}$)	Form and size	Supplier
Matrix		Thermoplastic copolyester elastomer (commercially Arnitel EM740)—TCE	1.29	Pellets—$1 \times 4\,mm^2$	Gargi Enterprises, Bengaluru, India
Reinforcement		Short E-glass fiber—SGF	2.50	$12\,\mu m \times 4\,mm$	Fine Organics, Mumbai, India
Particulate fillers	Functional	Short carbon fiber—SCF	1.74	$7 \times 100\,\mu m^2$	
		Silicon carbide—SiC	3.21	5–10 μm	Carborundum Ltd., Chennai, India
		Micro alumina—mAl_2O_3	3.95	5–10 μm	Triveni Chemicals, Gujarat, India
		Nano alumina—nAl_2O_3	3.60	70 nm	
	Lubricating	Polytetrafluoroethylene—PTFE	2.16	Powder—5 μm	Du Pont Co. Ltd.
		Molybdenum di sulfide—MoS_2	4.70	5–10 μm	Omkar chemicals, Pune, India
		Perflouropolyether—PFPE	1.88	200 nm	Nye Lubricants, Fairhaven, USA

Table 3 Details of composition of the composites under present investigation.

Composite designation	Matrix (wt.%)	Fiber (wt.%)	Filler (wt.%)
T_0	TCE (100)	–	–
T_1	TCE (80)	–	PTFE (20)
T_2	TCE (60)	SGF (17.5)	PTFE (10) + SCF (2.5) + SiC (5) + Al_2O_3 (5)
T_3	TCE (60)	SGF (17.5)	PTFE (10) + SCF (2.5) + SiC (4) + Al_2O_3 (4) + MoS_2 (2)
T_4	TCE (60)	SGF (17.5)	PTFE (10) + SCF (2.5) + SiC (8) + nano-Al_2O_3 (1) + PFPE (1)

Fig. 3 displays the line diagram of the low speed reciprocating wear test equipment. The subtleties of the test conditions are displayed in Table 4. The friction and wear test procedure pursued as mentioned about somewhere else [41].

2. Dry sand three-body abrasive wear (3-BAW) was done as per ASTM G-65 [42] utilizing dry sand/rubber wheel abrasion equipment (RWAT) (Make: Magnum Engineers, Bangalore) norms. Fig. 4 introduces the schematic portrayal of the test set up. The subtleties of the test conditions are exhibited in Table 5. The 3-BAW test method pursued as mentioned somewhere else [43].
3. An adhesive wear test was conducted on POD assessment equipment (Make: DUCOM, Bangalore) as per ASTM G99 [38]. The schematic portrayal of the test rig is delineated in Fig. 5. The subtleties of the test situations are exhibited in Table 6. The test procedure pursued as discussed about somewhere else [44].

2.1 Two-body abrasive wear performance of thermoplastic copolyester elastomer composites

The abrasive wear conduct of TCE compounds in multipass (MP) and (SP) single-pass situations has been examined. The K_s as a component of TCE compounds scratched opposite to 165 μm scaled SiC grating paper beneath MP and SP situations is delineated in Fig. 6. It is seen that the K_s of the unfilled TCE is greater than the TCE compounds loaded with 20 wt.% PTFE grout. A decrease in the K_s of the T_1 compound is because of the inclusion of lubricant grout (PTFE), after scratching creates a slender skin that impels a slicker gliding activity of the scratchy grains, thus

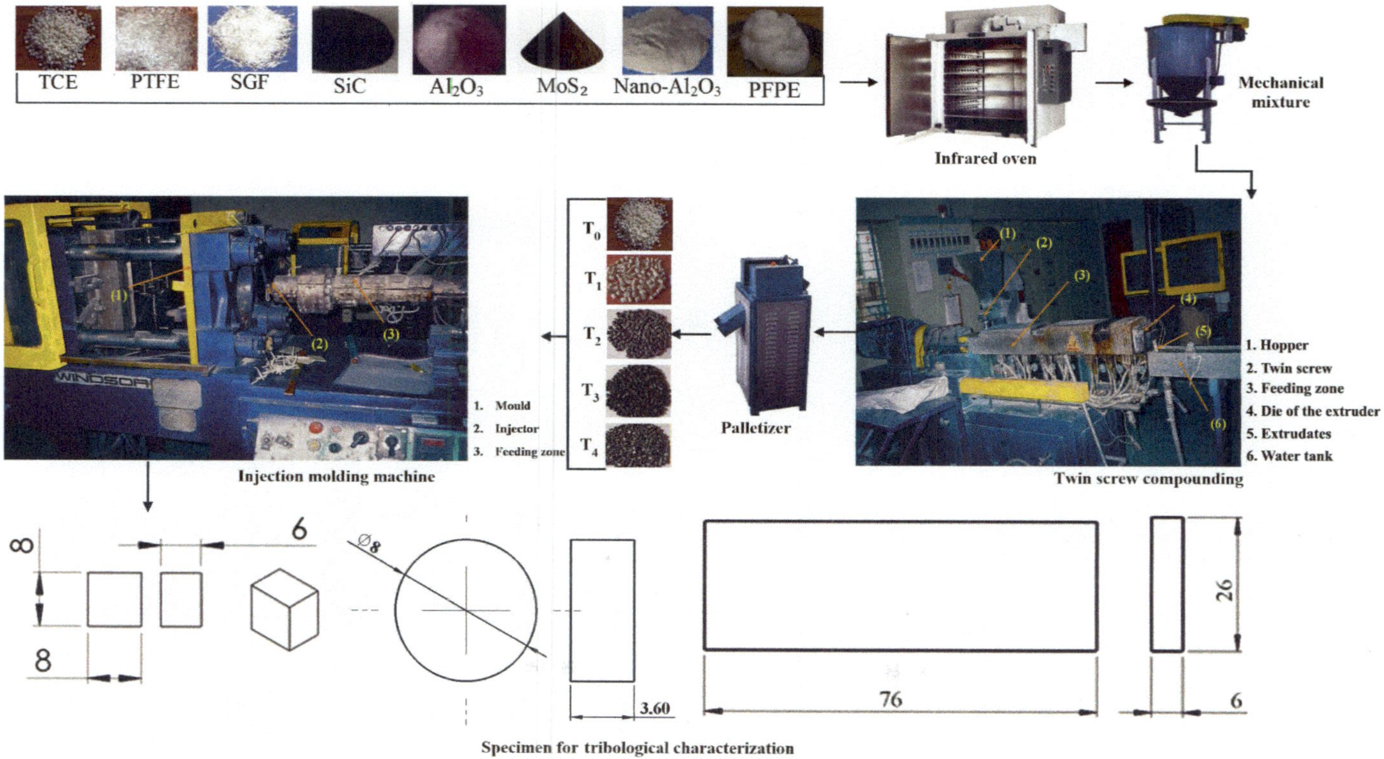

Fig. 1 Stages in composite fabrication.

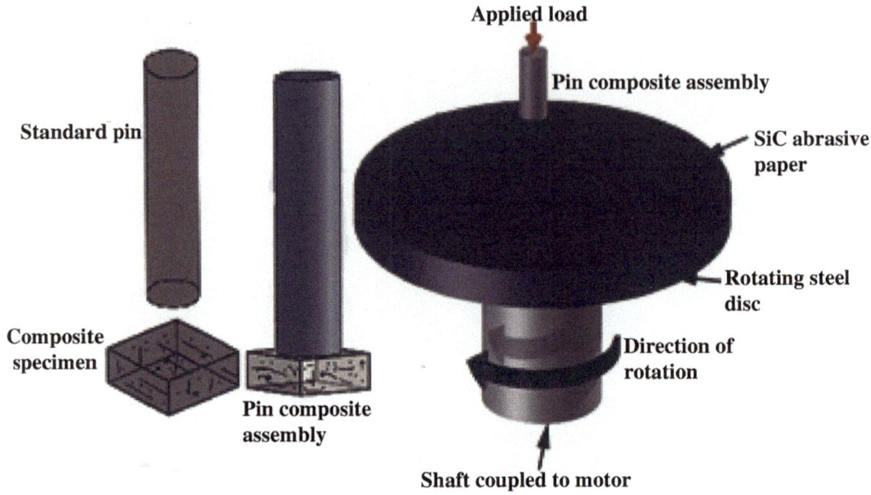

Fig. 2 Schematic of 2-BAW investigation equipment below MP situation.

diminishing the cutting activity confronting at the gradient of the harsh grains. These effects in elevated confrontation to wear of T_1 compound in the investigation group. It is seen that the K_s improved with an increment in the filament/grout intensity in the TCE compounds. Still, the accumulation of SiC, micro-Al_2O_3, MoS_2, nano-Al_2O_3, and PFPE into SGF-fortified and PTFE-filled TCE (T_2, T_3, and T_4) crossbreed compounds

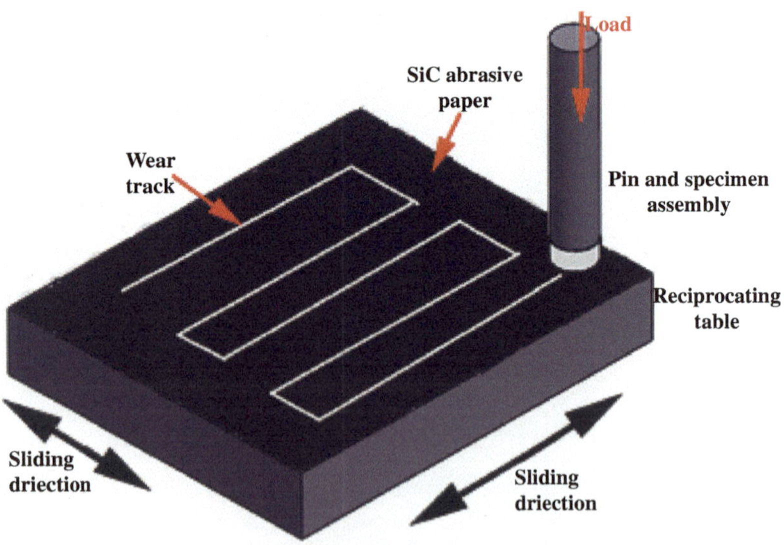

Fig. 3 Graphic of 2-BAW investigation equipment beneath SP situation.

Table 4 Test condition used in 2-BAW under MP and SP conditions.

Parameters	Test conditions (multipass)	Test conditions (single-pass)
Load	10, 15, 20 N	5, 10, 15, 20 N
Speed of disk/table	200 rpm (v=0.31 m s^{-1})	v=0.008 m s^{-1}
Track diameter	30 mm	–
Abrasive particles	Silica sand abrasive paper	Silica sand abrasive paper
Abrasive grain size	165 μm, 70 μm, 31 μm	165 μm, 70 μm, 31 μm
Abrading distance	20 m, 40 m, 60 m	1.5, 2.0, 2.5, 3.0, 3.5, 4.0, 4.5

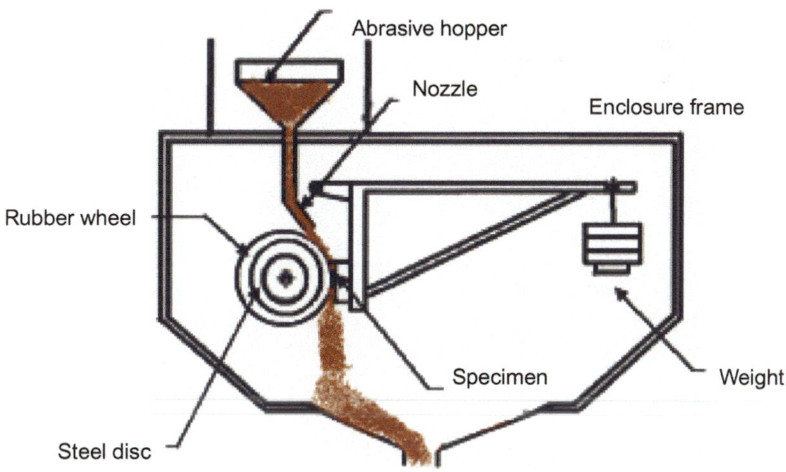

Fig. 4 Schematic of 3-BAW test rig.

Table 5 Test parameters considered for 3-BAW test.

Parameters	Test conditions
Load	40 N
Rotational speed of rubber wheel	200 rpm
Diameter of rubber wheel	228.6 mm
Abrasive particles	Silica sand, angular, 212, 425, 600 μm
Size of the specimen	76 mm × 26 mm × 6 mm
Total test duration	5.25, 10.25, 15.25 min
Abrading distance	752 m, 1468 m, 2185 m
Sand flow rate	255 ± 5 g m^{-1}

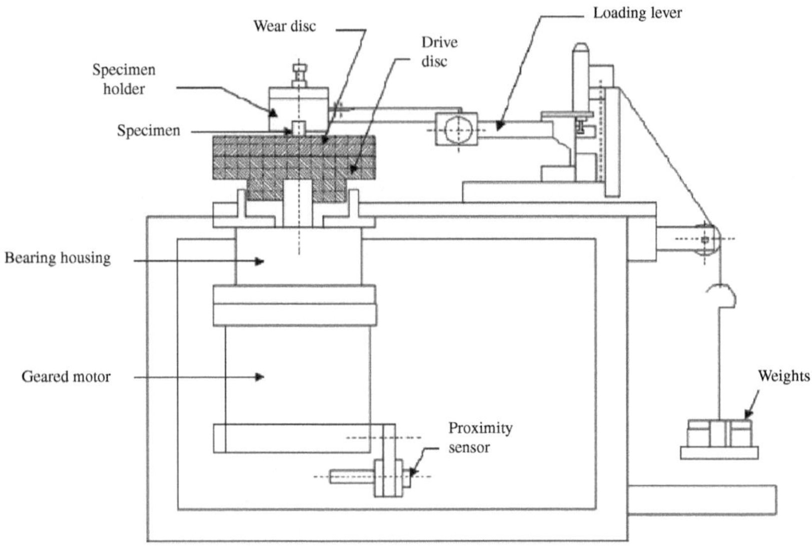

Fig. 5 Graphic of pin-on-disc equipment.

Table 6 Test parameters considered for a dry sliding wear test.

Parameters	Test conditions
Applied normal load (N)	20 and 40
Sliding velocity ($m s^{-1}$)	0.5, 1.0, 1.5, 2.0, 2.5
Sliding distance (m)	2000, 4000, 6000

showed a further increment in K_s by 271.4%, 209.5%, and 159.7% individually, contrasted with the former TCE + PTFE composite in the MP situation. The matrix substance is scratched by the gruelling SiC grains with shrill trims, prompting the development of wear trash. An SGF fortified with matrix substance splits up and aggregates the gap in the middle of the asperities of grating paper in the wear pathway alongside the wreckages. The fragmented filament in the wear pathway sets up a discontinuous interaction with the severities and encounters the applied normal load, prompting its breakup. These broken strands in the middle of the severities at the counter face go about as grainy substances and thereby expand the K_s of the T_2 compound. Likewise, splitting the ceramic grouts from the matrix in the course of the wear procedure and the fragmented strands consume the gap among the severities and turns as grainy specks and enhance the K_s of the TCE hybrid compound than the others in the investigation group. Bijwe et al. [12, 19] revealed that the filament and

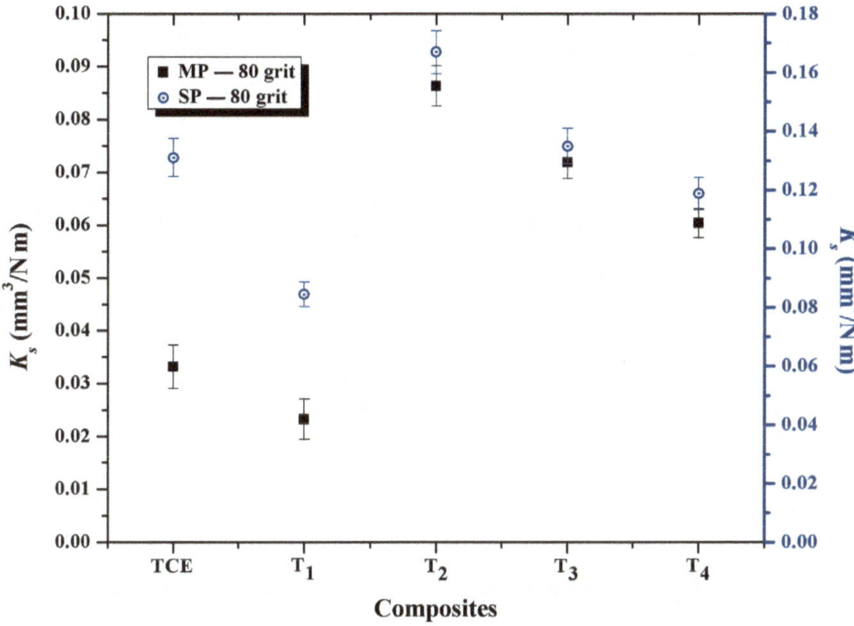

Fig. 6 K_s as a function of TCE composites.

grout content in the PEI is ominous to the scratchy wear conduct of the PMCs. Harsha and Tewari [17] likewise detailed that the incorporation of glass strands into the PSU disintegrated the scratchy wear conduct of the PSU compounds. These are in accordance with the current examination outcomes. In any case, T_3 and T_4 compounds involve a higher wt.% of lubricants (12 and 11 wt.%) contrasted with the T_2 crossbred compound, prompting a decrease of the K_s of the TCE crossbred compounds. A comparative pattern has been seen in the SP situation with an expansion in K_s by around 55% for unfilled TCE contrasted with the TCE+PTFE compound. This is trailed by the T_4 (40.5%), T_3 (59.6%), and T_2 (97.5%) crossbred compounds, correspondingly. A greater decrease in the K_s in the T_4 and T_3 compounds than that of the T_2 compound is because of the existence of salves. Pettarin et al. [24] announced the least-scratchy wear assessment in the HMWHDPE/MoS_2 compound, which is because of the existence of MoS_2, a solid salve. The expanded extent of K_s in the T_2 crossbred compound is because of the way that the increased focus of the strand/grout in the matrix lessens the interfacial holding forte, henceforth prompting strand/grout debonding during the scratchy procedure. Nonetheless, the SP situation witnessed greater K_s than the MP wear situation.

Unfilled TCE showed an expanded K_s in the SP wear situation by around 294.5%, although the T_1 compound displayed expanded K_s by around 264%, trailed by T_4, T_2, and T_3 with relating estimations of 97.1%, 93.4%, and 87.6% correspondingly. This expansion in K_s beneath the SP contrasted with the MP wear procedure can be credited to the consistent commitment of compounds with new SiC grainy paper through the wear procedure. A huge diminishing pattern in the degree of K_s in the TCE + PTFE crossbred compound in the relative examination with different methods of wear is an intriguing component with regard to the examination. The wear procedure in the SP situation is simply by the grainy specks and not with the joined consequence of shattered and detached strands and grouts collected in the gap among the severities. Consequently, the TCE crossbred compounds have shown bounteous confrontation to abrasion than that of the TCE + PTFE compound in the SP situation contrasted with the MP wear procedure.

2.1.1 Relative wear functioning

It is realized that a confrontation to the abrasion of plastics fortified with diminutive strands and particulate grouts will decrease. However, in several actual appliances such as gears, bushes, wheels, etc., strand/filler fortified polymers will be utilized. Consequently, it is necessary to think about the attributes of the material with various sorts and measures of strands and fillers utilized under different test circumstances before it is put to use in a particular application. It was seen that the K_s beneath the SP situation spans 1.25–3.9 times more noteworthy than the MP situation. With this, it is evident that the SP wear process is profoundly successful in segregating the substance contrasted with the MP situation wear procedure. From the above discussion, it tends to be noticed that an expansion in the strand/grout intensity expands the K_s, and henceforth the strand/grout particulates (particularly ceramic grouts) are viewed as detrimental to 2-BAW performance of the TCE compounds under examination. Increments in the scratchy speck dimension expanded the K_s. Likewise, increments in the grazing distance lessened the K_s of the TCE and their compounds. The discoveries of the current examination were in better concurrence with prior explored results revealed by Harsha and Tewari [17, 45, 46]; Bijwe et al. [19]; Friedrich [16]; and Sole and Ball [23]. It is apparent that the T_1 compound had lesser K_s and the T_2 compound displayed elevated K_s in the investigation.

2.1.2 Worn surface morphology

Figs. 7A and 7B show the SEM pictures of the tatty faces of the TCE compounds exposed to 2-BAW under MP circumstances, grazed opposite to 165 μm SiC grainy paper, and a 60 m grazing distance. The dart point determines the grazing direction. The infiltration of SiC hard severities from the grainy paper into the TCE compound surface during the graze would bring about slicing, furrowing, plastic distortion, and face cracks. Fig. 7A delineates the furrows (showed as DG) bringing about the arrangement of lengthy fibrils (demonstrated as F) that are delivered by slicing and furrowing because of gliding activity. Proceeded with graze outcomes in the detachment of fibrils prompting the fragmentation of film (indicated as LF). The matrix substance that is isolated from the test sample in the shape of shards (indicated as WD) can be watched, clinging to the tatty face of the test sample toward the finish of the grazing distance. These effects in the distortion of the fibrils (demonstrated as DF). The lengthy fibrils formed throughout the wear procedure demonstrate the ductility of the matrix substance. Fig. 7B portrays the tatty face morphology of the TCE+PTFE crossbred compound. The glass detachment (demonstrated as SGF and FD) is apparent from the SEM picture. Different highlights incorporate the haul out of the PTFE (shown as P), SiC,

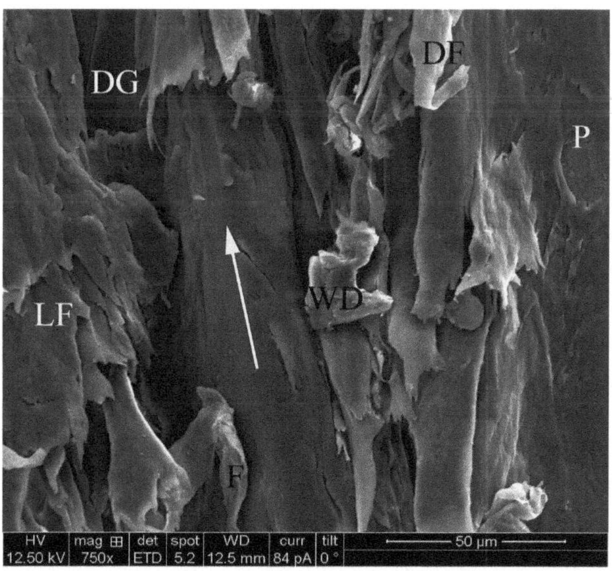

Fig. 7A SEM micrographs of impaired exterior of T_1 compound.

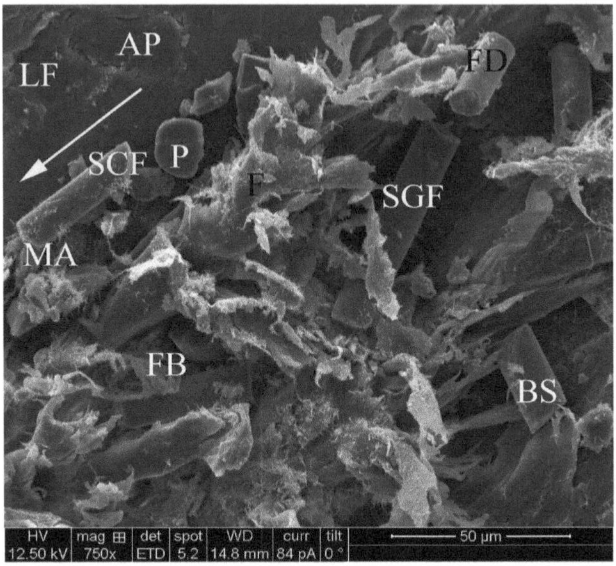

Fig. 7B SEM micrographs of impaired exterior of T_2 crossbred compound.

and Al_2O_3 grains from the matrix substance throughout the wear procedure. The debonded strand was hauled out and uprooted (showed as BS) through the wear procedure. Good bonding among strand and matrix (showed as FB) and furthermore with grout (demonstrated as AP); and diminutive fibrils sticking to the strands (showed as MA), supports for the evidence for effective interfacial holding amid the strand/grout and the matrix material.

The conceivable purpose behind the elevated K_s of the TCE centered crossbred compounds was credited to the existence of larger grainy specks (165 μm) that drastically affected the TCE centered crossbred compounds. That is, the better hardness of the crossbred compound (72.80 and 71.0 Shore D contrasted with 65.0 Shore D for the T_1 compound) appeared to be insufficiently hostile to the impact of hard SiC specks existing on the grainy paper. For example, the substance seemed to be essentially evacuated by the rigid SiC severities in the two-body slicing or furrowing wear mechanism.

Figs. 8A and 8B delineate the wreckage stored on the 165 μm SiC grainy paper grazed with the T_1 and T_2 crossbred compounds correspondingly, grazed to a length of 60 m. The dart assigns the abrading track. The SiC grating (shown as SA) grips the bleeding verge (demonstrated as E).

Figs. 9A and 9B show the SEM pictures of the tatty faces of the TCE compounds exposed to the SP situation, grazed opposed to 165 μm SiC paper to a grazing length of 4.5 m. The dart demonstrates the grazing route.

Fig. 8A SEM picture of T_1 compound wreckage dumped on SiC paper.

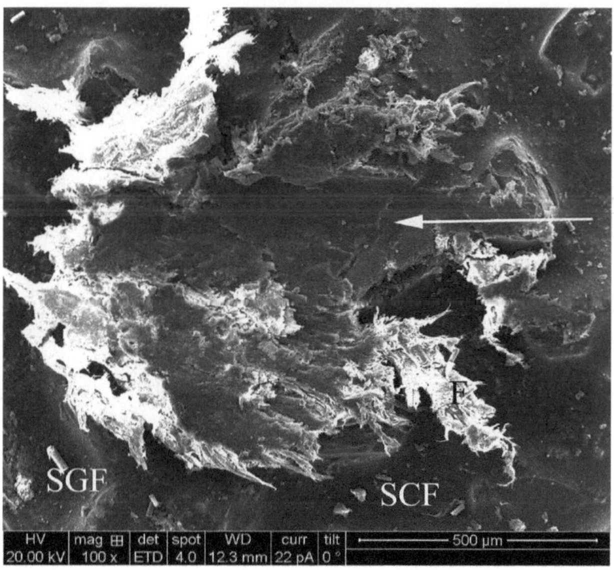

Fig. 8B SEM picture of T_2 compound wreckage dumped on SiC paper.

The fibrils (appeared as F) are shaped because of the furrowing and slicing action of the grating specks. The PTFE (demonstrated as P) grout and filaments (showed as FB and MA) can be observed clinging to the matrix substance in Fig. 9B, affirming the existence of a noble interfacial attachment among the matrix and the grout substance.

Fig. 9A SEM picture of impaired surface of T_1 composite.

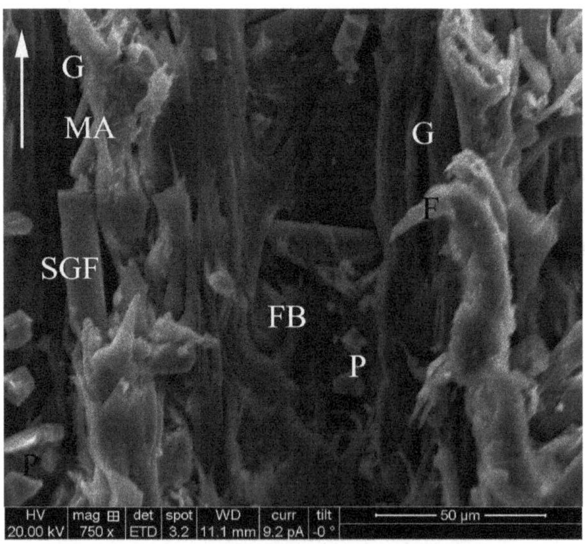

Fig. 9B SEM picture of impaired surface of T_2 crossbred compound.

2.2 Three-body grainy wear performance of thermoplastic copolyester elastomer compounds

The 3-BAW conduct of the TCE compounds has been considered. K_s as a component of the TCE compounds grazed opposed to 212 and 600 μm dimensioned SiC grainy specks, presented to an operational weight of 40 N and abraded to a graze length of 752 m, are shown in Fig. 10. It is seen

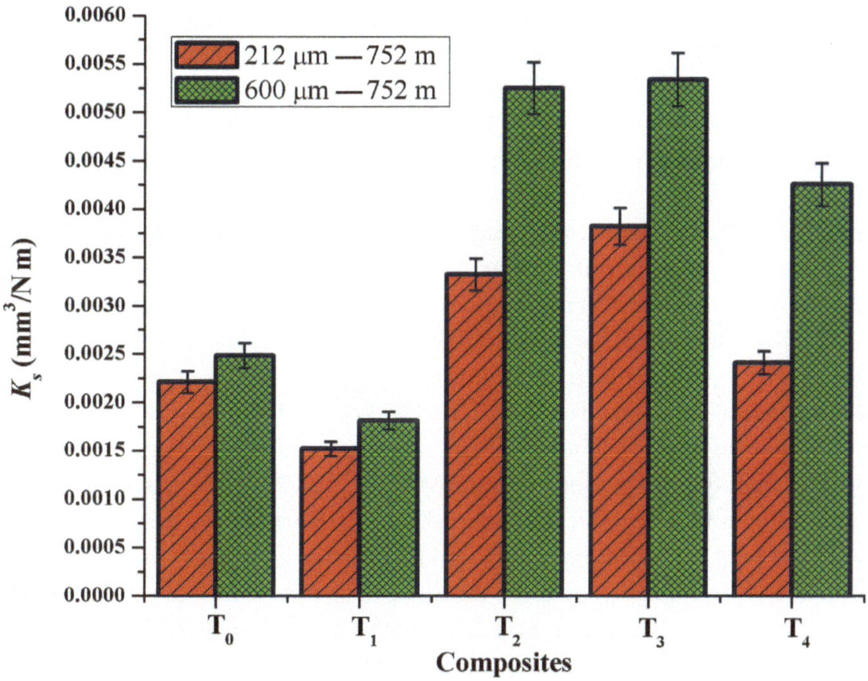

Fig. 10 Effect of strand/filler filling on K_s of TCE and their compounds.

that the K_s of the unfilled TCE loaded with 20 wt.% PTFE filler (T_1) is lower, displaying the most elevated resistance to wear in the investigation. Unfilled TCE (T_0) has demonstrated the K_s in the span of $2.21E^{-03}$ to $2.49E^{-03}\,mm^3\,N^{-1}\,m^{-1}$, which is higher than the T_1 compound. The insertion of PTFE alone upgraded the resistance to wear of the T_0 compound. A decrease in the K_s of the T_1 compound is because of the inclusion of the PTFE particulate filler, a solid lubricant, which prompts the smoother sliding activity of the grating particles. This diminishes the cutting activity experienced at the tip of the abrasives. Another explanation may be because of the better grip among the TCE and PTFE. The inclusion of SGF and hard particles, namely SiC, Al_2O_3, and MoS_2, is less powerful for resistance to abrasion in the T_2 and T_3 compounds.

The insertion of lubricants such as PFPE (T_4) alongside the nano-Al_2O_3 decreased the K_s contrasted with the T_2 and T_3 crossbred compounds. The unfilled TCE and TCE crossbred compounds (T_2, T_3, and T_4, correspondingly) indicated an expanded K_s in the span of 102%–190%, 163%–350%, 160%–350%, and 140%–310%, respectively, than that of the T_1 compound. The insertion of SGF alone in the T_1 compound indicated better resistance to wear contrasted with the other TCE crossbred compounds (T_2, T_3, and

T_4). This is in concurrence with the discoveries of Voss and Friedrich [47]. They explored the gliding and grainy wear conduct of SGF- and SCF-fortified PEEK compounds at the accommodation hotness. They reported that the inclusion of SGF to PEEK demonstrated negligible enhancement in the confrontation to abrasion. It ought to be seen that the expansion in K_s was more when SGFs, SiC, and miniaturized-scale Al_2O_3 were utilized together. Clearly, there is an antagonistic impact of ceramic grouts on the improvement of confrontation to abrasion of the PTFE-packed TCE. This may be because of the grainy conduct of the SGFs and the rigid specks in the delicate fragment, for example, the TCE-filled PTFE compound. Still, the K_s has been somewhat diminished with the insertion of lubricating fillers such as MoS_2 and PFPE. This has been upheld by the discoveries of Pettarin et al. [24]. They detailed the wear conduct of high molecular weight, high density polyethylene (HMWHDPE)/MoS_2 compounds under dry gliding and grainy wear situations. They reported that the insertion of MoS_2 to HMWHDPE expands its gliding and grainy wear conduct with a 10 wt.% MoS_2 substance to get a minimal wear rate. Consequently, it is presumed that the expansion of SGF, SiC, micro/nano-Al_2O_3, MoS_2, and PFPE expanded the focus of strands and filler stuffing in the TCE compounds, which is unfavorable to grainy wear conduct.

2.2.1 Worn surface morphology

By and large, grainy attrition happens by three distinct processes: micro-ploughing, microcutting. and microcracking. as uncovered by Chand and Neogi [48]. Worn surfaces were analyzed to discover the foremost mechanism during the wear procedure. SEM pictures of the impaired surface for the T_1 and T_2 compounds abraded opposed to a shingle speck magnitude of 212 μm to a distance of 2185 m and at an applied load of 40 N are portrayed in Figs. 11 and 12, respectively. The extreme depressions in the direction of abrasion (shown with an *arrow*) were shaped because of the microcutting activity by the shrill grainy specks and viewed on the face. Uprooting of the substance happened by the activity of trundling and gliding of grainy specks, which led to the development of wrinkles (F) and shaping edges (R) in the crosswise route, like a furrowing procedure.

The substance was distorted beneath the weight across the furrowing and slicing with the appliance of weight. The distorted substance (M) can be viewed holding fast to the grazed face (A) on account of the pliable type of the matrix substance. The PTFE specks (T) can be viewed on the face. The worn-out face gives proof of t the rundling activity of the grainy specks

Fig. 11 SEM picture of impaired exterior of T_1 compound.

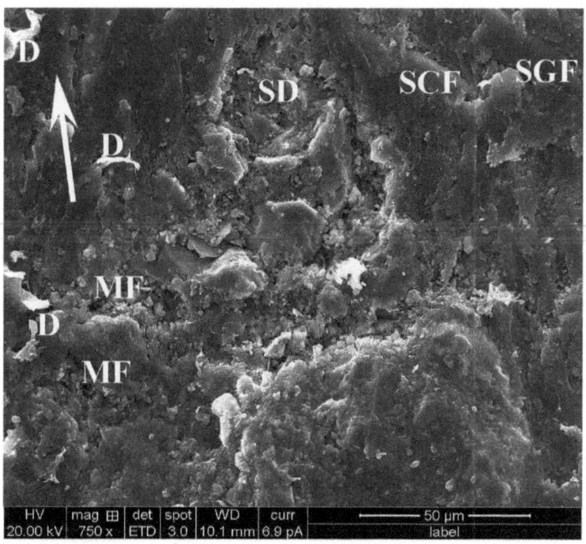

Fig. 12 SEM picture of impaired exterior of T_2 crossbred compound.

being predominant in the grazing procedure contrasted with the gliding activity. The profound wrinkles and wear impairment were because of the rolling activity of the abrasive particles. The development of edges and shallow notches was the least, and this might be because of the decreased sliding activity of abrasive particles.

Fig. 12 shows the impaired surface of the T_2 crossbred compound. Extreme impairment (SD) can be found in the SEM image, which brought about elevated attrition volume impairment. Fiber damage and removal can also be seen which has resulted in expanded K_s of the compound in the investigation. Because of the furrowing and cutting activity of the abrasive particles, breaks, wrinkles, and edges can be seen. Microspecks (MF) are freely detached to the matrix substance, which brought about the creation of the nadirs. The shrill edges of the microspecks ended up being abrasive particles and henceforth augmented the wear volume. The wear flotsam and jetsam as a slim film holding fast to the faces shows the sturdiness of the matrix substance.

2.3 Dry sliding wear performance of thermoplastic copolyester elastomer compounds

2.3.1 Specific wear rate

Fig. 13 portrays the intrigue of specific wear rate (K_s) as a role of strands and grout fortification in TCE at a distance of 6000 m and a speed of 2.5 m/s for a practical weight of 40 N.

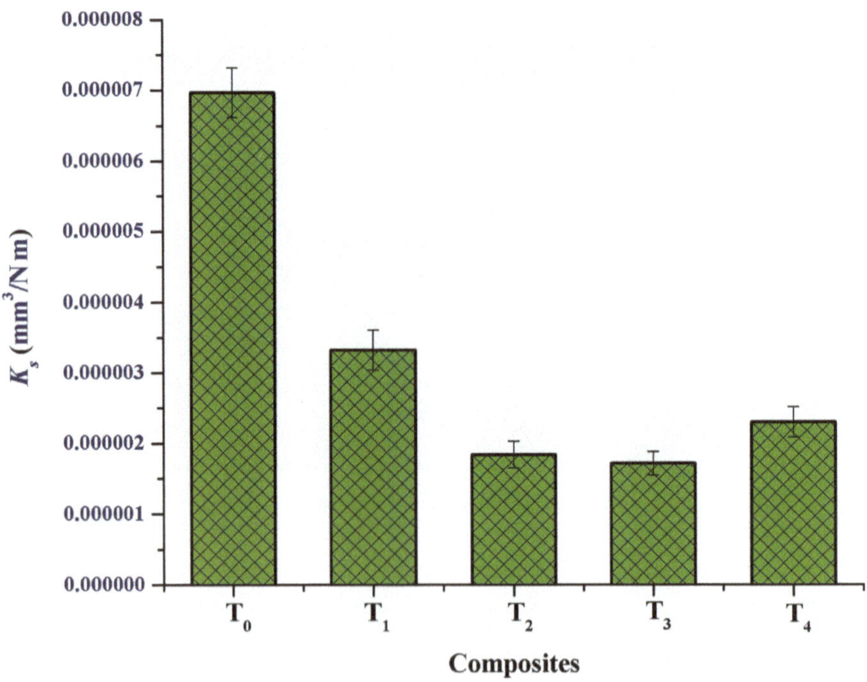

Fig. 13 K_s as a function of strand and filler fortification in TCE.

The strife and attrition conduct of TCEs and their crossbred compounds were estimated as far as K_s. It is evident from the intrigue that the K_s of the unfilled TCE (T_0) was diminished extensively by the addition of grouts and strand fortifications. The unfilled TCE (T_0) showed a K_s of $6.97\times10^{-6}\,mm^3N^{-1}m^{-1}$ with a minimum confrontation to wear. In contrast with the unfilled TCE, the K_s of the T_1 compound was diminished by 108.8%. This is because of the development of transmit layers among the tatty faces of the sample and the steel counter face as noted while assessing. The matrix substance TCE and the micron-scaled PTFE specks plug the lumpiness or severities of the counter face, ensuing in the development of consistent and compacted turnover layers on the counter face. This justifies the lessening in the K_s of the T_1 compound, which is in great concurrence with the discoveries of Lu and Friedrich [49]. They contemplated the strife and attrition qualities of PEEK + PTFE compounds. They prescribed a 10–20 vol.% of PTFE in PEEK as the ideal range, and showed improved strife and attrition attributes over that of the unfilled PEEK. The incorporation of SGF, PTFE, and ceramic grouts in TCE (T_2), a multistaged compound, decreased the K_s by 45% more than that of the T_1 compound. This is attributed to the existence of stiff filaments and ceramic grouts, which are encompassed by the matrix substance and bestow strengthening solidarity to the compound. The micron-scaled SiC and Al_2O_3 specks, being utilized in amalgamation with considerably higher SGF and SCF, were not just furrowed out of the matrix substance by counter face protrusions. Rather, they were consistently freed and expected to begin to proceed as a lurching element among the interfaces, as revealed by Jiang et al. [50]. They reported that the potential collaboration of SCF and TiO_2 may possibly be because of the affirmative lurching activity of the TiO_2 specks, shielding the SCF from being hauled out from the PPS matrix by the counter face protrusion, bringing about upgradation in the resistance to wear of the compounds.

Nevertheless, the multistaged T_3 compound encompassing strands, ceramic grouts, and greasing grout (comprising PTFE) MoS_2 displayed a lesser K_s of $1.72\times10^{-6}\,mm^3N^{-1}m^{-1}$ and it expanded by about 306%, 93%, and 7% for unfilled TCE, T_1, and T_2 compounds, correspondingly. Crystal-like polymers are more attrition-safe than amorphous polymers, as stated by Rabinowicz [51]. The restricted upsurge in crystallinity (35.1%, the most extreme in the investigation) endorsed by MoS_2 in the TCE multistaged compound can prove the augmentation in confrontation to attrition. Further, Bahadur [52] uncovered that PTFE, graphite, and MoS_2 incorporation in compounds augments the development of the transmit layer.

The attrition conduct of the multistage substances is progressively entangled. This is because of the incidence of chemical responses and tangible communications among multilevels during the attrition, which regulates the attrition conduct of PMCs. It has been recommended by Bahadur and Tabor [53] that the uncovered grout ought to chemically respond throughout the attrition procedure with the counter face to produce a better attachment of the transmit layer, prompting diminished attrition. Hu et al. [54] recognized both Mo and S existence on the counter face as MoO_3 and FeS while gliding MoS_2 specks opposed to a steel counter face. These mixes aided the counter attrition activity in the course of gliding. Along these lines, it appears that MoS_2 diminishes K_s by altering the fundamental attrition process of TCE from serious softening attrition to bland adhesive attrition. The existing study results are in accordance with the exploration results of Pettarin et al. [24]. They contemplated the friction and wear performance of MoS_2 specks stuffed high molecular weight high density polyethylene (HMWHDPE) matrix and revealed that the existence of MoS_2 improved the attrition conduct of the compound. This is due to the development of an even and adherent transfer film on the counter face.

The multistaged TCE compounds fortified with filaments and grouts of both micro- and nanoscaled specks (T_4) exhibited an expanded K_s of around 34%, contrasted with that of the T_3 compound. Nevertheless, the K_s of the T_4 compounds was higher by around 25%, contrasted with that of the T_2 compound. This can be ascribed to the incorporation of nanogrouts that have a lofty surface-to-volume fraction and the inclusion of microgrouts jointly, ensuing the feeble joining strength of the matrix substance and thus showing the inferior hardness (Shore D 70) of the compounds. Nonetheless, the K_s of the T_4 compounds is inferior to that of the T_0 and T_1 compounds. This is because of the presence of nanogrouts. The detached nanospecks at the crossing point were allowed to move, proceeding as a metal roller component. This implied that they roll rather than slide between the interfaces, and they would scatter equally above the transmit layer through gliding. This would bring about an additional steady interaction stress among the crossing points, and hence a diminished stress intensity on the distinct filaments. Thus, the thermal letdown of the matrix in the crossing point zone among the strand, grout, and matrix was dodged, and a progressively steady expulsion procedure of the short filaments happened. This affirmed that the K_s of the nanocompound was lower. Sawyer et al. [55] contemplated the strife and attrition execution of a PTFE-occupied nanoalumina compound. They uncovered that the confrontation to wear of these compounds expanded with the expansion of grouts and no ideal grout division was discovered up to 20 wt.%. Chang and Friedrich [56] researched the strife and attrition conduct of SCF-fortified graphite and nano-TiO_2 loaded with different vol.% in PA 66 and epoxy-centered compounds. They stated that the nano-TiO_2 stuffed compounds showed enhanced strife and attrition conduct. These examinations uncovered that crossbred compounds have shown improved strife and attrition conduct. Notwithstanding in the existing examination, TCE compounds involving the blend of micro- and nanoscaled grouts have given synergistic impacts, showing the resistance to attrition in the accompanying edict $T_3 > T_2 > T_4 > T_1 > T_0$ in the investigation.

2.3.2 Strife coefficient

Fig. 14 delineates the graph of strife coefficients (μ) of TCEs and their compounds as opposed to the gliding speed. The assessment was directed to a gliding distance of 6000 m underneath the functional usual weight of 40 N. At a gliding speed of 2.5 m s^{-1}, the unfilled TCE (T_0) displayed elevated μ of 0.664 in the investigation. The addition of 20 wt.% PTFE

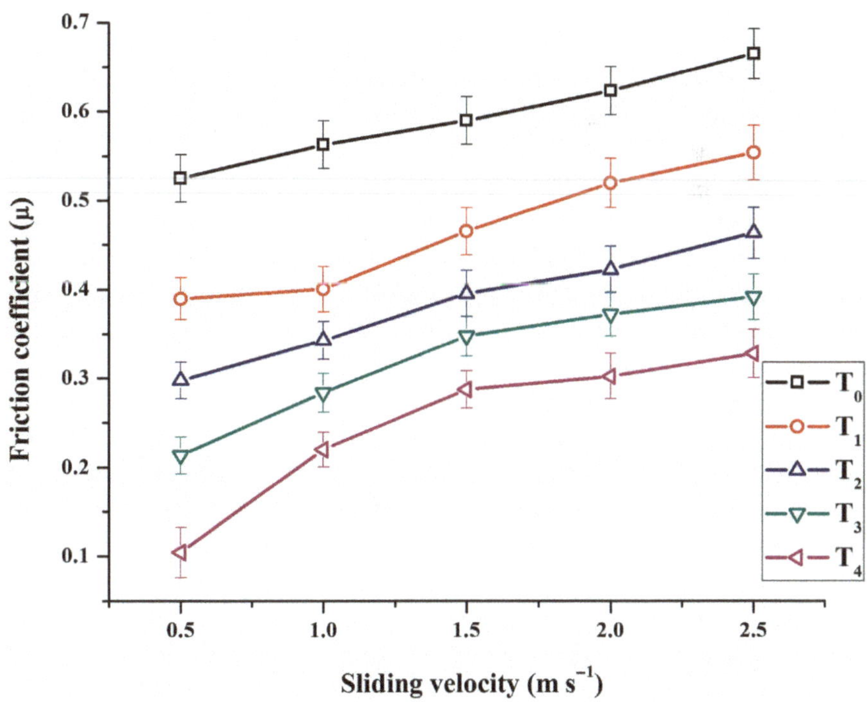

Fig. 14 Friction coefficients of TCE and their composites vs. sliding velocity.

in TCE was helpful in diminishing the μ of the T_1 compound by 16.5%. Further, inclusion of SGF and ceramic fillers in the T_2 composite and solid lubricant such as MoS_2 in the T_3 hybrid composite diminished the μ by 30.1% and 40.8%, correspondingly. The SGFs are surrounded by the matrix substance, which was worn out upon the application of weight opposed to the rigid steel counter face. The damaged matrix substance suits the severities of the counter face surface, shaping a transmit layer all through the running-in phase. This effectually diminishes the immediate interaction of the SGF-fortified compound with the severities of the rigid metallic counter face as stated by Bahadur [52], Myshkin et al. [57], and Chang et al. [58], in this way diminishing the μ. Numerous examinations by Reinicke et al. [59], Palabiyik and Bahadur [60], Voss and Friedrich [61], and Bahadur et al. [62] uncovered that the incorporation of strands as fortification in the compounds showed enhanced confrontation to attrition and diminished the strife constant. The incorporation of PFPE in T_4 nano-crossbred compounds has radically diminished the μ by a measure of 50%, contrasted with that of the unfilled TCE. This is attributed to the presence of nanoparticles at the interface may wrap the micron sized particles and the transfer film, and there by act as ball-bearing component. These specks roll among the two pair-off faces, bringing about decreased cut-off pressure, strife constant, and interaction heat. Wang et al. [63] explored the dry gliding attrition execution of Nylon 1010 loaded with MoS_2 and fortified with SCF compounds. They led the tribo-chemical investigation by X-ray photoelectron spectroscopy, and uncovered that MoS_2 particles disintegrated into MoO_3, FeS, $FeSO_4$, and $Fe_2\ (SO_4)_3$ throughout gliding, which expanded the sticking togetherness among the transmit layer and the counter face surface. The development of a slender, even, and persistent transmit layer would add to the reduction in μ of the Nylon 1010 compounds.

Henceforth, the present examination is in-accordance with the past exploration. Additionally, the prior explorations of Pettarin et al. [24] and Wang et al. [63] have uncovered that the incorporation of frozen lard grouts diminished the strife constant. As a result of this, the frozen lard grout stacking in wt.% was chosen as one of the noteworthy aspects affecting the adhesive attrition conduct of the compounds. Nevertheless, in the existing examination, TCE compounds involving the combination of micro- and nanoscaled grouts have given synergistic impacts, showing the strife constant in the accompanying edict $T_4 < T_3 < T_2 < T_1 < T_0$, respectively, in the investigation.

2.3.3 Worn surface morphology

The SEM micrographs appear in Figs. 15A and 15B, 16A and 16B, 17A and 17B correspondingly glided to a distance of 6000 m, with a speed of 2.5 $m s^{-1}$ and at a normal applied weight of 40 N. The darts indicate the route of the

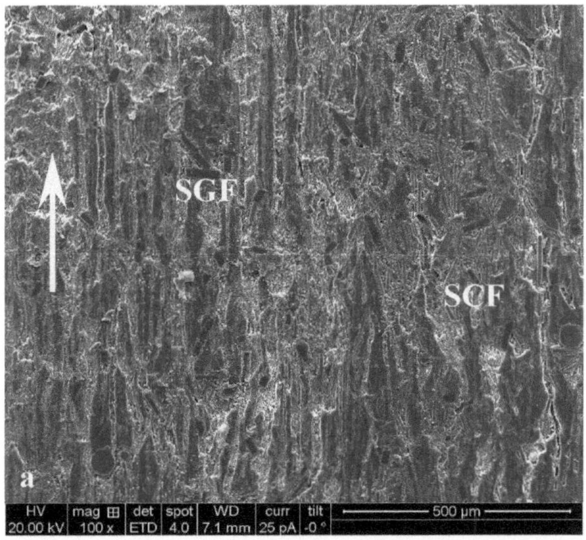

Fig. 15A SEM picture of impaired surface of T_2 crossbred PMC (magnification: 100×).

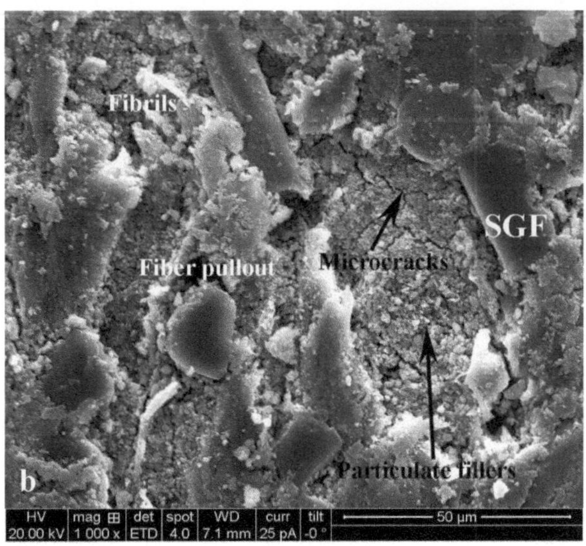

Fig. 15B SEM picture of impaired surface of T_2 crossbred PMC (magnification: 1000×).

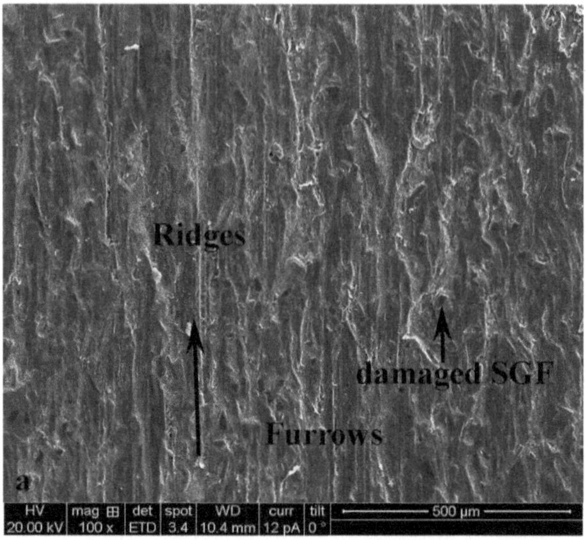

Fig. 16A SEM picture of impaired surface of T_3 crossbred PMC (magnification: 100×).

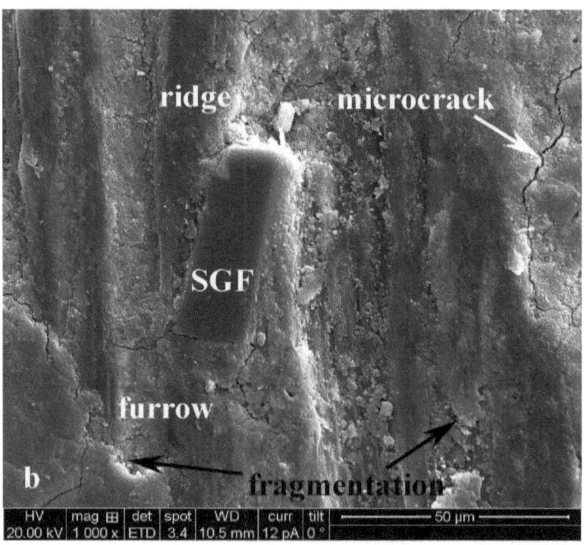

Fig. 16B SEM picture of impaired surface of T_3 crossbred PMC (magnification: 1000×).

adhesive attrition. Figs. 15A and 15B delineate the tatty face micrograph of the T_2 crossbred compound; Figs. 16A and 16B exhibit the tatty face micrograph of the T_3 crossbred compound; and Figs. 17A and 17B portray the worn surface micrograph of the T_4 crossbred compounds, correspondingly with various amplifications. The SGF represents the short glass fiber.

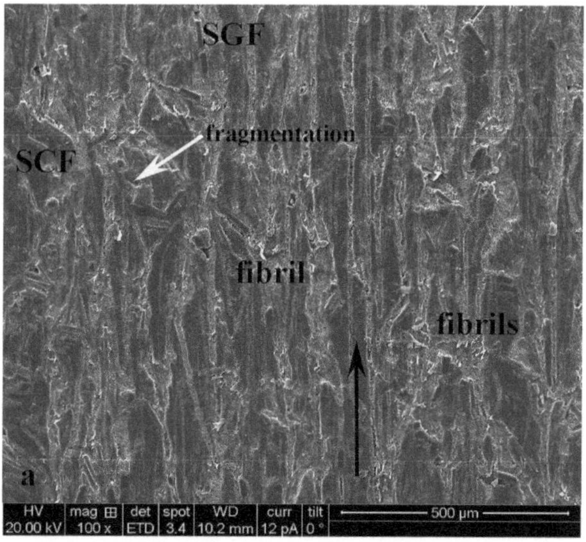

Fig. 17A SEM picture of impaired surface of T_4 crossbred PMC (magnification: 100×).

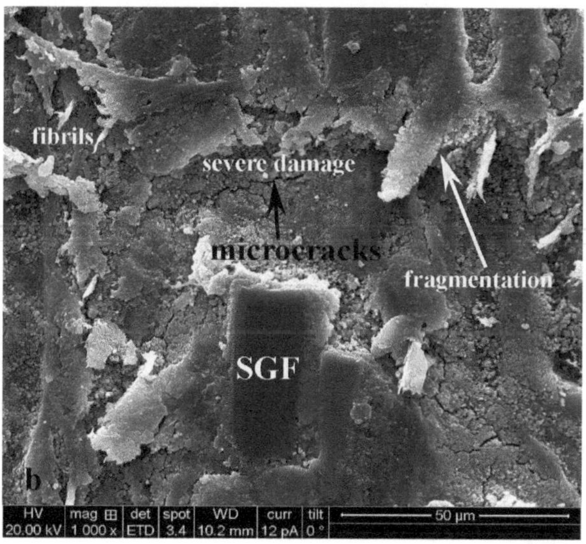

Fig. 17B SEM picture of impaired surface of T_4 crossbred PMC (magnification: 1000×).

From the SEM images, it is derived that the wrinkles and crests are made on the tatty faces of the samples under examination, and these are framed because of the existence of the severities on the counter face above which the test sample is glided. Fig. 15A obviously shows that the filaments are haphazardly arranged in the TCE grid. The perception of the tatty face at greater amplification (Fig. 15B) uncovers the detachment of the filaments and speck grouts (10 wt.%) alongside the grid discontinuity, prompting microscale splits and tiny fibrils clinging to the face. This demonstrates the flexible conduct of the TCE substance.

The existence of 10 wt.% ceramic/speck grouts has upsurged the hardness of the T_2 compound. Still, this has diminished the clinging potency among the matrix and filaments/grouts in the compound (Fig. 15B), resulting in transitional wear resistance in the examination. The existence of the inorganic filaments in the T_3 compound is apparent from Fig. 16A. The filaments limited by the matrix and existence of ceramic grouts (8 wt.%) nearly slighter than the T_2 and T_4 compounds means the presence of noble clinging potency, which is apparent from Fig. 16B. After gliding, the filaments that are covered by the matrix substance interact with the steel disc. Filaments experiences cracking leading to fracture into smaller fragments. This procedure expends more energy with the dissipation of heat, resulting in the development of microcrevices. This upsurges the attrition confrontation of the T_3 crossbred compound, steering to the diminished attrition of the substance. Fibrils and fragmentation of the matrix material with severe damage of the T_4 crossbred compound can be experiential in Figs. 17A and 17B. The existence of microceramic grout (8 wt.%) and nanoceramic grout (1 wt.%) fills a bigger volume, ensuing the deprived attachment strength of the compound substance. As discussed, with the advancement in gliding of the compound sample, the minute broken strands and detached ceramic grouts turn out to be a third party among the polymer crossbred compound and the counter face of the steel disc. The microscopic broken strands and detached ceramic grouts produce harsh impairment of the polymer crossbred compound (T_4) in the research group. From this, it is apparent that the attrition process comprised with TCE crossbred compounds is governed by the detachment of particulate grouts and the detaching and pullout of filaments from the matrix.

3 Conclusions

- The 2-BAW (beneath single-pass and multipass situation) conduct of TCE compounds was studied and compared. The K_s under SP condition ranges from 1.3 to 4 times greater than the MP condition. From this, it is clear that the SP wear mode is highly effective in detaching the material compared to the MP condition wear mode. A further increase in abrading distance resulted in the reduction in K_s; however, an increase in the abrasive grain size resulted in an increase in K_s. Wear mechanisms such as penetration, ploughing, and cutting were identified using SEM images of worn surfaces of TCE composites.

- The 3-BAW conduct facts (attrition volume) of TCEs and their compounds were discussed. The consequences of abrasive grain size and abrading distance were studied and the trend in terms of K_s was found to be similar with that of 2-BAW. The wear mechanisms such as ploughing and cutting were recognized and conferred using SEM pictures of worn surfaces of TCE compounds.
- The strife and attrition conduct (K_s) of TCE crossbred compounds was reported. TCE compounds encompassing the blend of micro- and nanosized fillers have offered synergistic impact, demonstrating the resistance to wear in the subsequent edict $T_3>T_2>T_4>T_1>T_0$ and the friction coefficients in the subsequent edict $T_4<T_3<T_2<T_1<T_0$ correspondingly in the study group. The wear mechanisms such as debonding of fillers and debonding and withdrawal of filaments from the matrix are recognized and conversed utilizing SEM pictures of worn surfaces of TCE compounds.
- The particulate reinforcements in TCE were found to be detrimental for abrasive wear; however, the same was found to be beneficial in dry sliding wear behavior.

References

[1] B.D. Agarwal, L.J. Broutman, Analysis and Performance of Fiber Composites, second ed., John Wiley & Sons, Inc., New Jersey, 1990, pp. 2–16.

[2] B. Jiang, C. Liu, C. Zhang, B. Wang, Z. Wang, The effect of non-symmetric distribution of fiber orientation and aspect ratio on elastic properties of composites, Compos. Part B 38 (2007) 24–34.

[3] J.A. Brydson, Plastics Materials, seventh ed., Elsevier, Oxford, UK, 2011, pp. 737–739.

[4] S. Amin, M. Amin, Thermoplastic elastomeric (TPE) materials and their use in outdoor electrical insulation, Adv. Mater. Sci. 29 (2011) 15–30.

[5] T.A. Stölarski, Tribology in Machine Design, Heiman Newnes, United Kingdom, 1990.

[6] K.G. Budinski, Surface Engineering for Wear Resistance, Prentice-Hall, New Jersey, 1998.

[7] K.R. Päivi, Correlation of Material Characteristics and Wear of Powder Metallurgical Metal Matrix Composites (Doctoral Thesis), Materials and Earth Sciences, Helsinki University of Technology, Laboratory of Materials Science, Espoo, 2006.

[8] T.P. Sathishkumar, J. Naveen, S. Satheeshkumar, Hybrid fiber reinforced polymer composites—a review, J. Reinf. Plast. Compos. 33 (5) (2014) 454–471.

[9] N. Kistaiah, C. Udaya Kiran, G. Ramachandra Reddy, M. Sreenivasa Rao, Mechanical characterization of hybrid composites: a review, J. Reinf. Plast. Compos. 33 (14) (2014) 1364–1372.

[10] J. Kuljanin, M. Vuckovic, M.I. Comor, N. Bibic, V. Djokovic, J.M. Nedeljkovic, Influence of CdS-filler on the thermal properties of polystyrene, Eur. Polym. J. 38 (8) (2002) 1659–1662.

[11] B. Weidenfeller, M. Hofer, F. Schilling, Thermal conductivity, thermal diffusivity and specific heat of particle filled polypropylene, Compos. Part A: Appl. Sci. Manuf. 35 (4) (2004) 423–429.

[12] J. Bijwe, J. Indumathi, Influence of fibers and solid lubricants on low amplitude oscillating wear of polyetherimide composites, Wear 257 (2004) 562–572.

[13] R. Hemanth, B. Suresha, M. Sekar, Physico-mechanical behaviour of thermoplastic co-polyester elastomer/polytetrafluroethylene composite with short fibers and micro-fillers, J. Compos. Mater. 49 (18) (2015) 2217–2229.

[14] H.H. Parikh, P.P. Gohil, Tribology of fiber reinforced polymer matrix composites—a review, J. Reinf. Plast. Compos. 34 (16) (2015) 1340–1346.

[15] A.D.A. Lucas, D.J. Ambrosio, H. Otaguro, L.C. Costa, J.A.M. Agnelli, Abrasive wear of HDPE/UHMWPE blends, Wear 270 (2011) 576–583.

[16] K. Friedrich, On the wear of reinforced thermoplastics by different abrasive papers, Wear 103 (1985) 333–344.

[17] A.P. Harsha, U.S. Tewari, Abrasive wear of glass fiber reinforced polysulfone composites, Indian J. Eng. Mater. Sci. 09 (2002) 203–208.

[18] B.N. Ravikumar, M. Venkataramareddy, B. Suresha, Two-body abrasive wear behaviour of nano-clay filled LDPE/EVA composites, J. Reinf. Plast. Compos. 28 (24) (2009) 3001–3007.

[19] J. Bijwe, J. Indumathi, J.J. Rajesh, M. Fahim, Friction and wear behavior of polyetherimide composites in various wear modes, Wear 249 (8) (2001) 715–726.

[20] B. Suresha, G. Chandramohan, N.D. Jawali, Siddaramaiah, Effect of short glass fiber content on three-body abrasive wear behaviour of polyurethane composites, J. Compos. Mater. 41 (22) (2007) 2701–2713.

[21] B.N. Ravikumar, B. Suresha, M. Venkataramareddy, Effect of particulate fillers on mechanical and abrasive wear behaviour of polyamide 66/polypropylene nanocomposites, Mater. Des. 30 (9) (2009) 3852–3858.

[22] Poomali, Siddaramaiah, B. Suresha, J.-H. Lee, Mechanical and three-body abrasive wear behaviour of PMMA/TPU blends, Mater. Sci. Eng. A 492 (2008) 486–490.

[23] B.M. Sole, A. Ball, On the abrasive wear behaviour of mineral filled polypropylene, Tribol. Int. 29 (6) (1996) 457–465.

[24] V. Pettarin, M.J. Churruca, D. Felhos, K.K. Joseph, M.F. Patricia, Changes in tribological performance of high molecular weight high density polyethylene induced by the addition of molybdenum disulphide particles, Wear 269 (1–2) (2010) 31–45.

[25] X. Zhaohong, L. Zhenhua, L. Jian, F.F. Yuan, The effect of CF and nano-SiO_2 modification on the flexural and tribological properties of POM composites, J. Thermoplast. Compos. Mater. 27 (3) (2014) 287–296.

[26] L. Jian, S. Tao, The mechanical and tribological properties of CF/PPS composite filled with PA6, J. Thermoplast. Compos. Mater. 27 (5) (2014) 594–602.

[27] S. Bahadur, D. Gong, The role of copper compounds as fillers in the transfer and wear behavior of polyetheretherketone, Wear 154 (1) (1992) 151–165.

[28] S. Bahadur, G. Deli, The transfer and wear of nylon and CuS-nylon composites: filler proportion and counterface characteristics, Wear 162–164 (1993) 397–406.

[29] S. Bahadur, D. Tabor, The wear of filled polytetrafluoroethylene, Wear 98 (1984) 1–13.

[30] S. Bahadur, A. Kapoor, The effect of ZnF_2, ZnS and PbS fillers on the tribological behavior of nylon 11, Wear 155 (1) (1992) 49–61.

[31] S. Bahadur, D. Gong, J.W. Anderegg, The role of copper compounds as fillers in transfer film formation and wear of nylon, Wear 154 (2) (1992) 207–223.

[32] S. Bahadur, D. Gong, J. Anderegg, Investigation of the influence of CaS, CaO and CaF_2 fillers on the transfer and wear of nylon by microscopy and XPS analysis, Wear 197 (1996) 271–279.

[33] J. Li, Y.C. Xia, The friction and wear properties of thermoplastic PA6 composites filled with carbon fiber, J. Thermoplast. Compos. Mater. 23 (2010) 337–349.

[34] J. Li, The impact and tribological properties of PTFE composites filled with PA6, J. Thermoplast. Compos. Mater. 23 (2010) 807–816.
[35] R. Wei, Tribological properties of short carbon/PA/PTFE hybrid composites under dry sliding conditions, J. Thermoplast. Compos. Mater. 27 (4) (2014) 518–527.
[36] C. Chilan, W. Wang, J. Zhang, The effect of surface functionality of carbon nanotubes on mechanical and tribological behaviours of carbon fibre-filled styrene-butadiene rubber composite, J. Thermoplast. Compos. Mater. 29 (09) (2016) 1167–1175.
[37] A. Hausberger, Z. Major, G. Theiler, T. Gradt, Observation of the adhesive and deformation contribution to the friction and wear behaviour of thermoplastic polyurethanes, Wear (2018), https://doi.org/10.1016/j.wear.2018.07.006.
[38] ASTM G99-05, Standard Test Method for Wear Testing with a Pin-on-Disk Apparatus, 2002.
[39] R. Hemanth, B. Suresha, S.-H. Yang, M. Sekar, Abrasive wear behaviour of thermoplastic copolyester elastomer composite: a statistical approach, Int. J. Precis. Eng. Manuf. 17 (06) (2016) 755–763.
[40] ASTM G133, Standard Test Method for Wear Testing with a Reciprocating Apparatus, 2002.
[41] R. Hemanth, M. Sekar, B. Suresha, Abrasive wear behaviour of thermoplastic copolyester elastomer composites, Emerg. Mater. Res. (2015), https://doi.org/10.1680/emr.15.00008.
[42] ASTM G65-04, Standard Test Method for Measuring Abrasion Using the Dry Sand/Rubber Wheel Apparatus, 2010.
[43] R. Hemanth, M. Sekar, P.K. Pallathadka, B. Suresha, Dynamic mechanical analysis and three-body abrasive wear behaviour of thermoplastic copolyester elastomer composites, Adv. Tribol. (2014), https://doi.org/10.1155/2014/210187.
[44] R. Hemanth, P.S. Thomas, B. Suresha, M. Sekar, Effect of short glass fiber and fillers on dry sliding wear behaviour of thermoplastic copolyester elastomer composites, Int. J. Mech. Eng. Technol. 05 (09) (2014) 71–80.
[45] A.P. Harsha, U.S. Tewari, Tribo performance of polyaryletherketone composites, Polym. Test. 21 (2002) 697–709.
[46] A.P. Harsha, U.S. Tewari, Two-body and three-body abrasive wear behaviour of polyaryletherketone composites, Polym. Test. 22 (2003) 403–418.
[47] H. Voss, K. Friedrich, On the wear behaviour of short-fibre reinforced PEEK composites, Wear 116 (1) (1987) 1–18.
[48] N. Chand, S. Neogi, Mechanism of material removal during three-body abrasion of FRP composite, Tribol. Lett. 04 (1998) 81–85.
[49] Z.P. Lu, K. Friedrich, On sliding friction and wear of PEEK and its composites, Wear 181–183 (1995) 624–631.
[50] Z. Jiang, L.A. Gyurova, A.K. Schlarb, K. Friedrich, Z. Zhang, Study on friction and wear behaviour of polyphenylene sulphide composites reinforced by short carbon fibers and sub-micro TiO_2 particles, Compos. Sci. Technol. 68 (2008) 734–742.
[51] E. Rabinowicz, Friction and Wear of Materials, John Wiley & Sons, New York, 1995.
[52] S. Bahadur, The development of transfer layers and their role in polymer tribology, Wear 245 (2000) 92–99.
[53] S. Bahadur, D. Tabor, Role of fillers in the friction and wear behaviour of high density polyethylene, in: Polymer Wear and Its Control, American Chemical Society Publications, 1985, pp. 253–268.
[54] X.G. Hu, S.L. Hu, Y.S. Zhao, Synthesis of nanometric molybdenum disulphide particles and evaluation of friction and wear properties, Lubr. Sci. 17 (2005) 295–308.
[55] W.G. Sawyer, K.D. Freudenberg, P. Bhimaraj, L.S. Schadler, A study on the friction and wear behaviour of PTFE filled with alumina nanoparticles, Wear 254 (2003) 573–580.

[56] L. Chang, K. Friedrich, Enhancement effect of nanoparticles on the sliding wear of short fiber-reinforced polymer composites: a critical discussion of wear mechanisms, Tribol. Int. 43 (2010) 2355–2364.
[57] N.K. Myshkin, M.I. Petrokovets, A.V. Kovalev, Tribology of polymers: adhesion, friction, wear, and mass-transfer, Tribol. Int. 38 (2005) 910–921.
[58] L. Chang, Z. Zhang, L. Ye, K. Friedrich, Tribological properties of epoxy nanocomposites III. Characteristics of transfer films, Wear 262 (2007) 699–706.
[59] R. Reinicke, F. Haupert, K. Friedrich, On the tribological behavior of selected, injection moulded thermoplastic composites, Composites 29A (1998) 763–771.
[60] M. Palabiyik, S. Bahadur, Tribological studies of polyamide 6 and high-density polyethylene blends filled with PTFE and copper oxide and reinforced with short glass fibers, Wear 253 (2002) 369–376.
[61] H. Voss, K. Friedrich, Wear performance of a bulk liquid crystal polymer and its short-fibre composites, Tribol. Int. 19 (3) (1986) 145–156.
[62] S. Bahadur, Q. Fu, D. Dong, The effect of reinforcement and the synergism between CuS and carbon fiber on the wear of nylon, Wear 178 (1994) 123–130.
[63] J. Wang, M. Gua, B. Songhao, S. Gec, Investigation of the Influence of MoS_2 filler on the tribological properties of carbon fiber reinforced nylon 1010 composites, Wear 255 (2003) 774–779.

CHAPTER 17

Tribological applications of polymer composites

T.G. Yashas Gowda[a], M.R. Sanjay[a], Jyotishkumar Parameswaranpillai[a], Suchart Siengchin[a], and Klaus Friedrich[b]
[a]Department of Mechanical and Process Engineering, The Sirindhorn International Thai-German Graduate School of Engineering (TGGS), King Mongkut's University of Technology North Bangkok, Bangkok, Thailand
[b]Institute for Composite Materials (IVW GmbH), Technical University of Kaiserslautern, Kaiserslautern, Germany

1 Introduction

Tribology is a Greek word, with "tribo" meaning "rubbing" and "logy" meaning "knowledge." Therefore, tribology is the study of the friction mechanism between two solid surfaces that have relative motion between them. The fundamental study of tribology deals with the nature of contact surfaces, the type of contact materials, the wear and frictional properties, and lubrication. Tribological studies are significant in terms of economy, reliability, maintenance, and wear behavior in different applications such as aircraft, automotive, and other industries. The understanding of the basic principles of the tribological behavior of a material is very complex. It requires fundamental knowledge in various subjects such as chemistry, solid mechanics, thermodynamics, machine design, lubrication, etc. The research in tribology is mainly focused on minimizing the loss of materials (wear) occurring during the frictional contact between the two rubbing surfaces. This becomes crucial to machinery that involves sliding and rolling contacts such as clutches, brakes, wheels of trains, automobiles, IC engines, gears, bearings, cams, and aircraft engines [1–3].

The development of new materials for various industries has become important for technological advancements. Polymers have good elasticity, low friction, and wear resistance, making them suitable for tribological applications. Polymers are generally reinforced with synthetic or natural fibers that enhance the strength of the composites, making them compatible for different applications where they can replace traditional metal materials. Other advantages of polymer composites are low cost, light weight, high strength, corrosion resistance, and easy manufacturing [4–6].

Tribology of Polymer Composites
https://doi.org/10.1016/B978-0-12-819767-7.00017-7

Besides, the properties can be tailor-made for different applications such as sliding elements, bearing, and other components that require higher or lower wear and friction. However, the wear mechanisms of polymer composites are very complex, and it is quite difficult to predict the wear rates of the composites due to different variables. Therefore, researchers are studying various factors involved in the tribology of polymer composites.

2 Polymer composites for tribological applications

Thermosetting polymers such as phenol-formaldehyde, epoxy resin, polyester, polyamide, and polyurethane and thermoplastics such as polypropylene (PP), high-density polyethylene (HDPE), polyether-ether ketone (PEEK), etc., are generally used as matrices for composites as they transfer stress to the reinforcements [7–8]. Thermoplastics can be used as the matrix for easy manufacturing of different components that can operate under heavy loads, high speed, high temperature, and extreme conditions.

The advantages of the polymers such as low density, better wear resistance, ease of manufacture, and self-lubrication encourage design engineers to use them as an important material for components that involve friction such as gears, piston parts, brakes, etc.. Among the various polymers, PEEK is widely used for tribological applications as it exhibits excellent friction and wear characteristics [9–10]. Chen et al. [11] studied the wear behavior of PEEK reinforced with carbon fiber composites under seawater lubrication. The authors reported that the incorporation of the CF fibers resulted in an improvement in wear properties under seawater lubrication. The composites can be used for many marine applications where tribology is involved and they can replace existing metallic materials and have better wear and corrosion properties.

Friction and bearing parts are the main tribological components. The friction parts used in industries are elevator brakes, wind turbine brakes, winch brakes, etc., and they are also used in various components in agriculture equipment as well as the automotive, marine, and aerospace industries. Polymer components are extensively used for making roller bearings due to their low density, better wear properties, lesser noise, and excellent corrosion resistance, which leads to better service [12–15]. PTFE-based composite cages and their functions in ball bearings are shown in Fig. 1.

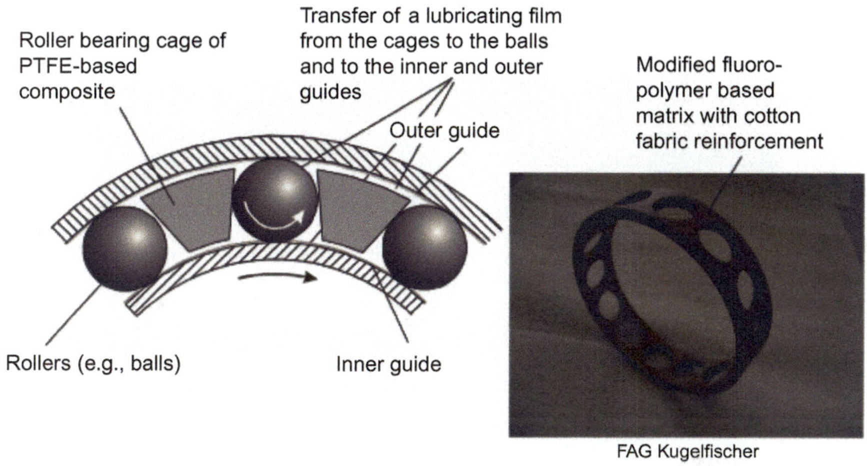

Fig. 1 PTFE-based composite cages and their functions in ball bearings [14].

Polymer gears are nowadays used in many industrial applications such as paper mills, copiers, and printers. The use of polymer gears is limited to some extent due to their low properties compared to metallic gears. Polymer gears are usually manufactured by injection molding, using materials such as polyoxymethylene (POM), ABS, HDPE, etc. The torque and rotational speed have significant effects on the tribological performance. To improve polymer gears, they are often reinforced with glass and carbon fibers, resulting in higher strength and impact resistance, that is, improved performance and durability [16]. Fig. 2 shows the polymer gears made of different polymers.

Fig. 2 Gears made of different polymers [16]. *(Reused with permission from Elsevier, License number 4743671127293.)*

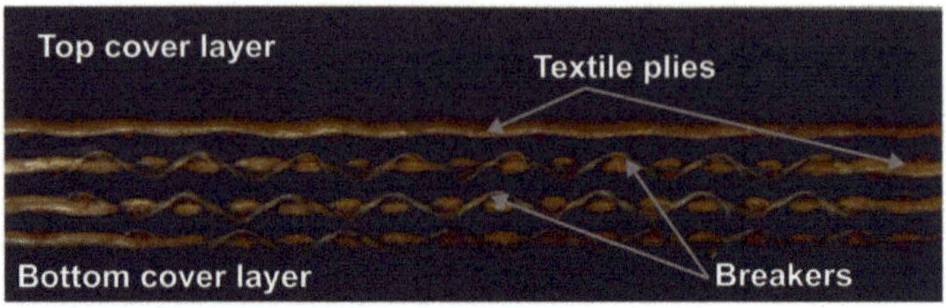

Fig. 3 Schematic representation of a rubber-textile conveyor belt [17]. *(Reused with permission from Elsevier, License number 4743680224599.)*

Fig. 3 shows the schematic of a rubber-textile conveyor belt used in many industries such as construction and mining equipment, where the materials are transported in very harsh environments under different operating conditions. It is made of different layers of textile piles. The piles are made up of polyamide fibers or polyester fibers because they are resistant to the chemicals, humidity, and acids while also having better mechanical and tribological properties [17]. The piles are bonded with a rubber mixture that acts as an adhesive between the piles. A photograph of textile-reinforced rubber conveyor belts used in civil engineering applications is shown in Fig. 4.

Hydraulic power plants are a major power generation source in many countries. Even though hydropower plants minimize the environmental impact, the use of stainless steel has caused a relatively heavy weight and greater

Fig. 4 Textile-reinforced rubber conveyor belts in civil engineering [13]. *(Reused with permission from Elsevier, License number 4743891395291.)*

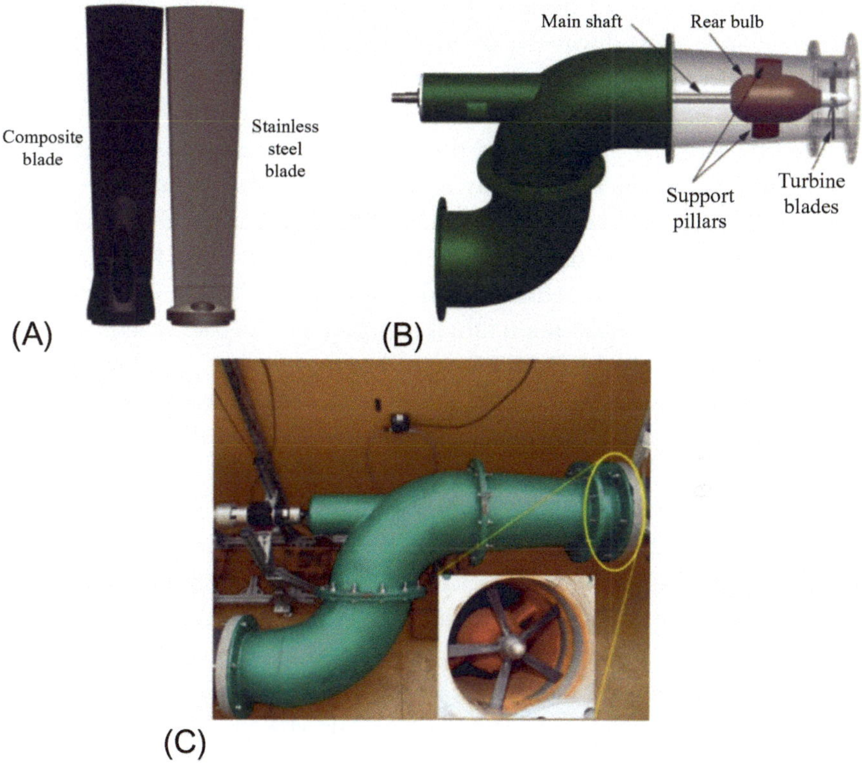

Fig. 5 Designs of the composite blade (A) and the turbine housing (B, C) [19]. *(Reused with permission from Elsevier, License number 4743680831303.)*

cost with a lower output. Recently, polymer-reinforced fiber composites have been used in water turbines, for example, a carbon and glass fiber-reinforced polymer composite material was used to make blades for a 1.2 MW SeaGen tidal turbine in Northern Ireland [18]. In a recent work, Li et al. [19] compared the performance of turbine blades made of stainless steel with those made of fiber-reinforced polymer composites. The turbine blades made of either stainless steel or carbon fiber reinforced-polymers were fabricated for an ultralow head (ULH) turbine of about a 1 m head for a comparative study (Fig. 5). The density of the stainless steel was about 7500–8000 kg/m^3, whereas that of the composites was only about 1500 kg/m^3. Because of the low elastic modulus of the composite blade, the base thickness of the blade was increased to reduce the bending stress and tip deflection during operation, as shown in Fig. 5A. The authors observed that the composite blade was more effective than the stainless steel blade and recommend it as a good material for the development of hydroelectric turbines.

The degeneration of human joints in bone due to aging is referred to as degenerative joint disease or osteoarthritis. This is caused by damage to the articular cartilage, mainly due to wear and tear that leads to pain in the knee joints. Osteoarthritis is treated by surgeries where the knee joints are replaced by artificial ones that aid the patients in recovering and leading a normal life. Polymers are used for human bone replacements such as the joints in the hips and knees. The UHMWPE (ultrahigh molecular weight polyethylene) is one of the most important polymers used for total knee replacement and the procedure called arthroplasty [20–21]. The knee replacement unit consists of the femoral component, the patellar component, the tibial tray, and the UHMWPE tibial inserts. Similarly, the hip replacement is called hip arthroplasty. The hip replacement unit consists of the metal femoral head and an acetabular cup connected with liners made up of UHMWPE [22]. The schematics of the knee and hip joint replacement units are shown in Fig. 6.

Huang et al. [23] studied the effect of temperature on the friction behavior of water-lubricated materials, namely nitrile rubber (NBR), ultrahigh molecular weight polyethylene (UHMWPE), UHMWPE/NBR composites (NU), and polymer resin (PR) against a copper alloy shaft. The friction coefficient of NBR and UHMWPE/NBR increases with temperature while the friction coefficient of UHMWPE and PR decreases with temperature. However, at 80°C, the friction and wear are maximum for PR. Thus, water-lubricated bearing materials may have different behavior at different temperatures. Hence, one should select the water-lubricated materials based on the environmental conditions. Fig. 7 shows the water-lubricated stern bearing and shaft damage due to poor lubrication. The tribological applications of some of the important polymers are listed in Table 1.

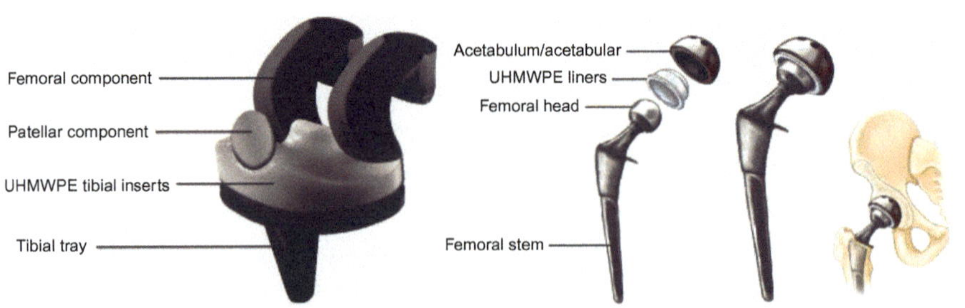

Fig. 6 UHMWPE inserts used in knee and hip joint replacements [22]. *(Reused with permission from Elsevier, License number 4743681427139.)*

Fig. 7 Damaged water lubricated bearing (A) and shaft (B) due to poor lubrication [23]. *(Reused with permission from Elsevier, License number 4743690212937.)*

Table 1 Application of polymers in tribology.

Polymer	Applications	References
PEEK	Seat adjustment parts, transmission parts, oil-pump parts.	[24]
UHMWPE	Bearings, pulleys, a lining of trucks or dock guards, articulating surfaces in orthopedic implants, bearing components for human joint replacements.	[25–26]
High-density polyethylene graphite-based nanocomposites	Hip joint replacements	[27]
PTFE/HDPE	Label/wiper in labeling machine	[28]
PTFE	TENMAT FEROFORM T814 self-lubricating bearing	[29]
PTFE	Marine bearing	[30]
Aromatic thermosetting polyester (ATSP)/ PEEK/PTFE	Tilting pad bearings	[31]

3 Tribological applications of natural fiber-based polymer composites

The demand for materials with high structural strength along with better wear resistance and lubricating properties has led to searches for newer materials for components such as gears, bearings, seals, etc., used in various industrial applications such as automobiles, chemical industries, mining, etc. [32]. Natural fibers are an alternative to replace synthetic fibers. They have advantages such as biodegradability, low density, low cost, low toxicity, and good mechanical and tribological properties. Plant fibers are abundantly available in nature and a good alternative for synthetic fibers. Nowadays, natural fibers are used in the interiors of automobiles and other components that are subjected to repeated wear and tear. Therefore, the study of the behavior of different natural fibers that can be utilized in different tribological applications is very important [33–34].

Commercially available automotive brake pads are made up of different polymers along with asbestos. Asbestos, which is widely used as a fiber along with polymers for brake pads, brake liners, couplings, etc., has harmful effects on the environment and the health of human beings. Therefore, its use has to be minimized with alternative environmentally friendly materials [35]. Natural fibers are one of the best alternatives. Finding a potentially sustainable friction material brings greater value to its performance, economy, and ecofriendliness [36]. Recently, electric buses, hybrid cars, bicycles, etc., were introduced in the market with high performance. The future emphasis in all vehicles is high fuel efficiency and zero emissions. This means that brakes should be lighter and should not release any toxic or carcinogenic substances during usage.

The design of brake friction materials mainly depends on the composition and fabrication process involved. Liu et al. [37] developed a brake friction material with the use of a wet granulation method. The friction material was made up of nitrile rubber (NBR)-modified phenolic resin, $BaSO_4$, CF, abaca fiber, Al_2O_3, graphite, shale, purple sands, porous iron powder, tire powder, water granulated slag, vermiculite, Sb_2S_3, husk powder, and frictional powder. It was found that the wet granulation method enhanced the friction and wear properties. The natural fibers are abundantly available and biodegradable. Because polymers are self-lubricating materials, the combination of natural fibers and polymer provides the best suitable material for tribological applications. Rashid et al. [38] used sugar palm fiber (SPF) as a reinforcement for phenolic resin-based composites. The fibers were treated with alkali and seawater to study the wear behavior of treated and untreated composites under the dry condition using a pin-on-disk apparatus. The study on wear behavior showed that the treatment of natural fibers led to better matrix fiber adhesion and improved the wear properties of the composites by avoiding delamination. The alkali-treated composites showed the best results, followed

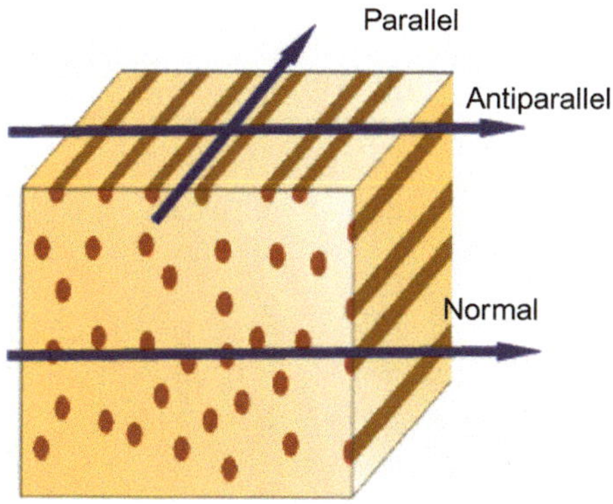

Fig. 8 Fiber orientation with respect to sliding direction [39]. *(Reused with permission from Elsevier, License number 4743690392911.)*

by seawater-treated composites and untreated composites. Chin et al. [39] attempted to introduce a kenaf fiber-reinforced epoxy composite for bearing applications. The test was carried out on a pin-on-disk apparatus for three different orientations: parallel orientation, antiparallel orientation, and normal orientation. Fig. 8 shows the fiber orientation with respect to sliding direction. The kenaf fibers in normal orientation enhanced the wear resistance of the composites by 85%, and the values were higher than under parallel orientation or antiparallel orientation.

Yousif et al. [40] studied the tribological behavior of oil palm fiber-reinforced polyester composites. The effect of speed, load, and sliding distance using a pin-on-disk apparatus was of special interest. The oil palm-reinforced composites showed greater improvement in the tribological properties by approximately 3–4 times over the neat polyester. Wear was observed in the form of debonding, bending, and deformation of the polyester. Krishnan et al. [41] used areca sheath fiber as the main ingredient in a brake pad application to develop an ecofriendly material. Similarly, waste banana tree powder, aramid pulp, ramie, corn stalk fiber, etc., were used as the friction components in brake pads, liners, and other applications [42–45]. Some of the important works on tribology are given in Table 2. Both synthetic and natural fiber-reinforced polymer composites are listed.

4 Conclusion

Advancements in technology have made polymer and polymer composites more popular in various tribological applications. This is due to their important characteristics such as low density, low friction, high wear resistance,

Table 2 Tribology studies of fiber-reinforced polymer composites.

Synthetic/natural/ hybrid fiber	Polymer matrix	Manufacturing method	Test apparatus/method	References
E-glass fibers	Epoxy	Hand lay-up method	Pin-on-disc apparatus	[46]
Glass fabric/graphite filler	Epoxy resin (LY556)	Hand layup followed by compression molding	Pin-on-disc apparatus	[47]
Basalt fibers	Epoxy	Resin transfer molding (RTM) technique	Pin-on-disc apparatus	[48]
Oil palm fiber/kenaf fiber	Epoxy	Casting method	Pin-on-disc apparatus	[49]
E-glass fibers	Epoxy	Hand-layup technique	Three-body abrasive wear	[50]
Sugarcane fibers and chopped glass fibers	Unsaturated polyester resin of type Reservol P9509	Hand-layup and open mold techniques	Adhesive wear test using pin on disc	[51]
Musaceae fibers	Vinyl ester resin	BMC fabrication technique	Pin-on-disc apparatus	[52]
Polyester fabric	Unsaturated polyester resin	–	Cylinder-on-plate testing configuration under reciprocating sliding	[53]
Jute fibers	Epoxy	Hand layup	Block-on-ring apparatus under sliding environment	[54]
Chopped glass fibers	Polyester	Hand lay-up technique	Three-body abrasive test using a sand/steel wheel (novel method)	[55]
Glass fibers	High-density polyethylene	Injection molding	Disc abrasive abrasion tester	[56]
Short glass fibers	Polyaryl ether ketone	Injection molding	Pin-on-disc apparatus	[57]
Flax fibers	Polypropylene	–	Nanoindentation and scratch tests	[58]
Cow dung fibers	Phenolic resin	Hot press molding	Constant-speed friction testing instrument	[59]
Cyperus pangorei fiber	Polyester	compression molding	Pin-on-disc apparatus	[60]

self-lubrication, and good dimensional stability. The increase in the use of polymers in various tribological applications is mainly to replace the existing materials such as metals. The understanding of the wear behavior of materials in various environments has become very important, especially to solve the fundamental problems related to friction and to design the components in a favorable way. The tribology of neat polymer can be altered by the incorporation of fibers and fillers. The need for environmentally friendly materials due to an increase in global demand has led to the use of natural fiber-based materials for tribology. This has resulted in the development of green tribology in various industries. Natural fibers are cost-effective and ecofriendly. Therefore, the study on natural fiber-based polymer composites is interesting due to the sustainability of natural fibers and the self-lubricating behavior of polymers.

References

[1] J. Gegner (Ed.), Tribology: Fundamentals and Advancements, BoD—Books on Demand, 2013.
[2] B. Bhushan, S. Sundararajan, Micro/nanoscale friction and wear mechanisms of thin films using atomic force and friction force microscopy, Acta Mater. 46 (11) (1998) 3793–3804.
[3] P.L. Menezes, M. Nosonovsky, S.P. Ingole, S.V. Kailas, M.R. Lovell (Eds.), Tribology for Scientists and Engineers, Springer, New York, 2013, pp. 295–389.
[4] W. Brostow, V. Kovačevic, D. Vrsaljko, J. Whitworth, Tribology of polymers and polymer-based composites, J. Mater. Educ. 32 (5) (2010) 273.
[5] S.W. Zhang, State-of-the-art of polymer tribology, Tribol. Int. 31 (1–3) (1998) 49–60.
[6] M. Cirino, K. Friedrich, R.B. Pipes, Evaluation of polymer composites for sliding and abrasive wear applications, Composites 19 (5) (1988) 383–392.
[7] M. Avalle, E. Romanello, Tribological characterization of modified polymeric blends, Procedia Struct. Integr. 8 (2018) 239–255.
[8] B.J. Briscoe, S.K. Sinha, Tribological applications of polymers and their composites: past, present and future prospects, in: Tribology and Interface Engineering Series, vol. 55, Elsevier, 2008, pp. 1–14.
[9] G.Y. Xie, G.S. Zhuang, G.X. Sui, R. Yang, Tribological behavior of PEEK/PTFE composites reinforced with potassium titanate whiskers, Wear 268 (3–4) (2010) 424–430.
[10] A.C. Greco, R. Erck, O. Ajayi, G. Fenske, Effect of reinforcement morphology on high-speed sliding friction and wear of PEEK polymers, Wear 271 (9–10) (2011) 2222–2229.
[11] B. Chen, J. Wang, F. Yan, Comparative investigation on the tribological behaviors of CF/PEEK composites under sea water lubrication, Tribol. Int. 52 (2012) 170–177.
[12] S. Karthikeyan, N. Rajini, M. Jawaid, J.T. Winowlin Jappes, M.T.H. Thariq, S. Siengchin, J. Sukumaran, A review on tribological properties of natural fiber based sustainable hybrid composite, Proc. Inst. Mech. Eng. Part J: J. Eng. Tribol. 231 (12) (2017) 1616–1634.
[13] K. Friedrich, L. Chang, F. Haupert, Current and future applications of polymer composites in the field of tribology, in: Composite Materials, Springer, London, 2011, pp. 129–167.

[14] K. Friedrich, Polymer composites for tribological applications, Adv. Ind. Eng. Polym. Res. 1 (1) (2018) 3–39.
[15] Broschure No. 5187 E, The Maintenance-Free Heavy DutyBushing, 2002, www.skf.com.
[16] P.K. Singh, A.K. Singh, An investigation on the thermal and wear behavior of polymer based spur gears, Tribol. Int. 118 (2018) 264–272.
[17] M. Andrejiova, A. Grincova, D. Marasova, Failure analysis of the rubber-textile conveyor belts using classification models, Eng. Fail. Anal. 101 (2019) 407–417.
[18] W. Marc, A. Roberto, How composite materials can be used for small hydro turbines, in: Turbines and Mechanical Components, New Development, vol. 34(2), Voith Hydro, North America, 2015.
[19] H. Li, D. Zhou, J.J. Martinez, Z.D. Deng, K.I. Johnson, M.P. Westman, Design and performance of composite runner blades for ultra-low head turbines, Renew. Energy 132 (2019) 1280–1289.
[20] B.C. Carr, T. Goswami, Knee implants—review of models and biomechanics, Mater. Des. 30 (2) (2009) 398–413.
[21] G.B. Cornwall, J.T. Bryant, C.M. Hansson, J. Rudan, L.A. Kennedy, T.D.V. Cooke, A quantitative technique for reporting surface degradation patterns of UHMWPE components of retrieved total knee replacements, J. Appl. Biomater. 6 (1) (1995) 9–18.
[22] N.S. Sharip, H. Ariffin, Polymeric composites for joint replacement, in: Nanostructured Polymer Composites for Biomedical Applications, Elsevier, 2019, pp. 385–404.
[23] J. Huang, X. Zhou, J. Wang, X. Tang, F. Kuang, Influence of temperature on friction of polymeric materials in water, Wear 426 (2019) 868–876.
[24] H. Koike, T. Honda, K. Kida, E.C. Santos, Y. Kashima, K. Kanemasu, Influence of radial load on PEEK plastic bearings life cycle, in: Advanced Materials Research, vol. 154, Trans Tech Publications, 2011, pp. 1288–1291.
[25] S.M. Kurtz, UHMWPE Biomaterials Handbook: Ultra High Molecular Weight Polyethylene in Total Joint Replacement and Medical Devices, Academic Press, 2009.
[26] B.P. Chang, H.M. Akil, R.B.M. Nasir, Dry sliding wear behaviour of talc-reinforced UHMWPE composite for implant application, Sains Malays. 44 (6) (2015) 819–825.
[27] H. Fouad, R. Elleithy, High density polyethylene/graphite nano-composites for total hip joint replacements: processing and in vitro characterization, J. Mech. Behav. Biomed. Mater. 4 (7) (2011) 1376–1383.
[28] D.H. Cho, B. Bhushan, Friction and wear of various polymer pairs used for label and wiper in labeling machine, Tribol. Int. 98 (2016) 10–19.
[29] Material data from: http://apexprecision.co.in/damadv.html.
[30] http://www.orkotmarine.us/pdf/orkot.txm.marine.material.data.sheet.2002.us.pdf.
[31] P. Lan, J.L. Meyer, B. Vaezian, A.A. Polycarpou, Advanced polymeric coatings for tilting pad bearings with application in the oil and gas industry, Wear 354 (2016) 10–20.
[32] P. Sampathkumaran, S. Seetharamu, A. Murali, R.K. Kumar, Dry sliding wear behavior of glass-epoxy composite, J. Reinf. Plast. Compos. 18 (1) (1999) 55–62.
[33] M.R. Sanjay, S. Siengchin, J. Parameswaranpillai, M. Jawaid, C.I. Pruncu, A. Khan, A comprehensive review of techniques for natural fibers as reinforcement in composites: preparation, processing and characterization, Carbohydr. Polym. 207 (2019) 108–121.
[34] M. Fahim, N. Chand (Eds.), Tribology of Natural Fiber Polymer Composites, Elsevier, 2008.
[35] S. Venkatesh, K. Murugapoopathiraja, Scoping review of brake friction material for automotive, Mater. Today: Proc. 16 (2019) 927–933.
[36] J. Kukutschova, V. Roubíček, K. Malachová, Z. Pavlíčková, R. Holuša, J. Kubačková, V. Mička, D. MacCrimmon, P. Filip, Wear mechanism in automotive brake materials, wear debris and its potential environmental impact, Wear 267 (5–8) (2009) 807–817.
[37] Y. Liu, L. Wang, D. Liu, Y. Ma, Y. Tian, J. Tong, P. Senthamaraikannan, S. Saravanakumar, Evaluation of wear resistance of corn stalk fiber reinforced brake friction materials prepared by wet granulation, Wear 432–433 (2019) 102918. https://doi.org/10.1016/j.wear.2019.05.033.

[38] B. Rashid, Z. Leman, M. Jawaid, M.J. Ghazali, M.R. Ishak, M.A. Abdelgnei, Dry sliding wear behavior of untreated and treated sugar palm fiber filled phenolic composites using factorial technique, Wear 380 (2017) 26–35.
[39] C.W. Chin, B.F. Yousif, Potential of kenaf fibres as reinforcement for tribological applications, Wear 267 (9–10) (2009) 1550–1557.
[40] B.F. Yousif, N.S.M. El-Tayeb, The effect of oil palm fibers as reinforcement on tribological performance of polyester composite, Surf. Rev. Lett. 14 (06) (2007) 1095–1102.
[41] G.S. Krishnan, L.S. Jayakumari, L.G. Babu, G. Suresh, Investigation on the physical, mechanical and tribological properties of areca sheath fibers for brake pad applications, Mater. Res. Express 6 (8) (2019) 085109.
[42] T. Singh, N. Kumar, J.S. Grewal, A. Patnaik, G. Fekete, Natural fiber reinforced non-asbestos brake friction composites: influence of ramie fiber on physico-mechanical and tribological properties, Mater. Res. Express 6 (11) (2019) 115701.
[43] Y. Ma, S. Wu, J. Zhuang, Y. Tian, H. Qi, J. Tong, The effect of lignin on the physicomechanical, tribological, and morphological performance indicators of corn stalk fiber-reinforced friction materials, Mater. Res. Express 6 (10) (2019) 105325.
[44] R. Yun, P. Filip, Y. Lu, Performance and evaluation of eco-friendly brake friction materials, Tribol. Int. 43 (11) (2010).
[45] İ. Sugözü, M. Güdük, Experimental investigation of a novel friction modifier in composite materials, Turk. J. Eng. 3 (4) (2019) 197–200.
[46] G.V. Kumar, R. Mageshvar, R. Rejath, S. Karthik, R. Pramod, C.S.P. Rao, Characterization of glass fiber bituminous coal tar reinforced polymer matrix composites for high performance applications, Compos. Part B: Eng. 175 (2019) 107156.
[47] B. Shivamurthy, K.U. Bhat, S. Anandhan, Mechanical and sliding wear properties of multi-layered laminates from glass fabric/graphite/epoxy composites, Mater. Des. 44 (2013) 136–143.
[48] N. Mohan, R.A. Kumar, K. Rajesh, R. Prabhu, S. Preetham, A.N. Rakshith, M.A. Prasad, Investigation on sliding wear behaviour of UHMWPE filled basalt epoxy composites, in: AIP Conference Proceedings, vol. 2057(1), AIP Publishing, 2019, p. 020048.
[49] F.F. Shuhimi, M.F.B. Abdollah, M.A. Kalam, M. Hassan, H. Amiruddin, Tribological characteristics comparison for oil palm fibre/epoxy and kenaf fibre/epoxy composites under dry sliding conditions, Tribol. Int. 101 (2016) 247–254.
[50] K. Gupta, Mechanical and abrasive wear characterization of bidirectional and chopped E-glass fiber reinforced composite materials, Mater. Des. 35 (2012) 467–479.
[51] N.S.M. El-Tayeb, A study on the potential of sugarcane fibers/polyester composite for tribological applications, Wear 265 (1–2) (2008) 223–235.
[52] C.E. Correa, S. Betancourt, A. Vázquez, P. Gañan, Wear performance of vinyl ester reinforced with Musaceae fiber bundles sliding against different metallic surfaces, Tribol. Int. 109 (2017) 447–459.
[53] P. Samyn, Tribological properties and thermomechanical analysis of unsaturated polyester fabric composite in oscillating line-contact sliding, Tribol. Int. 99 (2016) 127–139.
[54] F.Z. Alshammari, K.H. Saleh, B.F. Yousif, A. Alajmi, A. Shalwan, J.G. Alotaibi, The influence of fibre orientation on tribological performance of jute fibre reinforced epoxy composites considering different mat orientations, Tribol. Ind. 40 (3) (2018).
[55] B.F. Yousif, N.S.M. El-Tayeb, Wear characteristics of thermoset composite under high stress three-body abrasive, Tribol. Int. 43 (12) (2010) 2365–2371.
[56] X. Liang, Z. Guo, T. Jun, C. Yuan, Effect of modified glass fiber on tribological performance of water-lubricated bearing, Polym. Test. (2019) 106153.
[57] J.N. Panda, J. Bijwe, R.K. Pandey, Optimization of the amount of short glass fibers for superior wear performance of PAEK composites, Compos. A: Appl. Sci. Manuf. 116 (2019) 158–168.
[58] F. Chegdani, Z. Wang, M. El Mansori, S.T. Bukkapatnam, Multiscale tribo-mechanical analysis of natural fiber composites for manufacturing applications, Tribol. Int. 122 (2018) 143–150.

[59] Y. Ma, S. Wu, J. Zhuang, J. Tong, H. Qi, Tribological and physio-mechanical characterization of cow dung fibers reinforced friction composites: an effective utilization of cow dung waste, Tribol. Int. 131 (2019) 200–211.
[60] N. Rajini, K. Mayandi, S. Siengchin, N. Ayrilmis, C. Bennet, S.O. Ismail, Tribological properties of *Cyperus pangorei* fiber-reinforced polyester composites (friction and wear behavior of *Cyperus pangorei* fiber/polyester composites), J. Nat. Fibers (2019) 1–13.

Index

Note: Page numbers followed by *f* indicate figures and *t* indicate tables.

T